Arctic Oil and Gas

The Arctic is predicted to hold a large part of the world's undiscovered oil and gas fields. With the rise in global energy demand and the focus upon energy security, the Arctic environment is likely to experience increasing pressure from the push to uncover these reserves.

This book analyzes the expanding oil and gas activities in the Arctic from the perspective of sustainable development and corporate social responsibility. The focus is on the territories of the Arctic rim where the current and future oil and gas activities in the Arctic are and will be located. The book raises a number of questions including how sustainable development has been framed in the Arctic and the interaction between indigenous peoples, governments and oil and gas companies.

The book is divided into three parts. In the first part of the book, oil and gas are approached through the concepts of sustainable development and corporate social responsibility together with the challenge of climate change. The second part consists of case studies from Alaska, Canada, Norway and Russia, where the discourses on oil and gas in the Arctic are explored, and the final part of the book draws together the material from the country studies in a comparative manner.

The book is primarily aimed at academics, universities, public authorities and the oil and gas industry. It is useful, more specifically, for postgraduate students who are undertaking research in energy and gas industries, environmental economics and ecological economics.

Aslaug Mikkelsen is Rector of the University of Stavanger in Norway.

Oluf Langhelle is Professor at the Faculty of Social Sciences at the University of Stavanger in Norway.

Routledge Explorations in Environmental Economics
Edited by Nick Hanley
University of Stirling, UK

Greenhouse Economics
Value and ethics
Clive L. Spash

Oil Wealth and the Fate of Tropical Rainforests
Sven Wunder

The Economics of Climate Change
Edited by Anthony D. Owen and Nick Hanley

Alternatives for Environmental Valuation
Edited by Michael Getzner, Clive Spash and Sigrid Stagl

Environmental Sustainability
A consumption approach
Raghbendra Jha and K.V. Bhanu Murthy

Cost-Effective Control of Urban Smog
The significance of the chicago cap-and-trade approach
Richard F. Kosobud, Houston H. Stokes, Carol D. Tallarico and Brian L. Scott

Ecological Economics and Industrial Ecology
Jakub Kronenberg

Environmental Economics, Experimental Methods
Edited by Todd L. Cherry, Stephan Kroll and Jason F. Shogren

Game Theory and Policy Making in Natural Resources and the Environment
Edited by Ariel Dinar, José Albiac and Joaquín Sánchez-Soriano

Arctic Oil and Gas
Sustainability at risk?
Edited by Aslaug Mikkelsen and Oluf Langhelle

Arctic Oil and Gas
Sustainability at risk?

Edited by Aslaug Mikkelsen
and Oluf Langhelle

Routledge
Taylor & Francis Group
LONDON AND NEW YORK

HD9574.A68 A73 2008

0134111317662

Arctic oil and gas :
sustainability at risk?
2008.

2009 01 12

First published 2008 by Routledge
2 Park Square, Milton Park, Abingdon, Oxon OX14 4RN

Simultaneously published in the USA and Canada
by Routledge
270 Madison Ave, New York, NY 10016

*Routledge is an imprint of the Taylor & Francis Group,
an informa business*

© 2008 Editorial matter and selection, Aslaug Mikkelsen and Oluf
Langhelle; individual chapters, the contributors

Typeset in Times New Roman by Keyword Group Ltd
Printed and bound in Great Britain by MPG Books Ltd, Bodmin Cornwall

All rights reserved. No part of this book may be reprinted or reproduced
or utilised in any form or by any electronic, mechanical, or other means,
now known or hereafter invented, including photocopying and recording,
or in any information storage or retrieval system, without permission in
writing from the publishers.

British Library Cataloguing in Publication Data
A catalogue record for this book is available from the British Library

Library of Congress Cataloging in Publication Data
A catalog record for this book has been requested

ISBN 10: 0-415-44330-X (hbk)
ISBN 10: 0-203-89374-3 (ebk)

ISBN 13: 978-0-415-44330-2 (hbk)
ISBN 13: 978-0-203-89374-6 (ebk)

Contents

List of figures		vii
List of tables		viii
List of contributors		ix
Preface		xii
Acknowledgements		xiii

1 Introduction — 1
ASLAUG MIKKELSEN AND OLUF LANGHELLE

Part I
The Arctic: context, framework and methodology — 13

2 Framing oil and gas in the Arctic from a sustainable development perspective — 15
OLUF LANGHELLE, BJØRN-TORE BLINDHEIM AND OLAUG ØYGARDEN

3 Climate change and consequences for the Arctic — 45
TORLEIV BILSTAD

4 Corporate social responsibility: the economic and institutional responsibility of business in society — 57
BJØRN-TORE BLINDHEIM

5 Framework and methodology: regulation and discourse analysis as a research strategy — 87
OVE HEITMANN HANSEN, OLUF LANGHELLE AND ROBERT ANDERSON

Part II
Legal and institutional framework: case studies — 109

6	Legal and institutional framework: a comparative analysis NIGEL BANKES	111
7	Expanding oil and gas activities on the North Slope of Alaska ASLAUG MIKKELSEN, SHARMAN HALEY AND OLAUG ØYGARDEN	139
8	Oil and gas activities at the Mackenzie Delta, in Canada's Northwest Territories ALDENE MEIS MASON, ROBERT ANDERSON AND LEO-PAUL DANA	173
9	Going North: the new petroleum province of Norway OVE HEITMANN HANSEN AND METTE RAVN MIDTGARD	200
10	The Russian model: merging profit and sustainability ELENA N. ANDREYEVA AND VALERY A. KRYUKOV	240

Part III
Comparisons and managerial implications 289

11	Human rights and indigenous peoples in the Arctic: what are the implications for the oil and gas industry? KETIL FRED HANSEN AND NIGEL BANKES	291
12	Perceptions of Arctic challenges: Alaska, Canada, Norway and Russia compared OLUF LANGHELLE AND KETIL FRED HANSEN	317
13	Managerial implications ASLAUG MIKKELSEN, RONALD D. CAMP II AND ROBERT E. ANDERSON	350

Index 381

Figures

1.1	The Arctic's undiscovered oil and gas resources	3
3.1	Maximum oil exploration travel activities on the tundra (ACIA, 2004)	49
3.2	Methane concentration doubling over the last 150 years (Petit, Jouzel, Raynaud et al., 1999)	50
3.3	Arctic ice cover (Norsk Romsenter/National Snow and Ice Center, 2007)	52
3.4	Concentrations of CO_2 and temperature from 400,000 years Vostok Ice Core data set (Corell, 2006)	53
4.1	The degrees of responsibility assigned to business in different approaches to CSR	60
5.1	Major actors associated with the oil and gas industry in the Arctic (Anderson, MacAulay, Kayseas et al., 2007)	88
7.1	Allocation of gross Alaskan North Slope (ANS) production value 2004 (US$13.3 billion total)	148
10.1	Dynamics of oil and gas output in the Northern regions of Russia compared to the total output for the country	244
10.2	Dynamics of receipts of the budget of the Yamal-Nenets Autonomous Okrug (YaNAO) v. gas production dynamics	267
12.1	Arctic share of global petroleum production, 2002	319
13.1	Major actors associated with the oil and gas industry in the Arctic	351
13.2	Corporate social responsibility and long-run survival	354
13.3	Corporate approaches to social responsibility	355
13.4	Corporate approaches to social responsibility	363

Tables

4.1	The social responsibility of business	76
5.1	Interview matrix	102
6.1	Canadian land claim agreements	124
10.2	Distribution of oil and gas reserves and resources in oil- and gas-bearing provinces, as a percentage of Russia's total	243
10.1	Russian oil and gas revenue flows (federal/regional as of 2006): primary distribution of revenues	247
10.3	Development indicators for the key oil and gas regions of the Russian North	259
10.4	Development indicators for the key oil and gas regions of the Russian North	260
10.5	Complex ecological assessment in the Arctic regions of RF, 2003	278
11.1	Ratifications of selected human rights instruments	292
12.1	Kyoto Protocol targets for US, Canada, Norway and Russia	327
12.2	Increase in CO_2 emissions from 1990 (baseline year in the Kyoto Protocol) to 2005 in percentage (without land use, land-use change and forestry), and from 2000 to 2005	329
12.3	Energy industries' sector share of total greenhouse gas emissions (without LULUCF) in 1990 and 2005 in Canada, Norway, Russia and US	329
12.4	Protected areas in Arctic Canada, Norway, Russia and Alaska, classified in International Union for Conservation of Nature categories I-V, plus Ramsar international wetland sites as of 2000. Areas smaller than 10 km^2 are not included	334

Contributors

Robert Anderson is a Professor in the Faculty of Business Administration at the University of Regina, in Saskatchewan, Canada. He is the Editor of the *Journal of Small Business and Entrepreneurship* and Co-editor, with Leo Paul Dana, of the *Journal of Enterprising Communities*. He is former President of the Canadian Council for Small Business and Entrepreneurship and is a Director of the Canadian Council for Small Business as well as a member of the Small Business Research Advisory Committee of the Department of Industry and Commerce Canada.

Elena N. Andreyeva is Chief of the Laboratory of Arctic Studies, Institute for System Analysis, Moscow, Russian Academy of Sciences; leader of the Russian and International research projects on social and economic issues of Arctic development; and an expert on governmental and regional structures. Most of her projects are connected with the use of non-renewable natural resources and the interactions of stakeholders, including indigenous people and industrial companies. She works directly with regional administrations on questions of social policy, programs of economic development and legal issues of resource use conflicts. Her main focus is on oil and gas areas, particularly the coastal zone and the Arctic shelf.

Nigel Bankes is a Professor of Law at the University of Calgary, Alberta, Canada, where he has taught since 1984. He was seconded to Canada's Department of Foreign Affairs and International Trade as Professor in Residence for the 1999–2000 academic year. His principal research interests are in the areas of indigenous peoples' law, water law, oil and gas law and international environmental law. He was the lead author of the 'Legal Systems' chapter of the Arctic Human Development Report of the Arctic Council and is a former chair of the Canadian Arctic Resources Committee, a Canadian non-governmental organization.

Torleiv Bilstad is Professor of Environmental Engineering at the University of Stavanger, Norway. He received his academic degrees from the University of Wisconsin: Ph.D. (1977) and M.Sc. (1973) in Civil and Environmental Engineering, and B.Sc. in Civil Engineering (1972). His main research activities are concerned with water quality and climate change. Treatments for potable water as well as water produced from petroleum production are typically handled by membrane separation in his research. The development of a biomembrane for agriculture, with possibilities for controlling the water content, adjusting the soil temperature and reducing CO_2 emission, is presently a central research activity.

Bjørn-Tore Blindheim is a Ph.D. student in management at the Department of Media, Culture and Social Science, University of Stavanger, Norway. His main research area is corporate social responsibility. He attained a Masters degree in Political Science from the University of Oslo in 2000.

Ronald D. Camp II, Ph.D., is Associate Professor in International Business/Organizational Behaviour in the Faculty of Business Administration at the University of Regina, in Saskatchewan, Canada.

Leo-Paul Dana is Senior Advisor to the World Association for Small and Medium Enterprises, and tenured at the University of Canterbury, in New Zealand. He was formerly Deputy Director of the International Business M.B.A. Programme at Singapore's Nanyang Business School. He also served on the faculties of McGill University and INSEAD. He holds B.A. and M.B.A. degrees from McGill University, and a Ph.D. from the Ecole des Hautes Etudes Commercials. He is the author of several books in business and economics.

Sharman Haley is a Professor of Economics and Public Policy employed by the Institute of Social and Economic Research at the University of Alaska, Anchorage, USA. She attained a Ph.D. in Economics from the University of California, Berkeley, in 1994. Her focus area is interdisciplinary social science research on community and resource development in rural Alaska and the circumpolar Arctic.

Ketil Fred Hansen is a post-doctoral scholar at the Department of Media, Culture and Social Science, University of Stavanger, Norway. His main research interests are the development–security nexus, including issues of aid, poverty, marginalized groups, human rights and security. His international publications include 'Cutting aid to promote peace and democracy? Intentions and effectiveness of aid sanctions' in *The European Journal of Development Research* 2006 and 'The politics of personal relations' in *Africa* (Edinburgh University Press), Vol. 73, no 2, 2003.

Ove Heitmann Hansen is a professional Strategy and Change Management adviser at IBM Centre of Excellence for Oil & Gas and a Ph.D. student at the University of Stavanger, Norway. He holds an M.Sc. in Industrial Economics, a Masters degree in Change Management, a B.Sc. in Business and Information Systems, and in Education from the Royal Norwegian Navy Academy. The research fields at the University of Stavanger focus on stakeholder management and decisions dilemmas in the oil and gas industry.

Valery A. Kryukov is a Doctor of Sciences (Economics), and Professor, Head of the Research Laboratory 'Economic Development of the West Siberia Oil and Gas Complex' at the Institute of Economics and Industrial Engineering, Siberian Division of the Russian Academy of Sciences in Novosibirsk, Russia. He is one of the leading specialists in Russia in the following areas: institutional aspects of the oil and gas sector development (particularly concerning regional issues); natural resources management (relating to regional socio-economic development); regional project design and preparation of regional development programmes.

Oluf Langhelle is a Professor in Social Sciences at the University of Stavanger, Norway. He attained a Dr.Polit. in political science from the University of Oslo in 2000. His main research interests are sustainable development, environmental policy, corporate social responsibility and negotiations within the World Trade Organization. He has published several books and international papers on sustainable development.

Aldene Meis Mason is a Doctoral Candidate in Management at the University of Canterbury, New Zealand, and a Lecturer with the Faculty of Business Administration, University of Regina, Canada. Her interests are sustainable indigenous entrepreneurship, resilience, capacity building, innovation and strategic partnerships. After completing an interdisciplinary BSc. (Kin) and a MBA, she held progressively more senior management and consulting positions in industry and government for over 20 years. Aldene is a Director of the Canadian Council on Small Business and Entrepreneurship. She is a Fellow of the Institute of Certified Management Consultants and a Certified Human Resource Professional.

Mette Ravn Midtgard is a Senior Researcher at NORUT (Northern Research Institute in Tromsø), Norway. She was educated as an Economic Geographer at the University of Oslo, and has worked as a researcher in Northern Norway for over 12 years. In 2007 she started her Ph.D. fellowship at the University in Tromsø focusing on recovery strategies subsequent to an oil potential spill.

Aslaug Mikkelsen is President of the University of Stavanger, Norway. She moved from a position as Professor of Change Management at the same institution and a professor II position at the Western Norway Regional Health Authorities, Stavanger. By education she has an M.Phil. in Sociology from the University of Oslo and a Ph.D. in Work and Organizational Psychology from the University of Bergen, Norway. Her main research areas are the management of organizational change, knowledge transfer, occupational health and corporate social responsibility. She has published several books and international papers in her field.

Olaug Øygarden is a Masters student in Change Management at the University of Stavanger, Norway, and has a Bachelors degree in Development Studies from the University of Oslo. She has worked as a research assistant on this project, and also on a project about corporate social responsibility in the textile industry.

Preface

The Arctic contains huge amounts of oil and gas ready to be exploited. Is it possible to exploit these resources in a sustainable manner? Several stakeholders have expressed concern that exploration and production activities in these sensitive areas may not be performed in an environmental sustainable manner nor be consistent with the interests of indigenous peoples.

This book addresses these issues in the following way. In the first part of the book, oil and gas are approached through the concepts of sustainable development and corporate social responsibility together with the challenge of climate change. The second part consists of case studies from Alaska, Canada, Norway and Russia, where the discourses on oil and gas in the Arctic are explored. The third part of the book draws together the material from the country studies in a comparative manner.

The book is the result of a collaborative effort between scientists from Russia, Norway, Canada and the United States. Together we travelled in 2005 and 2006 to remote places in Russia, Norway, Canada and Alaska with the purpose of observing the physical environment, interviewing people as well as screening literature, searching for effects from, and future challenges for, oil and gas activities in the fragile Arctic. It is our pleasure to extend gratitude to our colleagues who agreed to participate in such a challenging project.

Stavanger 15 November, 2007

Professor Aslaug Mikkelsen	Professor Oluf Langhelle
University of Stavanger	University of Stavanger

Acknowledgements

The research on which this volume is based was supported by the Research Council of Norway (175086/D15) and Shell International Exploration and Production BV. Thanks are due to a number of people from Shell who enthusiastically supported the project with their time. In particular, Mr Robert Blaauw and Dr James Parker initially asked us to design and carry out the studies on which this book is based. Dr Parker helped guide the project through all its stages. Mr Joppe Cramwinckel provided useful insights on industry approaches to sustainable development. The manuscript also benefited from comments by Dr Kim Cartwright and Ms Rebecca Nadel.

We gratefully acknowledge the assistance of all those who took the time to comment on earlier drafts on sections of the manuscript, including Aleksander A. Arbatov, Turid Arnegaard, Martin H. Belsky, Per Christian Brodschöll, Olga Gjerald, Robert T. Hamilton, Lassi Heininen, James Meadowcroft, Richard Otis, James Parker and Arild Aarskog. In addition, we would like to thank the individual authors of the chapters included in this book. Special thanks go to Sharman Haley and Bob Anderson who have commented on most of these chapters.

We also extend thanks to Suzanne Sharp at the Institute of Social and Economic Research (ISER), Anchorage, for facilitating the data collection in Alaska. Likewise, we are grateful for assistance from Dr Vera Toskunina who, together with Dr Elena Andreyeva, facilitated our second network meeting in Archangelsk. We should also like to acknowledge Senior Advisor Solveig Andresen at the Norwegian Petroleum Directorate for her contribution to data collection, discussions and encouragement. Professor Gordon Pullar was an active contributor in the first phase of this project and gave important contributions to the perspectives of this project. The University of Stavanger (UiS) has supported this project through involvement of staff as well as covering publishing expenses.

In addition, we would like to thank Clare Freeman for doing wonders with the manuscripts, and Imran Shahnawaz, Keyword Project Manager, who handled the production process for the book on behalf of James Rabson, Routledge. Last, but not least, we would like to thank Rob Langham and Sarah Hastings at Routledge for guiding us through the whole process.

A.M.
O.L.

1 Introduction

Aslaug Mikkelsen and Oluf Langhelle

We like to think of the Arctic as pristine, untouched nature, but the Arctic has become part of the global economy. Large natural resources such as oil, minerals and fish are already depleted from the region and sent to the global market. In the modern Arctic, practically every aspect of life has been and continues to be influenced and shaped by both developmental needs and market demands for natural resources elsewhere. Simultaneously, the Arctic is suffering the consequences of pollution and human-created global warming owing to carbon dioxide (CO_2), other gas emissions influencing the climate, and environmental damage and the improper dismantling of military installations from the Cold War period.

On this background, the Arctic states (Canada, Denmark, Finland, Iceland, Norway, the Russian Federation, Sweden and the United States) established the Arctic Council[1] in 1996, which focused on sustainable development from the very start. In the Ottawa Declaration, they committed themselves to: 'sustainable development in the Arctic region, including economic and social development, improved health conditions and cultural well-being'; '... the protection of the Arctic environment, including the health of Arctic ecosystems, maintenance of biodiversity in the Arctic region and conservation and sustainable use of natural resources'; and 'the well-being of the inhabitants of the Arctic, including special recognition of the special relationship and unique contributions to the Arctic of indigenous people and their communities'.[2]

The programme of the Russian Federation chairmanship of the Arctic Council in 2004–2006 states that the issues of social and economic development are lagging behind in the Arctic Council activities compared to the environmental component. The programme calls for the Arctic Council to take actions that will create a more balanced contribution to resolving the problems of sustainable development of the Arctic region. In 2006, Norway took over the chairmanship of the Arctic Council. The new programme statement confirms that there has been a greater emphasis on the environmental issues and that we now have a good knowledge base on this. However, the programme focuses on the fact that it will not be possible to maintain settlement patterns and to ensure growth and welfare in the Arctic without further economic activities. The programme suggests that protection of the environment, combined with sustainable utilization of natural resources, both renewable and non-renewable, should be the core areas of cooperation under the auspices of the Arctic Council in the years ahead. This statement might be seen as a door opener for further expansion of exploration and production of oil and gas in the area in line with worldwide *oil dependency* and for the economic, *geopolitical* and developmental interests of the main Arctic states, in spite of the *vulnerability* of the area.

The presence of hydrocarbons in the Arctic has already caused large-scale developments in the Arctic, such as: in the Alpine and Prudhoe Bay at the North Slope, Alaska; in the North West territories, Nunavut and Nunavik in Canada; at Melkøya in the Northern Norway; and in the Yamalo-Nenets and Khantyt-Mansii Autonomous Okrugs in Russia. The fact that there are few signals of a decline in either the demand for oil and gas, or for security of the energy supply has become one of the top political issues in the European Union (EU), the United States (US) and other nations in recent years, creating an expectation of a further expansion of oil and gas activities worldwide.

For many years, the International Energy Agency (IEA) has systematically over-estimated how much new oil will come out on the market outside the Organization of the Petroleum Exporting Countries (OPEC).[3] Most oil reserves are found in only a few countries, but estimates of their size vary widely. The most available and economical oil and gas fields in the world are already exploited. The main future oil and gas fields are situated in remote areas with vulnerable environments, where the costs of exploration and production are high, or within politically unstable countries. In contrast to the other areas, most Arctic countries are politically stable, but from a developmental perspective, there is uncertainty surrounding the requirements of technology that is currently (2008) undeveloped or untested.

The US geological survey (USGS) world petroleum assesment 2000 estimates, that approximately 25 per cent of the undiscovered petroleum reserves are in the Arctic, makes Arctic oil and gas an important economic and geopolitical issue. Map 1.1 shows assessed Arctic oil and gas provinces and potential Arctic oil and gas provinces. The yellow areas constitute the areas where USGS estimates that as much as 25 per cent of the world's remaining undiscovered conventional oil and gas fields may be found. The areas marked in blue are some other potential Arctic areas. In addition to the four countries with operating fields today, Alaska/US, Canada, Norway, Russia, Greenland/Denmark may also have huge oil and gas reserves.

Figure 1.1 presents the volume of the undiscovered resources in the Arctic presented by the USGS 2000 assessment.[4] The total estimate for the world is 268 billion Sm^3 o.e., or 1690 billion barrels o.e. of the 70 billion Sm^3 that is stipulated to be in the Arctic. The major reserves seem to be in Western Siberia on land, North Slope of Alaska offshore and East Greenland offshore. There has been no exploration in East Greenland: only a few seismic lines have been acquired, basically due to year-round sea-ice cover. Explorations conducted in West Greenland for several years have made no exploitable discoveries. The USGS estimates for the Lofoten area in Norway, which is equivalent to the Norwegian Sea area, are considerably higher than the Norwegian Petroleum Directorate's estimate.

This shows that there is considerable uncertainty regarding how big the Arctic reserves are, which the USGS has also confirmed. In fact, the estimate included

Introduction 3

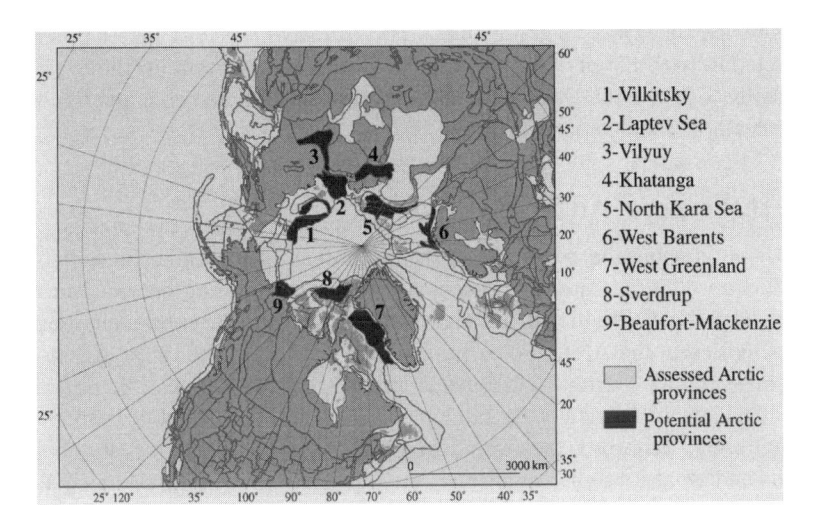

Map 1.1 Assessed Arctic oil and gas provinces, and potential Arctic oil and gas provinces. Source: The USGS World Assessment 2000.

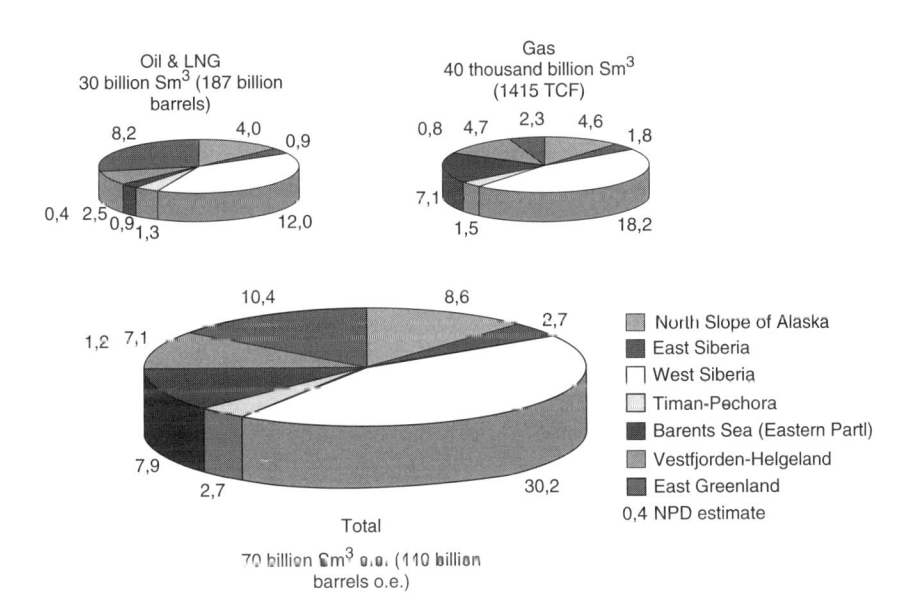

Figure 1.1 The Arctic's undiscovered oil and gas resources. Source: The USGS World Assessment 2000.

areas which lie entirely south of the Arctic Circle. If the East Siberian basin is deducted from the 25 per cent, the number is closer to 14 per cent (Bailey, 2007). Currently, USGS is undertaking a study of several of these basin areas and will publish a new estimate later in 2008.

The challenges of Arctic oil and gas

This book is about the expansion of the oil and gas activities in the Arctic. Our focus is on Alaska, Canada, Norway and Russia, where the current and near-future oil and gas activities in the Arctic will be located. The expansion of oil and gas activities in the Arctic poses tremendous challenges. We believe, however, that many of the current and potential future conflicts in the Arctic can be addressed in a meaningful way within the framework of sustainable development. In this book, therefore, the overall question is what sustainable development entails and what it might imply for oil and gas activities in the Arctic. Will the desire for economic progress in both developing and developed countries put *the sustainability of the Arctic*, as well as the globe, *at risk*? The answers to the question are neither straightforward, as they depend upon a number of factors and additional information, nor are the challenges identical in different regions/countries in the Arctic.

This study is structured along three steps to answer the research question: The first step is to provide an initial understanding of the concept of sustainable development. Our approach places sustainable development within the processes of the United Nations Conference on Environment and Development (UNCED) and of the report from the World Commission on Environment and Development (WCED), *Our Common Future* (1987). Sustainable development is an ethical development goal that frames some global issues in a particular way and defines a global sustainability agenda in which the prominent issues are, among others, developmental concerns, climate change, biodiversity and indigenous peoples (Langhelle, 2000).

Although *Our Common Future* and the following UNCED process barely mention the Arctic, it is from these international processes that the concept of sustainable development was placed – and has further evolved – on the international political agenda (Lafferty and Langhelle, 1999). It is also from these processes that sustainable development has been transferred to the Arctic context, primarily through the auspices of the Arctic Council. The questions we attempt to answer are:

- how do the Arctic Council and its Working Groups understand the concept of sustainable development?
- what are the implications drawn from sustainable development for oil and gas in the Arctic? and
- how are the challenges from oil and gas conceived in the Arctic?

The second step is to focus upon the responsibilities of the oil and gas industry to address challenges of sustainable development in the Arctic. This is done theoretically by exploring the nature of the concept of corporate social responsibility (CSR), and by discussing the potential relationship between CSR and sustainable development. Most basically, business has an economic responsibility towards society, that is, to enhance value for shareholders, employees, local communities and society in general. However, business also has an institutional responsibility towards society, that is, to comply with both national and international laws and regulations, and politically sanctioned standards and guidelines aimed at promoting responsible business practice. Both Arctic public policy documents, and international standards and guidelines aimed at promoting responsible business practice in general, suggests that the oil and gas industry should support the goals of sustainable development through their decisions and actions. They should take into consideration not only the economic aspects of their operations, but also the social and environmental issues.

Alongside national governments and supranational bodies such as the EU, the multinational oil and gas companies are the main actors in the competition to gain control over the Arctic region's natural resources. Although the oil and gas companies have financial power, they are dependent on access to oil and gas reserves and licenses to operate in order to satisfy their stockholders' present and future targets for value creation. Managers have been willing to fake the size of the available reserves[5] to prevent negative development in shareholder values.

Oil and gas companies seem to adjust their behaviour and definitions of corporate social responsibility in response to unexpected criticism of their activities by the media and non-governmental organizations (NGOs). When these activities develop into critical incidents and become the sustained focus of public attention, they also attract the attention of corporate management and can influence their strategic decisions, corporate activities and institutional transformation (Flanagan, 1954; Dutton and Dukerich, 1991; Hoffman and Ocasio, 2001), as well as a collective definition or redefinition of social problems (Pride, 1995). How vulnerable the same companies are to damage to their reputation is illustrated by such critical incidents as the Brent Spar, the Ken Saro Wiwa execution in Nigeria and the *Exxon Valdez* accident (Mikkelsen, Engen and Grønhaug, 2008).

The third step is to map out and analyse the main disputes and conflicts within and between different Arctic regions/countries – in our case, Alaska, Canada, Norway and Russia – from a discourse perspective. How do stakeholders in the oil and gas industry perceive and frame the expansion of oil and gas activities in the Arctic? What are the main conflicts? How do these conflicts eventually relate to ideas about sustainable development and corporate social responsibility? How disputed are Arctic oil and gas activities, and what are the main concerns of – and conflicts between – indigenous peoples, local people, governments and oil and gas companies? Can the interests of these different groups be reconciled and combined with an expansion of oil and gas activities? Or do some stakeholders perceive sustainable development in the Arctic to be at risk from further expansion of oil and gas activities? And finally, what are the managerial implications of the defined sustainable development issues for oil and gas companies that want to expand their activities in the Arctic?

The content of the book

The book has three parts. Part I outlines the Arctic context, the background for the research questions, the theoretical approach of the project and the methodology. Part II consists of a comparative chapter on the institutional framework across countries and four case studies of Alaska, Canada, Norway and Russia. Their aim is to identify the main conflicts and framings of oil and gas activities in the different countries. Part III consists of cross-country studies and comparisons. It includes a chapter on human rights, a comparative analysis based on the four case studies from Part II, and a chapter on the implications of the research conducted for managers of the oil and gas companies.

This chapter has addressed the Arctic context and the background for the research questions. In Chapter 2, *Langhelle, Blindheim* and *Øygarden* present an understanding of the concept of sustainable development, based on *Our Common Future* (WCED, 1987), also referred to as 'the Brundtland report' after the Chairman of the Commission, former Prime Minister of Norway, Gro Harlem Brundtland. The first part of the chapter explores the concept and its core elements. It also answers the question of whether it makes sense to talk about 'sustainable' extraction of non-renewable resources. The second part explores how sustainable development has been interpreted and conceived in an Arctic context by the Arctic states in the Arctic Council and its Working Groups. The last section explores the specific framing of oil and gas within the Working Groups of the Arctic Council.

In Chapter 3, *Bilstad* focuses on the issue of climate change in general and climate change in the Arctic, and discusses both the causes and effects of these changes. Climate change and global warming are considered major environmental and economic threats to the global community. Commitments to prevent and minimize causes and to mitigate adverse effects are continuously on the political and scientific agenda. Strategies are implemented to achieve reductions in CO_2, considered to be the major human-induced greenhouse gas. While the Arctic is rich in hydrocarbons (HCs), and production of oil and gas will improve global energy security, burning of HCs irrespective of where they are mined advances global warming. Changes in climate influence not only snow and ice, but also the way of life in the Arctic.

In Chapter 4, *Blindheim* discusses both the nature of the concept of CSR and its potential relationship with and effect on sustainable development. The chapter starts out with presenting four broad approaches to CSR: the classical, stakeholder, social demanding, and social activist approaches. Based on an analysis of the contemporary discourse and agenda of CSR, it is argued that the contemporary conceptualization of CSR largely falls within what may be called a 'social activist' approach to CSR, suggesting corporations as complementing and sometimes replacing states as the primary structures and shapers of the world. Such an understanding of CSR, it is argued, implying that corporate managers take on the roles as moral, social and political leaders, will not necessarily promote a sustainable development path. A concept of CSR – heavily emphasizing the advantage of voluntary action instead of political solutions to developmental questions and societal challenges – may do the opposite, that is, to undermine a sustainable development path.

On this background, a limited conceptualization of CSR is proposed, and the companies' institutional responsibility towards society is stressed. This implies that corporations have a responsibility to adapt their activity to the constraints and opportunities set by political institutions and processes of public policy, and, as such, should not undermine the capacity of these institutions' to perform their fundamental mission in society (e.g. framing and regulating the activity of the oil and gas industry in the Arctic). Based on this conceptualization of CSR, some implications for the oil and gas industry operating in the Arctic are briefly discussed.

In Chapter 5, *Heitmann Hansen, Langhelle* and *Anderson* present the methodology used for the different case studies from Alaska, Canada, Norway and Russia. The first part introduces a 'framework', or an analytical tool based on regulation theory, which is used to identify the main actors influencing developments in the Arctic. The second part outlines discourse analyses as a research strategy and methodology and explains the core concepts used in our study. We define a discourse, in accordance with Hajer, as 'a specific ensemble of ideas, concepts, and categorizations that are produced, reproduced, and transformed in a particular set of practices and through which meaning is given to physical and social realities' (Hajer, 1995: 44). Discourse analysis is the main tool for identifying how actors construct and frame oil and gas activities in the Arctic.

Chapter 6 is the first chapter in Part II of the book. In this chapter, *Bankes* gives a comparative analysis of the legal and institutional framework of Alaska, Canada, Norway and Russia. The chapter serves as an introductory chapter to the case studies and is organized around three main themes or sets of questions. The first part poses a set of questions about the form of government in each of the four jurisdictions, where three are federal states (Canada, Russia and the United States), while Norway has a unitary form of government.

The second set of questions deals with the ownership of land and resources, particularly the division between public ownership and private ownership. The premise here is that the bundle of rights associated with ownership provides both an important source of power or authority within society, and a place where the owner may have not only a greater degree of protection from state authority, but also a space within which to exercise autonomy. The oil and gas sector will always need to know who the owner of the resource is and with whom the industry must negotiate for a lease or concession. The third set of questions deals with the legal status of indigenous peoples and their land and resource claims in each of the four jurisdictions. Here, the sorts of questions that are posed focus upon the recognition of indigenous rights to land and resources, and the implications of this for the oil and gas sector.

In Chapter 7, *Mikkelsen, Haley* and *Øygarden* present the Alaskan case study. It is argued that the oil and gas activities in Alaska are expanding *in* the Arctic, not *to* the Arctic. While the Alaska economy has always centred on resource extraction, the discovery of Prudhoe Bay oil in 1969 was worth seven times more than all previous resources combined. The main conflict lines in Alaska are on how oil and gas activities can be combined with the subsistence activities that the indigenous people have defined as the foundation of their identity and culture. At the same time, the communities are highly dependent on revenues from oil and gas. With the advent of oil wealth, the North Slope society was modernized and increasingly integrated with the global cash economy.

To maintain new infrastructure and to afford modern equipment and lifestyles – like using computers and satellites in bowhead whaling – declining revenues must be shored up with industrial expansion. In this situation, onshore oil and activities have broad support. The oil and gas industry now, however, want to expand their activities into offshore areas. As noise from this activity may deflect the whales and other mammals, and spills are a risk to the marine environment, expansion offshore is perceived as a threat to subsistence activities, and local opposition is strong. The regional authorities and corporations, however, need the revenues and business to maintain the present infrastructure and socio-economic standard, and to cope with modern lifestyle problems, such as diabetes, mental health problems, drug abuse and unemployment. The oil and gas companies are challenged to take responsibility for the situation and to mitigate the negative impacts and provide a positive future for the coming generations.

In Chapter 8, *Mason, Anderson* and *Dana* present the Canadian case study. Canada is a major oil and natural gas producer and exporter. Exploration and production have been centered in Alberta, with significant additional activity in the neighbouring province of Saskatchewan. More recently, the Hibernia field in the Atlantic Ocean offshore of Newfoundland has been brought into production. Exploration continues offshore on both the east and west coasts.

Oil and gas activity in the Northwest Territories and Nunavut, the portion of Canada north of 60 degrees, has centred on the Beaufort Sea in the Western Arctic, where vast proven reserves exist. The challenge is to get the product to market. The preferred solution has been a pipeline down the Mackenzie River Valley, a route that crosses the traditional land of four indigenous groups: the Inuvialuit, the Sahtu, the Gwich'in and the Deh Cho. In the context of the petroleum industry in Canada, this chapter details the events surrounding the Beaufort Sea field and the ongoing efforts to build the Mackenzie Valley pipeline. There are four interwoven themes finding expression in the ongoing pipeline saga. First is the oil and gas exploration development and supply agenda as part of the drive to increase North American supply for the US market for security and other reasons. Second is the issue of indigenous rights to land and resources, and the modern land claim and treaty-making process in Arctic Canada. Third is the ongoing struggle to improve the socio-economic circumstances of the people of Canada's Arctic through development. Finally, the fourth is the issue of sustainability, the traditional economy and alternative development strategies.

The Norwegian case is presented by *Heitmann Hansen* and *Midtgard* in Chapter 9. Norway has enormous ocean areas in the North where few petroleum resources have been detected, where the environmental tensions have been high, and where economic depression characterizes the coastal communities. The Snøhvit gas field is the first petroleum project in the Barents Sea. The Norwegian government has also opened up for exploration in the Barents South, and the oil project Goliat is in the phase of being developed. The discussions about expansion of the oil and gas activities to Arctic Norway can be separated into four main discourses: the environmental, the regional economic, the rights and situation of the Sámi minority and the international. The environmental issues have dominated historically, but lately this dimension has shifted increasingly to a combination of regional economic benefits and how to utilize renewable and non-renewable resources and opportunities in a sustainable way.

The different fishermen's federations have been relieved by, and seem confident with, the general opinion that oil activity is compatible with fishing. In this context, the Norwegian Sámi people have recently raised their voices to claim a right to participate in the decision-making process and a right to benefits from oil and gas revenues. In the debates about how to manage Norwegian oil and gas resources, we have a link to international obligations and responsibilities, and the foreign policy about oil and gas in Norway contains elements of cooperation with Russia and Europe. Through collaboration with Russia, Norway is seeking a comprehensive and coherent development of the Barents Sea as a petroleum province. In this region, Norway wishes to address both the sustainable development of living resources, including some of the world's most precious fish stock, and the impact of climate change. And it is here that Norway seeks to develop the new petroleum province of Europe.

In Chapter 10, *Andreyeva* and *Kryukov* present the Russian case. It is an embedded case study devoted to the large changes that the country is undergoing now: changes in administrative and resource policy; changes in the interaction between business and federal structures; changes in regional opportunities to use money from resource development; and changes in the attitudes of indigenous peoples regarding the struggle for their rights to land and natural resources. These processes are very similar to those that occurred in many countries over the last two centuries. For Russia, the point is that the timeframe for such changes has tightened, and all – from government and legislative bodies to social and business groups – must adapt quickly to the new rules and requirements while simultaneously building new relations and responding to all the changes.

The large-scale transformation of Russian society is not without contradictions. Business groups are the most active part of society, and they must develop their operations in a very complex and tense social environment. Oil and gas are seen as key strategic resources and play an extremely important role in modern Russia. These companies' activities are under the close scrutiny of different parties, as is their cooperation with foreign partners.

The legislative process on key questions of resource use is not completed yet, which creates some uncertainties for stakeholders. Issues of sustainable development are steadily being included in governmental policy and have become the key questions for regional authorities in their programmes of social and economic development. Unfortunately, there are huge obstacles to the actual implementation of sustainable approaches to resource development and to regional policies regarding aboriginal peoples and their vital interests. Two sub-case studies exemplify this development. One case is from the Murmansk region; the other case is from the Yamal-Nenets Okrug. A new state law, effective in 2007, that will concentrate all income from oil and gas activities to the federal level adds to the uncertainties in Russia. Although money is to be distributed among all regions and spheres of national economy, the unknown effects of this new law create new worries. In addition, the formation of a legislative basis for oil and gas activity is not completed yet, which causes many problems for all stakeholders, including foreign companies.

Chapter 11 is the first chapter in Part III of the book. In this chapter, *Fred Hansen* and *Bankes* give a broad introduction to human rights and the indigenous people of the Arctic. It examines the interplay between the human rights of indigenous peoples and the interests of the oil and gas industry in being able to carry out exploration, production and transportation of oil and gas within the traditional territories of those indigenous peoples. The chapter focuses on the human rights of indigenous peoples under international law. It also tries to assess how these rights in international law form part of the discourse within the domestic legal and political systems of each of the four states, Canada, Norway, Russia and Alaska (US), and how these discourses may influence the commitments of the oil and gas industry towards ideas of corporate social responsibility.

Within the field of international human rights law, the focus is on the rights of indigenous peoples in relation to the lands and resources within their traditional territories. As a starting point, it is acknowledged that there is a close relationship between the right of all peoples (including indigenous peoples) to self-determination, and that the two International Covenants also recognize the inherent right of all people to their natural wealth and resources and their right not be deprived of their own means of subsistence. While the focus is not on the right of self-determination and its implications for claims to autonomy within the nation state, it is recognized that states party to the International Covenant on Civil and Political Rights have a duty to report on their implementation of this right, especially as it applies to indigenous peoples.

Chapter 12 brings together the material from the four countries – Alaska, Canada, Norway and Russia – in a comparative context and structures the analysis from the perspective of sustainable development. This chapter by *Langhelle* and *Fred Hansen*, identifies, highlights and discusses differences and similarities in more detail within the overall research questions raised in the introduction: What is sustainable development believed to entail, and what does it imply for oil and gas activities in the Arctic?

From the country studies, it is evident that little of the debate is explicitly structured in terms of sustainable development. At the same time, it is clear that sustainable development has become fairly widespread and accepted in the political vocabulary in all the countries, although there are noticeable differences – and similarities – between the countries. Development concerns are strong in all countries, and the main story line identified is the perception that the key sustainability challenge seen from the Arctic is how to ensure that more of the profits from resource extraction actually remain in the Arctic.

The issues of biological diversity (conservation, sustainable use) dominate the environmental debate in all countries. Climate change is an issue in all countries, although to a varying degree, and primarily from the perspective of *impacts* from climate changes in the Arctic. In Canada and Norway, however, Arctic oil and gas production is also seen as a *cause* of climate change. Expanding oil and gas activities in the Arctic will only continue a development path based on fossil fuels. Fossil fuel production should therefore be constrained. Norway is probably the country with the most opposition towards oil and gas in the Arctic. In Canada and Alaska, opposition is primarily to be found amongst some indigenous groups and environmental NGOs. In Russia, there is hardly any opposition to oil and gas activities, except from some indigenous groups and some small NGOs. State control of oil and gas resources is a major concern and policy goal. The last section in the chapter discusses the implications from the comparative study for future Arctic oil and gas.

In the last chapter (Chapter 13), *Mikkelsen, Camp* and *Anderson* set out the major contextual forces that managers should consider when deciding on strategy and actions with respect to oil and gas operations in the Arctic and, more generally; it then makes suggestions as to effective responses to these contextual forces. This is done in three sections. After the introductory section, the second revisits corporate social responsibility and proposes a matrix of four possible approaches (classical, social demand, stakeholder and sustainable development) from the viewpoint of the strategic importance of socially responsible behaviour and the nature of that behaviour. Section three addresses the different and often conflicting expectations of the different actor groups about appropriate corporate behaviour from the perspective of the community, the state and the environmental actors. Then in section four, the managerial implications of the preceding sections on the oil and gas industry at the various geographic scales, with a primary focus on the Arctic, are discussed.

The chapter explores possible approaches to CSR and sustainable development, for example: (i) whether to lead or lag relative to competitors and demands from other actors; (ii) whether to treat CSR as a constraint or a potential source of competitive advantage; and (iii) whether to treat the demands of actors as elements of a key strategic issue or as tactical issues differing by geographic scale and actors. In general, we conclude that it is better to lead than lag, and by doing so, to treat CSR as a source of competitive advantage, while simultaneously encouraging collaboration across companies to allow for integrated management planning in vulnerable areas. To do this, CSR and sustainable development must be treated as a key strategic issue and not a tactical one, although the corporate approach must, of course, be in response to the particular issues at the national, regional and local levels.

Notes

1 The Arctic Council is an intergovernmental forum for addressing many of the common concerns and challenges faced by the Arctic States: Canada, Denmark (including Greenland and the Faroe Islands), Finland, Iceland, Norway, the Russian Federation, Sweden and the United States. The Council is a unique forum for cooperation between national governments and indigenous peoples. Six international organizations representing many Arctic indigenous communities have the status of Permanent Participants of the Arctic Council and are involved in the work of the Council in full consultation with governments.
2 The Declaration on the Establishment of the Arctic Council, signed at Ottawa, Ontario, Canada, on September 19, 1996 (the 'Ottawa Declaration').
3 Dagens Næringsliv, August 8, 2006.
4 The numbers should be used with caution. The figure is based only on areas that are assessed by the USGS, and several areas are not part of the analysis. This is the case for the Norwegian part of the Barents Sea, for example.
5 As Shell did in 2004. About one-fourth of the oil and gas reserves reported in Shell's books did not exist. Komp, L. (2004) Enron, Parmelat, Shell Oil: Who will be next? *Executive Intelligence Review*, May 7.

References

Arctic Council (2004) *Arctic Human Development Report*, Akureyri, Iceland: Stefansson Arctic Institute.
Bailey, A. (2007) 'USGS: 25% Arctic oil, gas estimate a reporter's mistake', *Petroleum News*, 12 (42). Online. Available HTTP: <http://www.petroleumnews.com/pntruncate/347702651.shtml> (Accessed 20 November 2007).
Dutton, J. E. and Dukerich, J. M. (1991) 'Keeping an eye on the mirror – image and identity in organizational adaptation', *Academy of Management Journal*, 34 (3): 517–554.
Flanagan, J. (1954) 'The critical incident technique', *Psychological Bulletin*, 51 (4): 76–90.
Hajer, M. A. (1995) *The Politics of Environmental Discourse: Ecological Modernization and the Policy Process*, Oxford: Oxford University Press.
Hoffmann A. J. and Ocasio W. (2001) 'Not all events are attended equally: toward a middle-range theory of industry attention to external events', *Organization Science*, 12 (4): 414–434.
Lafferty, W. M. and Langhelle, O. (1999) *Towards Sustainable Development. On the Goals of Development and the Conditions of Sustainability*, Basingstoke: Macmillan Press Ltd.
Langhelle, O. (2000) 'Sustainable development and social justice: expanding the Rawlsian framework of global justice', *Environmental Values*, 9 (3): 295–323.
Mikkelsen, A., Engen, O. A. and Grønhaug, K. (2008) 'Consequences of critical events for the social construction of corporate social responsibility: the case of oil and gas companies in Norway', *Scandinavian Journal of Management* (submitted).
Pride, R. A. (1995), 'How activists frame social problems: critical events versus performance trends for schools', *Political Communication*, 12: 5–26.
US Geological Survey (USGS) *World Petroleum Assessment 2000*. Available HTTP: <http://energy.cr.usgs.gov/> (Accessed 8 April 2008).
World Commission on Environment and Development (WCED; 1987) *Our Common Future*, Oxford: Oxford University Press.

Part I
The Arctic
Context, framework and methodology

2 Framing oil and gas in the Arctic from a sustainable development perspective

Oluf Langhelle, Bjørn-Tore Blindheim and Olaug Øygarden

Introduction

Sustainable development has become the core idiom framing international and national debates about environment and development. From the release of the report *Our Common Future* (1987) by the World Commission on Environment and Development (WCED), sustainable development has been endorsed by world leaders on several occasions, most importantly at the Rio Earth Summit in 1992, and at the Johannesburg Summit in 2002. Sustainable development has also become a part of the conceptual framework of international organizations like the Organisation for Economic Co-operation and Development (OECD), the World Trade Organization (WTO) and the World Bank, and achieved a near constitutional status in the European Union. In the countries we are studying, sustainable development has figured prominently on the political agenda, especially in Norway and Canada, and to some lesser extent in the United States (US) and Russia. However, even in the US and Russia, there are signs of an increasing focus on sustainable development.

The focus on sustainable development is especially visible in the cooperative efforts between the Arctic states in the Arctic Council. In fact, sustainable development constitutes *the* objective of the Arctic Council: 'The Arctic Council is a regional forum for sustainable development, mandated to address all three of its main pillars: the environmental, social and economic'. The Arctic Council established a Working Group on Sustainable Development (WGSD) in 2000, at the Ministerial Meeting in Barrow, Alaska. At the same meeting, the Arctic Council also approved a strategic framework document on sustainable development. In the 'Programme of the Russian Federation Chairmanship of the Arctic Council in 2004–2006' (Arctic Council, 2004c), the Arctic Council is seen as a mechanism of implementing sustainable development in the Arctic:

> ... the Arctic Council has demonstrated that as far as the Arctic region is concerned it is the main mechanism for implementing the principles of sustainable development set forth in the Program of Action on the Implementation of the Agenda 21 adopted by the UN Conference on Environment and Development in 1992 in Rio-de-Janeiro and the decisions

taken by the World Summit on Sustainable Development in 2002 in Johannesburg.

As such, sustainable development has been placed as the core political objective for the Arctic by the Arctic states. But what does the concept of sustainable development entail and what does it imply? And following from these questions: What are the implications of sustainable development for oil and gas activities in the Arctic? Is it possible to speak about 'sustainable' oil and gas exploration at all? What does it imply to 'frame' the Arctic from a sustainable development perspective?

This chapter provides some initial answers to these questions. To be able draw actual policy implications from sustainable development, there are a number of issues that have to be resolved. These issues can be organized along two axes, one horizontal and one vertical. The horizontal axis relates to the debates about what sustainable development means, what should be included in the concept, and how the different dimensions of sustainable development should be weighted. This we refer to as the horizontal framing of sustainable development. On the other hand, it relates to whether one operates at a global, regional or local level. *Our Common Future* first and foremost addressed the global environment and development concerns, and these global concerns have to be 'translated' to a regional, national or local setting. The challenges of sustainable development may vary across these levels, but they may also interact. The issue of climate change is a good example. Although it is first and foremost conceived as a global problem, it has regional and local effects, and it also requires regional and local efforts. How the global sustainable development agenda is 'translated' to regional, national and local levels is referred to as the vertical framing of sustainable development.

This chapter, therefore, addresses the above issues in two ways, along the vertical and the horizontal axes. In the horizontal, we explore the background of the concept of sustainable development. In the vertical, we present an overview of how sustainable development has been conceived within the Arctic, most notably from the political processes related to the Arctic Council. Hence, the point of departure here is not the global sustainable development agenda as such, but the problems that have warranted attention from a regional and to some degree, local perspective. Finally, we look at how oil and gas has been addressed within the Arctic Council.

The global framing of sustainable development

The report from the Brundtland Commission has been referred to as 'the key statement of sustainable development' (Kirkby, O'Keefe and Timberlake, 1995: 1). Although sustainable development is a contested concept, much of the international and also national follow-up has been firmly located within the context of the Brundtland report and what is usually referred to as 'the UNCED-process' (Lafferty, 1996; Jacobs, 1999; Lafferty and Langhelle, 1999; Lafferty and

Meadowcroft, 2000; Langhelle, 2000; Dresner, 2002). The decision to convene the United Nations Conference on Environment and Development (UNCED) was formally taken by the General Assembly, 22 December, 1989 (GA/ res. 44/228) (United Nations, 1989; Mugaas, 1997; Lafferty and Meadowcroft, 2000). Among the purposes of the conference was:

> To recommend measures to be taken at the national and international levels to protect and enhance the environment, taking into account the specific needs of developing countries, through the development and implementation of policies for sustainable and environmentally sound development with special emphasis on incorporating environmental concerns in the economic and social development process and of various sectoral policies and through, inter alia, preventive action at the sources of environmental degradation, clearly identifying the sources of such degradation and appropriate remedial measures, in all countries.
> (United Nations, General Assembly, GA/res. 44/228, 22 December, 1989)

The United Nations Conference on Environment and Development (UNCED), took place in Rio De Janeiro in June 1992 and resulted in five major documents, referred to as 'the Rio Accords': the Rio Declaration on Environment and Development; the Framework Convention on Climate Change; the Convention on Biodiversity; a Statement of Forest Principles; and Agenda 21. Agenda 21 was described as a 'Programme of Action for Sustainable Development' (Grubb, Koch, Thomson et al., 1993; Lafferty and Meadowcroft, 2000). In the 'Preamble' of Agenda 21, the opening paragraph describes the problems – and partly the solution – in the following way:

> Humanity stands at a defining moment in history. We are confronted with a perpetuation of disparities between and within nations, a worsening of poverty, hunger, ill health and illiteracy, and the continuing deterioration of the ecosystems on which we depend for our well-being. However, integration of environment and development concerns and greater attention to them will lead to the fulfilment of basic needs, improved living standards for all, better protected and managed ecosystems and a safer, more prosperous future. No nation can achieve this on its own; but together we can – in a global partnership for sustainable development.
> (United Nations, 1993: 12)

As pointed out by Lafferty and Meadowcroft (2000: 13), sustainable development was never formally defined in the UNCED outputs, and sustainable development is treated somewhat differently in the various UNCED documents. As Lafferty and Meadowcroft (2000: 13) argue, this was no doubt a reflection of the different forums in which the texts were negotiated. It could be argued, however, that the different treatment of sustainable development also reflects different opinions of how sustainable development should be understood. Nonetheless, the meaning of sustainable development was taken as essentially given and deriving from the Brundtland report (Lafferty and Meadowcroft, 2000: 13). To understand the concept of sustainable development within the UNCED process, therefore, it is necessary to turn to the Brundtland report.

Our Common Future

The Brundtland Commission was not the first to summon 'sustainable development', but there are different opinions concerning the origin of the concept (O'Riordan, 1993; Worster, 1993; Jacob, 1996; McManus, 1996; Langhelle, 2000; Dresner, 2002). The *World Conservation Strategy* (1980), published jointly by the International Union for Conservation of Nature and Natural Resources (IUCN), the World-Wide Fund for Nature (WWF) and the United Nations Environment Programme (UNEP), is often seen as one of the first to make use of the term, but the earliest expression, to our knowledge, relates to work done within the World Council of Churches in the early seventies. The following, which could have been a quotation from *Our Common Future* (WECD, 1987), is actually from a report made by a working group within the World Council of Churches in 1976:

> The twin issues around which the world's future revolves are justice and ecology. 'Justice' points to the necessity of correcting misdistribution of the products of the Earth and of bridging the gap between rich and poor countries. 'Ecology' points to humanity's dependence upon the Earth. Society must be so organised as to sustain the Earth so that a sufficient quality of material and cultural life for humanity may itself be sustained indefinitely. A sustainable society which is unjust can hardly be worth sustaining. A just society that is unsustainable is self-defeating. Humanity now has the responsibility to make a deliberate transition to a just and sustainable global society.
> (Cited from Birch et al., 1979; Langhelle, 2000)

Although this report speaks of a transition to a just and sustainable global society, not sustainable development, there are several stipulations made on the relationships between social justice and physical sustainability. On the one hand, justice is here not necessarily functional for physical sustainability: a just society can be unsustainable. On the other hand, a sustainable society that is unjust is not worth sustaining. In one sense, this can be seen as a contradiction, but it need not be. It implies a view that justice and physical sustainability in principle and practice can be reconciled.

What the Brundtland report accomplished was to 'relaunch' sustainable development by casting it in a form that could appeal to a wide range of political actors, and at the same time deriving legitimacy from the UN-sponsored process through which it had been formulated. As such, the report drew together diverse strands of the international discourses of environment and development. As we shall argue, however, in drawing these diverse strands together, the Commission made some crucial changes from the more traditional conservationist usage in the

World Conservation Strategy (IUCN/WWF/UNEP, 1980), changes which have implications for the framing of oil and gas extraction in the Arctic.

The World Commission: reframing environment and development concerns

The first reframing that was made by the World Commission was *the merging* of environmental and development concerns. As Gro Harlem Brundtland points out in the introduction of the report, when the terms of the mandate for the World Commission were being discussed, some wanted to limit its concerns to environmental issues only. This would, according to Brundtland, be 'a grave mistake' because

> The environment does not exist as a sphere separate from human actions, ambitions, and needs, and attempts to defend it in isolation from human concerns have given the very word 'environment' a connotation of naivety in some political circles ... But the 'environment' is where we all live; and development is what we all do in attempting to improve our lot within that abode. The two are inseparable.
> (WCED, 1987: xi)

The second reframing that was made was *the ordering* of environment and development. Sustainable development starts with people. Compared with the *World Conservation Strategy* (IUCN/WWF/UNEP, 1980), Adams (1990: 59) argues that '*Our Common Future* starts with people, and goes on to discuss what kind of environmental policies are required to achieve certain socio-economic goals', while the *World Conservation Strategy* 'started from the premise of the need to conserve ecosystems and sought to demonstrate why this made good economic sense (and although the point was underplayed – could promote equity)'. The orientation towards people obviously contributed to a more development-oriented approach within the Commission, leading a number of commentators to portray the report as strongly anthropocentric (Adams, 1990; Kirkby, O'Keefe and Timberlake, 1995; Lafferty and Langhelle, 1999; Reid, 1995).[1]

The World Commission defined sustainable development as 'development that meets the needs of the present without compromising the ability of future generations to meet their own needs'. This definition, according to the report, contains within it two key concepts:

– the concept of 'needs', in particular the essential needs of the world's poor to which over-riding priority should be given; and
– the idea of limitations imposed by the state of technology and social organization on the environment's ability to meet present and future needs. (WCED, 1987: 43)

Seen in the light of common usages of the word 'sustainable', this reframing has further important implications, also for how oil and gas activities *can be* viewed from a sustainable development perspective in the Arctic. The implications of the reframing are most easily seen when contrasted with other usages of the term 'sustainable'. The term itself is derived from the Latin *sus tenere*, meaning 'to uphold' (Redclift, 1993), which does not, of course, imply any specific normative content. Dixon and Fallon (1989) provide a basis for further clarification by identifying three types of usage.

1. As a *purely physical concept for a single resource*. Here the scope of sustainability is limited to particular renewable resources considered in isolation, with sustainability simply implying a usage no greater than the annual increase in the resource, without reducing the physical stock. Maximum sustainable yield, maximum sustainable cut, and so forth, are examples of the underlying logic of this approach. It is also in this usage that the exploitation of oil and gas by definition is unsustainable. Any consumption will reduce the stock and hence be unsustainable in the long run.
2. As a *physical concept for a group of resources, or an ecosystem*. With this usage, explicit attention is devoted to different aspects of the ecosystem. For example, forestry managed in accordance with maximum sustainable cut may create problems with increased soil erosion, changes in water yield, wildlife habitat and species diversity. As a result of system interaction, what may be considered as a sustainable usage of a single resource may actually be found to be unsustainable within the context of the entire system. From this usage, it is possible to argue that oil and gas activities *could have* profound negative impacts on the ecosystem as a whole. Hence, the activities *can be* unsustainable if the goal is to sustain the ecosystem as such.
3. As a *socio-economic physical concept*. Here the goal is not a sustained level of a physical stock or the physical production of a given ecosystem, but an unspecified 'sustained increase in the level of societal and individual welfare' (Dixon and Fallon, 1989: 6) or, more directly, in accord with the language of *Our Common Future*, a sustained level of need satisfaction and equal opportunities.

This third type is the usage developed by *Our Common Future* (Dixon and Fallon, 1989; Langhelle, 1999), and it is clear that oil and gas *can be* exploited in a meaningful sense within this usage of sustainable development, since what is to be sustained is not the resource itself but the level of need satisfaction and equal opportunities. A further consequence of the above framework, is that it is not, as Meadowcroft also notes, a particular institution, nor a specific pattern of activity, nor a given environmental asset which is supposed to be sustained, but rather *a process*, the process of development (Meadowcroft, 1997).

Sustainable development, understood as a process of development which is to be sustained, leads to the consequence that an activity which is not itself sustainable could be a part of an ongoing process which *is* sustainable (Meadowcroft, 1997). This applies not only to social behaviour, but also to activities like the consumption of renewable and non-renewable resources. Thus, contrary to the first two usages, the third usage of sustainability, identified by Dixon and Fallon (1989), may have implications for the use of renewable resources contradictory to the first two.

The environmental dimension of sustainable development

Where is the environment in this framing of sustainable development? The environment, so to speak, comes in the back door. The qualification that development (a sustained level of need satisfaction and equal opportunities) must also be sustainable is a constraint placed on this goal, meaning that each generation is permitted to pursue its interests only in ways that do not undermine the ability of future generations to meet their own needs. Hence, environmental sustainability (or physical sustainability) is not the primary goal of development, but a *precondition* for this goal in the long term and for justice between generations. Thus, physical sustainability becomes an inherent part of the *goal* of sustainable development: 'At a minimum, sustainable development must not endanger the natural systems that support life on Earth: the atmosphere, the waters, the soils, and the living beings' (WCED, 1987: 44–45).

This, however, also implies that not every environmental problem is necessarily a sustainable development issue. This is important because it partly determines *the scope* of sustainable development and how environmental effects are to be judged from a sustainable development perspective. The first is a reflection of the fact that the World Commission's primary concern was the *global* environmental (and developmental) problems. The second is a reflection of the minimum requirement and the notion of 'endanger'. The point is concisely put by Malnes:

> A policy that procures development by dint of damages to the environment violates the proviso only insofar as these damages are detrimental to future development. Policies are not to be judged on account of their environmental effects as such. This is the World Commission's point of departure.
> (Malnes, 1990: 7)

It is thus as a prerequisite for development that the injunction to conserve plants and animals in *Our Common Future* must initially be understood. It is because the environment is vulnerable to destruction through development itself that the constraint of sustainability is placed on the goal of development (WCED, 1987: 46; Malnes, 1990: 5). As such, an activity with negative environmental effects is not necessarily contradictory to sustainable development, but it can be.

But which environmental issues are not peripheral to sustainable development? Arguably, the environmental problems of sustainable development will differ along the vertical axes of sustainable development. At the regional and local level, environmental problems that threaten human need satisfaction today are no doubt sustainable development issues. They threaten the first priority of sustainable development, human need satisfaction and 'the essential needs of the world's poor to which overriding priority should be given'.

From the specific framing of sustainable development in *Our Common Future*, therefore, environmental problems that threaten human need satisfaction, today or in the future, are the subject of sustainable development. Environmental problems that endanger the natural systems that support life on Earth – the atmosphere, waters, soils and the living beings – relate to sustainable development. Accordingly, the environment, which is vulnerable to destruction through development, is itself part of the sustainable development agenda.

The global challenges of sustainable development

In the follow-up UNCED process, Agenda 21, adopted at the Rio Conference in 1992, addressed a number of issues under the umbrella of sustainable development. The first section of Agenda 21 addressed the social and economic dimensions of sustainable development under the headings of international cooperation, combating poverty, changing consumption patterns, demographic dynamics, protecting and promoting human health, promoting sustainable human settlement development and integrating environment and development in decision-making.

The second section was called 'Conservation and management of resources for development', and focused on the atmosphere, land resources, deforestation, fragile ecosystems, agriculture and rural development, conservation of biological diversity, biotechnology, protection of the oceans, freshwater resources, toxic chemicals, hazardous wastes, solid wastes and radioactive wastes (United Nations, 1993: Chapter 9–22).

All the above problems are no doubt important sustainable development issues. There are, however, two environmental problems in particular that were seen as crucial to sustainable development in *Our Common Future*: climate change and loss of biological diversity. Of the 'Rio Accords', these problems were the only problems that materialized as international Conventions and thus 'hard law'. These issues are not solely environmental issues but sustainable development issues. In both cases, development conflicts with the environment. Climate change and loss of biodiversity may threaten human need satisfaction, today or in the future, both problems may endanger the natural systems that support life on Earth, and both problems are closely associated with development itself.

Climate change implies the potential for human activities to alter the Earth's climate to an extent unprecedented in human history with severe but also unpredictable and unknown consequences for humans and ecosystems. Loss of biological diversity is perceived as a global problem that presents a conflict between the short-term economic interests of nations and the long-term interests of sustainable development (WCED, 1987: 160). The dilemma is, of course, that given the first priority of sustainable development (meeting the essential needs of the world's poor, to which overriding priority should be given) and expected population growth rates, 'a five- to tenfold increase in manufacturing output will be needed just in order to raise developing-world consumption of manufactured goods to industrialised world levels by the time population growth rates level off next century' (WCED, 1987: 15).

How is this to be done without compromising the needs and opportunities of present and future generations when it comes to climate change and biological diversity? How is the conflict between intra- and intergenerational justice to be solved? These questions illustrate what John Dryzek (1997) calls *the core story line* of sustainable development:

> The core story line of sustainable development begins with a recognition that the legitimate developmental aspirations of the world's peoples cannot be met by all countries following the growth path already taken by the industrialized countries, for *such action would over-burden the world's ecosystems* ... sustainable development is not just a strategy for the future of developing societies, but also for industrialized societies, which must reduce the excessive stress their past economic growth has imposed upon the earth.
> (Dryzek, 1997: 129)

This framing of the problem also goes to the heart of both the UN Framework Convention on Climate Change and the Convention on biological diversity (Langhelle, 2000).

The World Commission argues that different limits hold for the use of energy, materials, water and land (WCED, 1987: 45). The limits to be met first are the availability of energy and the biosphere's capacity to absorb the by-products of energy use. These limits may be approached far sooner than the limits imposed by other material resources, because of the depletion of oil reserves and carbon dioxide build-up leading to global warming (WCED, 1987: 58–59).[2] Indeed, this situation may already be imminent. Climate change thus constituted the first limit to global development in *Our Common Future*. Climate change, therefore, is fundamental to sustainable development (Langhelle, 1999).

The most radical changes proposed by the Commission in this context were based on climate change as perhaps the most pressing developmental constraint. To avoid an ecological disaster owing to the problem of climate change, *Our Common Future* recommends a low-energy scenario of 'a 50 per cent reduction in primary energy consumption per capita in industrial countries, to allow for a 30 per cent increase in developing countries within the next 50 years' (WCED, 1987: 173). This 'will require profound structural changes in socio-economic and institutional arrangements and it is an important challenge to global society' (WCED, 1987: 201). Further, 'the Commission believes that there is no other realistic option open to the world for the 21st century' (WCED, 1987: 174).

Our Common Future argued that 'the total expanse of protected areas needs to be at least tripled if it is to constitute a representative sample of Earth's ecosystems'. This was described as a 'prerequisite for sustainable development', with the further remark that 'failure to do so will not be forgiven by future generations' (WCED, 1987: 166). The current target set by the now 168 states that have signed the Convention is to have 'at least 10 per cent of each of the world's ecological regions effectively conserved', and to protect areas 'with particular importance to biological diversity' [Convention on Biological Diversity (CBD)/COP7,

Decision VII/30, 2004]. By 2010, terrestrially, and in 2012 in the marine area, 'a global network of comprehensive, representative and effectively managed national and regional protected area system' is to be established.

Sustainable development and indigenous people

Indigenous people were addressed in a general way in *Our Common Future*, mainly under the heading of 'Empowering Vulnerable Groups' (WCED, 1987: 114). Although the treatment of indigenous people only constituted three pages in the report, the policy prescriptions and conclusions were quite strong. Within the framework of sustainable development, it is not only the environment that was seen as vulnerable to destruction through development itself, but also indigenous people. Indigenous people had 'become victims of what could be described as cultural extinction' (WCED, 1987: 114), and also the victims of social discrimination, cultural barriers, exclusion from national political processes and exclusion from the processes of economic development. A more 'careful and sensitive consideration of their interests', therefore, was described as 'a touchstone of a sustainable development policy' as opposed to 'insensitive development'.

In addition to the recognition of traditional rights and empowering of indigenous people, *Our Common Future* argued that this should be accompanied 'by positive measures to enhance the well-being of the community in ways appropriate to the group's lifestyle' (WCED, 1987: 116). The challenge is to find the 'fine line between keeping them in artificial, perhaps unwanted isolation and wantonly destroying their life-styles' (WCED, 1987: 116). This did not, however, exclude broader measures of human development. The provision of health facilities, improvement of traditional practices, correction of nutritional deficiencies, the establishment of educational institutions and economic development where the local communities 'derive full benefit from the projects, particularly through jobs' were the main recommendations. Where to draw the 'the fine line' was by no means fully answered in *Our Common Future*. In the recognition of traditional knowledge, however, the report acknowledged a 'terrible irony that as formal development reaches more deeply into rain forests, deserts, and other isolated environments, it tends to destroy the only cultures that have proved able to thrive in these environments' (WCED, 1987: 115).

The Rio Summit further established and strengthened the concerns of indigenous people within the framework of sustainable development. Indigenous people's concerns were addressed in their own chapter in Agenda 21, and the Principle 22 in the Rio Declaration reads as follows:

> Indigenous people and their communities and other local communities have a vital role in environmental management and development because of their knowledge and traditional practices. States should recognize and duly support their identity, culture and interests and enable their effective participation in the achievement of sustainable development.
>
> (United Nations, 1993: 7)

The 'vital role of the indigenous peoples' was reaffirmed in the Johannesburg Declaration on Sustainable Development in 2002 (United Nations, 2002).

Sustainable development in the Arctic

What then about sustainable development in the Arctic? How should sustainable development be translated to the Arctic – what we have called the vertical framing? What are the implications from the global approach to sustainable development for the Arctic as a region? And if the question is posed the other way round, what are the specific challenges for sustainable development seen from the Arctic? Here we have to turn to other sources than the UNCED process. As Heininen remarked in the Arctic Human Development Report (AHDR), 'not all efforts to highlight Arctic concerns in international fora have been particularly effective' (Heininen, 2004: 211). The Arctic Council, however, has made several attempts to specify and translate sustainable development to the Arctic.

The official Intergovernmental Arctic cooperation started in the environmental sphere. It is often traced back to 1989 and the initiative from the Government in Finland to discuss cooperative measures to protect the Arctic environment. This again, was inspired by Mikhail Gorbachev's Murmansk speech two years earlier, October 2, 1987. Gorbachev's speech advanced a wide range of topics for dialogue and negotiation with other Arctic and North Atlantic states, including economic cooperation, environmental issues and indigenous people (Scrivener, 1989, 1996, 1999; Archer and Scrivener, 2000; Vanderzwaag, Huebert and Ferrara, 2003; AHDR, 2004; Koivurova and Vanderzwaag, 2007). The Finnish Government initiative culminated with the Arctic Environmental Protection Strategy (AEPS) established in 1991,[3] a strategy that was politically – but not legally – binding for the signatories (Haavisto, 2001; Koivurova and Vanderzwaag, 2007).

The AEPS strategy contained an explicit reference to *Our Common Future* and, although the strategy primarily addressed the environmental challenges, several of the paragraphs in the Strategy were highly influenced by the Brundtland report. It was explicitly stated that the Strategy 'should allow for sustainable economic development in the North so that such development does not have unacceptable ecological and cultural impacts' (AEPS, 1991: 6–7). The objectives of the Strategy were outlined as follows:

 i To protect the Arctic ecosystem including humans;
 ii To provide the protection, enhancement and restoration of environmental quality and the sustainable utilization of resources, including their use by local populations and indigenous people in the Arctic;
 iii To recognize and, to the extent possible, seek to accommodate the traditional and cultural needs, values and practices of the indigenous people as determined by themselves, related to the protection of the Arctic environment;
 iv To review regularly the state of the Arctic environment;
 v To identify, reduce and, as a final goal, eliminate pollution. (AEPS, 1991: 9)

In addition, the Strategy outlined a number of guiding principles which were highly influenced by *Our Common Future*. Among them the following general principle: 'Management, planning and development activities shall provide for the conservation, sustainable utilization and protection of Arctic ecosystems and natural resources for the benefit and enjoyment of present and future generations, including indigenous peoples' (AEPS, 1991: 9–10). Apart from general principles such as the above, however, the main focus in the AEPS Strategy was directed towards environmental problems. Since global warming and ozone depletion were being addressed in 'other fora' (within the UNCED process and the Montreal Protocol), these issues were not addressed in AEPS. The strategy identified and focused upon six issues: persistent organic contaminants, oil pollution, heavy metals, noise, radioactivity and acidification.

As part of the AEPS, the Arctic states also agreed to establish Working Groups in four programme areas, where member states, observers and indigenous groups could participate in the work.

- The Arctic Monitoring and Assessment Program (AMAP) was established to identify the levels and effects of selected anthropogenic pollutants in all compartments of the Arctic, including humans.
- A Working Group on the Conservation of Arctic Flora and Fauna (CAFF) was established to address species and habitat conservation.
- A Working Group on the Protection of the Arctic Marine Environment (PAME) was established to inventory the threats to the Arctic marine environment irrespective of origin and to identify gaps in the coverage of existing international Multilateral Environmental Agreements (MEAs) concerning the Arctic.
- A Working Group on Emergency Preparedness and Response (EPPR) was established to provide a framework for cooperation in response to the threat of environmental emergencies.

The Working Groups have moved from analysis and advice towards questions of implementation, and the scientific output from these groups has been huge (Archer and Scrivener, 2000; Haavisto, 2001; AHDR, 2004).[4]

The broadening of the AEPS environmental agenda began at the Nuuk Ministerial Meeting in 1993. Here, the Arctic states agreed to create a Task Force on Sustainable Development and Utilization (TFSDU). The Task Force was established as a response to the UNCED process and the World Summit in Rio de Janeiro 1992, and prompting by the Inuit Circumpolar Council (ICC), which also played an important role at the Rio conference (Reimer, 1993; Nuttall, 2000). In addition, The Nuuk Declaration affirmed Principle 22 of the Rio Declaration concerning indigenous people and contained statements with clear linkages to the UNCED process and sustainable development:

> **Determined**, individually and jointly, to conserve and protect the Arctic Environment for the benefit of present and future generations, as well as the global environment, **Noting** that in order to achieve sustainable development, environmental protection shall constitute an integral part of the development process and cannot be considered in isolation from it ... **We support** the achievements of the United Nations Conference on Environment and

Development, and state our beliefs that the Principles of the Rio Declaration on Environment and Development have particular relevance with respect to sustainable development in the Arctic.

(The Nuuk Declaration, 1993)

According to Vanderzwaag, Huebert and Ferrara (2003), the participants at the Nuuk meeting 'reached a general recognition that the AEPS needed to consider issues of sustainability and not only environmental protection'. In 1996, Foreign Ministers of the Arctic states agreed in the Ottawa Declaration to form the Arctic Council with a mandate to undertake a broad programme to include all dimensions of sustainable development.[5] The Arctic Council replaced the AEPS and took over the programmes already established under the AEPS. The Council was established as a 'means for promoting cooperation, coordination and interaction among the Arctic States', with the involvement of the Arctic indigenous communities and other Arctic inhabitants on common Arctic issues. Indigenous groups were accorded the status as Permanent Participants.[6] Decisions, however, are based on the 'consensus' of the Members states (Arctic Council, 1996).

The creation of the Arctic Council led several Environmental Ministries and environmental non-governmental organizations to worry that a shift from environmental concerns to the broader issues of sustainable development could weaken or downgrade environmental efforts in the Arctic (Archer and Scrivener, 2000: 615). According to Archer and Scrivener, the creation of the Arctic Council 'slowed the momentum of environmental cooperation in the AEPS':

> It revived mutual fears of hidden agendas behind the impetus to regional collaboration and also diplomatic caution about circumpolar institution-building, while bringing into sharper focus tension involved in reconciling pollution and conservation concerns with economic development under the elusive rubric of sustainable development in the Arctic.
>
> (Archer and Scrivener, 2000: 616).

The *WWF Arctic Bulletin* (1996: 6–8) contained comments on the creation of the Arctic Council from several respected scientists on Arctic issues. Oran R. Young commented that the creation of the Arctic Council was:

> likely to affect the contents of the Arctic policy agenda, redirecting interests from classic environmental protection issues ... to issues centred around the interaction between people and the environment (e.g. the sustainable use of living resources) and issues involving the impacts of airborne and winter-bourne pollutants.

According to Young, this was not necessarily bad news from the view of environmental protection. Monica Tennberg on the other hand, commented that 'the Arctic Council institutionalises the view that the Arctic is a rich region whose resources will be used increasingly in the future'. This was, according to

Tennberg, inconsistent with the goal of environmental protection, and the perception of the Arctic as 'a vulnerable and sensitive region in need of protection'.

At the first Ministerial Meeting of the Arctic Council in Iqaluit, Canada, September 17–18, 1998, the Council transformed the Task Force on Sustainable Development and Utilization to a Sustainable Development Working Group (SDWG).[7] The group was to work on the adopted *Terms of Reference for a Sustainable Development Program*. The goal of the programme was formulated as follows:

> The goal of the sustainable development program of the Arctic Council is to propose and adopt steps to be taken by the Arctic States to advance sustainable development in the Arctic, including opportunities to protect and enhance the environment, and the economies, cultures and health of indigenous communities and of other inhabitants of the Arctic, as well as to improve the environmental, economic and social conditions of Arctic communities as a whole.
>
> (Arctic Council, 1998b)

The SDWG was to facilitate completion of work on sustainable development proposals, propose possible priority areas in the further development of the sustainable development programme, and review specific proposals and prepare them for approval by the Ministers (Arctic Council, 1998a). At the second Ministerial Meeting in Barrow, Alaska, October 12–13, 2000, the Arctic Council adopted a *Framework Document (Chapeau) for the Sustainable Development Program*, which was an attempt to specify the meaning of sustainable development within the special circumstances of the Arctic. The Chapeau was to provide the Council with a capacity to identify projects, activities and priorities. Among the principles in the Chapeau was the definition of sustainable development from *Our Common Future*, and the ambition that

> The Sustainable Development Program should leave future generations in the North with expanded opportunities, and promote economic activity that creates wealth and human capital, while simultaneously safeguarding the natural capital of the Arctic.
>
> (Arctic Council, 2000)

The Chapeau also included a list of the subjects that the Arctic Council 'attaches special importance to' under the heading of sustainable development. These were described as 'critical to human health and sustainable development', and included health issues and the well-being of people living in the Arctic, prevention and control of disease and injuries, the long-term monitoring of the impact of pollution and climate change. In addition, the Chapeau focused on sustainable economic activities and the prosperity of Arctic communities, education and cultural heritage, children and youth, the management of natural, including living resources, and lastly, infrastructure development.

The third Ministerial Meeting in Inari, Finland, 2002, decided to develop the Chapeau further and to prepare a Sustainable Development Action Plan (SDAP). The plan was adopted at the Arctic Council fourth Ministerial Meeting, Iceland, 2004, 'as a tool for the practical realization of the Arctic Council's Sustainable Development Program and assessing the progress made by the Arctic Council in advancing sustainable development in the circumpolar region' (Arctic Council, 2004a). To a large extent, the Action Plan built on the previous documents addressing sustainable development. The Action Plan established three 'broad categories' as 'priorities for the activities of the Arctic Council on sustainable development' under the headings of economic, social and environmental dimensions of sustainable development. Among them were:

− Sustainable economic activity and increasing prosperity of Arctic communities.
− Sustainable use of natural, including living, resources.
− Health of the people living and working in the Arctic.
− Education and cultural heritage, including language.
− Prosperity and capacity building for the people of the Arctic, in particular for children and youth.
− Gender equality.
− Enhancing well-being, eradication of poverty among Arctic people.
− Prevention and elimination of environmental pollution in the Arctic.
− Arctic marine environment protection.
− Biodiversity conservation in the Arctic.
− Climate change impact assessment in the Arctic.
− Prevention and elimination of ecological emergencies in the Arctic, including those relating to climate change. (Arctic Council, 2004b)

While the SDAP contained an overview of the different ongoing projects under each of the above headings, the SDAP has not been especially operational within the Arctic Council. It is still only available in its draft form and the Action Plan seems to have had limited impact. This is partly due to the general nature of the document, and is also linked to the two contradicting views identified in AHDR (2004) on how sustainable development should be dealt with within the Arctic Council. One view is a project-based approach, where it is the projects that form the activities of sustainable development in the Arctic context. The other view stresses the importance of a wider programme of sustainable development to direct the selection and support of projects. As with the Chapeau on sustainable development, the Action Plan can be used to prioritize projects supported by SWDG, and as the Chapeau, the Action Plan can be seen as an attempt to overcome these differences (Haavisto, 2001). It is still unclear, however, how these strategies are to be balanced.

The SDWG Work Plan adopted at the Salekhard Ministerial in 2006 first and foremost represents the project-based approach. It is an operating document intended 'to provide a framework for the work and priorities of the SDWG during the period 2006–2008' that complements the existing Ministerial Declarations and other SDWG documents. The projects are placed under the headings of the following priority areas:

- *Economic dimension of sustainable development*: sustainable economic activity and increasing prosperity of Arctic communities; sustainable use of natural, including living, resources; development of transport infrastructure (including aviation, marine and surface transport), information technologies and modern telecommunications.
- *Social dimension of sustainable development*: health of the people living and working in the Arctic; education and cultural heritage, including language; prosperity and capacity building for the people of the Arctic, in particular for children and the youth; gender equality; enhancing well-being, poverty eradication in the Arctic.

Thus, the SDWG Working Group's agenda is broad, and the work within SDWG must be seen in relation to the other Working Groups (AMAP, CAFF, PAME, EPPR) of the Arctic Council. Even though they have a special focus on the environment, all Working Groups under the Arctic Council deal with sustainable development (Haavisto, 2001). While there have been some tensions between the different groups, and some questions over the ownership of 'sustainable development' (Haavisto, 2001), the social, economic and environmental dimensions are crucial parts of the sustainable development concept.

It seems, however, that there is still tension between the project-based approach and the wider programme approach to direct the selection and support of projects. To some extent, it is also difficult to detect clear priorities from a sustainable development perspective within the SDWG Working Group. It could also be argued that both the Arctic Council and the SDWG group are still struggling with ways to find a good and operational form to cope with sustainable development. There are also several people who voice the need for institutional reforms of the Arctic Council (Haavisto, 2001; Koivurova and Vanderzwaag, 2007). In a discussion paper distributed at the SDWG meeting in Tromsø, April 2007, the Arctic Athabaskan Council presented a number of suggestions for reforming the Arctic Council, including the establishment of a task force to look at the future direction, structure, priorities and communication of the Arctic Council, in order to improve its efficiency and effectiveness (Arctic Athabaskan Council, 2007).

Arctic oil and gas and sustainable development

Arctic oil and gas has been addressed in several of the Working Groups in the Arctic Council, both directly and indirectly. In the following, we present some of the major contributions and findings from the work conducted within the auspices of the Arctic Council, including the Arctic Human Development Report, and other projects and other Working Groups in relation to oil and gas. We start with the AHDR, which is perhaps the most important study done under the auspices of the Arctic Council in terms of trying to come to grips with the human dimensions of sustainable development. As the lead authors of the introductory chapter argued, the report was an explicit attempt to 'help to establish priorities for the activities of the Sustainable Development Working Group' within the Arctic Council, with the explicit aim of shedding light 'on the concept of human development itself', and to highlight 'dimensions of human well-being that are not prominent in mainstream discussions of this topic' (Young and Einarsson, 2004a: 15). In doing so, the report focused on a number of sustainable development challenges, although without using sustainable development as an explicit frame of reference.

Conclusions from the Arctic Human Development Report

The Arctic Human Development Report focused on a number of issues, such as demography, societies and cultures, indigenous people, economic systems, political and legal systems, resource governance, community viability, human health and well-being, education, gender issues, circumpolar international relations and geopolitics. In a way, these headings also reflect what is seen as essential parts of the sustainable development agenda for the Arctic.

The concluding chapter of the AHDR report, with Oran R. Young and Nils Einarsson as lead authors, singled out major conclusions from the report, which 'merit attention on the part of the Arctic Council and its Sustainable Development Working Group', and with a special focus upon 'policy-relevant conclusions' and 'gaps in knowledge' (Young and Einarsson, 2004b). Some of these conclusions are directly or indirectly linked to oil and gas activities in the Arctic. The first relates to the economic systems of the Arctic. Arctic economies are, according to Young and Einarsson, 'narrowly based and highly sensitive to outside forces, including market fluctuations and political interventions'. The narrow base typically consists of the extraction of non-renewable resources or industrial harvesting of living resources. The gap in knowledge relates to the 'understanding of the roles that modern industrial activities play in the pursuit of sustainable development at the regional level'. The challenge, or nature of the problem, is described in the following way:

> The Arctic today is the site of world-class industrial activities. Typically, these activities involve the extraction of non-renewable resources or industrial harvesting of living resources. The Prudhoe Bay oil field is the largest ever discovered in North America. The super giant gas fields of northwestern Siberia are critical to Russia's efforts to reconstruct the country's economy. Nickel and lead/zinc mines located in the Arctic are among the largest in the world. The industrial fisheries of the Bering, Norwegian, and Barents Seas figure prominently in the world harvest of marine living resources. Diamonds from Siberia and Canada's Northwest Territories now account for a sizable share of the world market. Taken together, these enterprises have become prominent symbols of the new North. They provide employment opportunities for Arctic residents, tax revenues for local governments like the North Slope Borough, and sources of income for regional governments like the State of Alaska and the Sakha Republic. At the same time, these industrial activities introduce more or less severe instabilities in many parts of the region.

> Non-renewable resources like oil and gas are not only exhaustible, but the income they produce is also subject to the volatility of world markets. Extractive industries are generally controlled by multinational corporations that are more responsive to global forces than to local concerns. The activities of industrial fishers can conflict with the needs of subsistence harvesters. We need to learn more about feasible responses to these sources of instability.
> (Young and Einarsson, 2004b: 239)

Moreover, Young and Einarsson argue that it is a misinterpretation to regard Arctic economic systems as 'backward' or 'marginal'. Although transfer payments are important in many parts of the Arctic:

> large quantities of profits and rents, arising mostly from the extraction of natural resources on a large scale, flow out of the Arctic, depriving public authorities in the region of potential sources of revenue. A comparison of outflows in the form of profits and rents and inflows in the form of transfer payments shows that the Arctic as a whole is a net exporter of wealth.
> (Young and Einarsson, 2004b: 231)

This issue is closely linked to another policy implication drawn by Young and Einarsson concerning the political systems. They argue that the 'devolution of political authority to regional and local governments in the Arctic has not been accompanied by significant reallocations of material resources' (Young and Einarsson, 2004b: 232). They point to some exceptions but, in general, they argue that the only alternative to continued dependence on transfer payments is to 'change the rules of the game to ensure that more of the profits remain in the Arctic'. Therefore, 'finding ways to address the imbalance of authority and resources is a matter of the utmost importance throughout the Arctic today'.

An associated challenge is to maintain the 'viability of Arctic communities'. This requires an 'enhanced ability to take advantage of the interactions among governmental, corporate, organizational and personal networks from the local to the global level', and 'a better understanding of the effects of cumulative changes on cultural and social well-being in the Arctic' (Young and Einarsson, 2004b: 234, 238). Although one of the main arguments put forward by Young and Einarsson is the high resilience of human societies in the Arctic, they argue that 'it would be a mistake to assume that Arctic societies and cultures can remain resilient in the face of all biophysical and social changes. Today, Arctic societies face an unusual combination of biophysical and socio-economic stresses' (Young and Einarsson, 2004b: 230), many of which were addressed in the AHDR report, and many of which can be linked to oil and gas.

Climate change and Arctic oil and gas

Climate change has been an important issue in the Arctic Council. In 2000, the Arctic Council endorsed, adopted and established the Arctic Climate Impact

Assessment (ACIA) at the Barrow Ministerial Meeting. In 2004, this Assessment was released at the ACIA International Scientific Symposium held in Reykjavik, Iceland, November 2004.[8] The main findings were presented in two different publications, a syntheses report (Impacts of a Warmer Arctic; ACIA, 2004) and a scientific report (ACIA, 2005). In addition, the Arctic Council endorsed a policy document based on the ACIA report at the Ministerial Meeting in Reykjavik, November 2004 (Arctic Council, 2004d).

The ACIA project represents a seminal effort on climate change impacts. In the Arctic Council Policy Document (Arctic Council, 2004d), ACIA is described as 'the world's most comprehensive and detailed regional climatic ... assessment to date'. No doubt, the ACIA report has played an important role in identifying impacts that affect human health, culture and well-being as well as risks to Arctic species and ecosystems. The effect of ACIA has been to increase attention towards and awareness of the problem of climate change in the Arctic.

In the ACIA project, climate change is conceived as a global problem with regional impacts, although in the case of the Arctic, it will in turn also have serious consequences for the global climate. The disappearance of Arctic ice and snow increases the absorption of heat from the sun, and thus further warms the planet. Glacial melt and river runoff add fresh water, which can slow down ocean circulation, affecting both the regional and global climate. It may alter the release and uptake of greenhouse gases in soils, vegetation and coastal oceans, and also have implications for biodiversity around the world affecting migratory species depending on the Arctic (ACIA, 2004: 10).

The focus of the ACIA project, however, was as the title indicates, explicitly on *impacts*. Mitigation efforts and the reduction of emissions of greenhouse gases were beyond the scope of the report.[9] For oil and gas activities in the Arctic, therefore, it is a striking feature of the ACIA framing of oil and gas that the focus is on the impacts of climate change on oil and gas activities, not the other way around. The main human-induced drivers of climate change in ACIA were identified to be the burning (consumption) of fossil fuels and clearing of land (ACIA, 2004: 2), activities that are mostly taking place outside the Arctic region.

Regarding the impacts of climate change on oil and gas activities, some of the ACIA conclusions were quite positive, especially regarding offshore extraction. Reduced and thinner sea ice would most likely increase marine access to oil and gas. It would also make marine transport easier, and thus also the transport of oil and gas to world markets (ACIA, 2004: 11; ACIA, 2005: 1001). In addition, reduced extent and thinner sea ice would also 'allow construction and operation of more economical offshore platforms' (ACIA, 2005: 1001).

On the negative side, it was argued increasing movement of ice in some areas initially could hamper oil and gas operations offshore, and that thawing ground could disrupt oil and gas extraction on land, affecting travel, buildings, pipelines and industrial facilities (ACIA 2004: 11; ACIA, 2005: 1001). Damage to oil and gas transmissions lines in the permafrost zone was said to present 'a particularly serious situation' (ACIA, 2004: 117). The cost of maintaining these structures would also increase (ACIA, 2004: 119; ACIA, 2005: 1001). Some offshore activities could also be hindered by increased wave action in sub-region II (Siberia and adjacent seas) (ACIA; 2004: 19), and reduced sea ice could 'hamper winter seismic work on shore-fast ice' (ACIA, 2005: 1001). The environmental challenges discussed were first and foremost linked to increased risk of oil spills. Referring to a study from the *Exxon Valdez* accident in the Alaskan Prince William Sound, the report worried that oil spills in high-latitude, cold ocean environments may last much longer and be far worse than first expected (ACIA, 2004: 85).

The Arctic Council Policy Document (Arctic Council, 2004d) addressed both mitigation and adaptation policies. In relation to mitigation, it was argued that even though 'overall emissions of greenhouse gases within the Arctic region are limited, there are important mitigation opportunities in the region that would contribute to sustainable development and global emission reduction efforts'. The document further requested action and 'climate change mitigation strategies across relevant sectors' and the promotion of 'appropriate energy sources, uses, technologies and efficiencies'. Relevant initiatives mentioned were the International Partnership for Hydrogen Economy (IPHE) and The Carbon Sequestration Leadership Forum (CSLF), and also initiatives to promote renewable energy production and more efficient energy use (Arctic Council, 2004d: 5). Apart from this, it did not address the issue of oil and gas production in the Arctic any further.

Arctic Offshore Oil and Gas Guidelines (PAME)

Greenhouse gas emissions from oil and gas production have, however, been addressed within the Working Group Protection of the Arctic Marine Environment. The first Offshore Oil and Gas Guidelines were developed by PAME in 1997 in cooperation with the other Working Groups. The guidelines undergo periodic review, and the Arctic Council endorsed a new version in 2002 (PAME, 2002). The guidelines are intended to be of use by the Arctic Nations central and regional authorities at:

> all stages during planning, exploration and development of offshore oil and gas activities. They should be used to secure common policy and practices. The target group for the Guidelines is thus primarily the national authorities, but the Guidelines may also be of help to the industry when planning for oil and gas activities and to the public in understanding environmental concerns and practices of Arctic offshore oil and gas activities. While recognizing the non-binding nature of these Guidelines, they are intended to encourage the highest standards currently available. They are not intended to prevent States from setting stricter standards, where appropriate.
> (PAME, 2002)

The stated goals for environmental protection during oil and gas activities are quite ambitious. Amongst other things, these operations should be planned and conducted so as to avoid: 'adverse effects on climate and weather patterns'; 'significant adverse effects on air and water quality'; 'significant changes in the atmospheric, terrestrial (including aquatic), glacial or marine environments in the Arctic; degradation of, or substantial risk to, areas of biological, cultural, scientific, historic, aesthetic or wilderness significance; and 'adverse effects on livelihoods, societies, cultures and traditional lifestyles for northern and indigenous peoples'. Moreover, the guidelines are directly linked to sustainable development:

> In permitting offshore oil and gas activities, Arctic governments should be mindful of their commitment to sustainable development, including, inter alia:
> – protection of biological diversity;
> – the duty not to transfer, directly or indirectly, damage or hazards from one area of the marine environment to another or transform one type of pollution into another;
> – promotion of the use of best available technology/techniques and best environmental practices ...;
> – the duty to cooperate on a regional basis for protection and preservation of the marine environment, taking into account characteristic regional features; and
> – the need to maintain hydrocarbon production rates in keeping with sound conservation practices as a means of minimizing environmental impacts. (PAME, 2002)

Furthermore, the guidelines suggest 'programs that emphasize energy efficiency and conservation in all activities, exploration (survey and exploratory drilling), development (construction and drilling), production, and transportation', and contain thorough descriptions and guidelines for the conduct of Environmental Impact Assessments. In 2004, PAME also published guidelines for transfer of refined oil and oil products in Arctic waters (PAME, 2004).

Oil and gas in the Arctic Monitoring and Assessment Programme

The first assessment report published by AMAP in 1997, included the collaborate efforts of over 400 scientists and administrators in monitoring the levels and effects of anthropogenic pollutants. The report looked at two sources of pollution: sources remote to the Arctic, and sources found within the Arctic. In short, many sources of persistent organic pollutants were located outside the Arctic. For other contaminants, like heavy metals, two-thirds were originating from industrial activities in the Kola Peninsula.

The assessment also included a chapter on petroleum hydrocarbons, which discussed several kinds of pollution associated with oil and gas activities. The chapter concluded that oil spills and chronic releases from poorly maintained pipelines and from ships pose the greatest threat to hydrocarbon contamination. The Arctic environment was also said to be more vulnerable to oil spills, since oil breaks down more slowly under cold, dark conditions and plants and animals in the Arctic need a longer time to recover. Remedial measures are equally difficult in these extreme conditions (AMAP, 1997: 157). It was also concluded that risks would increase but that:

> it is possible to limit the environmental impact of most routine operations. But as the exploitation of the huge resources of oil and gas increases, so does the risk of serious accidents. Although more stringent regulation will reduce the frequency of accidents, incidents due to human error and technical deficiencies over recent decades have shown that regulation alone cannot completely prevent spills.
>
> (AMAP, 1997: 145)

The 2002 AMAP report did not address oil and gas production (AMAP, 2002). The 2006 scientific report focusing on acidifying pollutants, Arctic haze[10] and acidification in the Arctic, however, discussed oil and gas activities and marine transport as Arctic sources of acidifying pollutants. The report argued that oil and gas activities are 'likely to result in increased pollutant emissions' (AMAP, 2006a: 4). In the policy report (AMAP, 2006b), it was argued that the increase in shipping and the expansion of the offshore oil and gas industry may enhance a number of acidifying pollutants. Acidifying pollutants are emitted at every stage, 'from exploration to the final closure of the field' (AMAP, 2006b) and, while the impact of the oil and gas industry on acidification is low, these emissions could have some impact on the vegetation, soil and surface waters near emission sites (AMAP, 2006b).

The scientific report, however, argued that the 'impact of oil and gas activities on climate in the Arctic areas is a less well-covered issue' than the impacts of a warmer climate for oil and gas activities, and that emissions to air are likely to increase' (AMAP, 2006a: 5). The scientific report stated further that a 'comprehensive coverage of all air emissions from the many activities related to oil and gas production may warrant a separate assessment at some point in the future (AMAP, 2006a: 4).

Oil and gas in the Conservation of Arctic Flora and Fauna Working Group

The CAFF Working Group addressed oil and gas activities in the assessment report *Arctic Flora and Fauna: Conservation and Status* (2001). Habitat disturbance is seen by CAFF as the most serious threat to biodiversity. This is a global trend, which is also reflected in the Arctic. It is linked to oil and gas exploration and development, roads and other infrastructure, off-road use of motorized vehicles, deforestation and so on, 'each of which has affected Arctic habitats in various areas' (CAFF, 2001: 254). Infrastructure poses a serious threat both to biodiversity and sustainable use in the Arctic:

> Between 1900 and 1950, less than 5 percent of the Arctic was affected by infrastructure. Between 1950 and 2000, the areas affected grew to 20–25 percent of the Arctic, mainly as a result of petroleum development in Alaska and Russia … . A recent study by UNEP predicts that continued development

of existing infrastructure at the current rate will leave few areas in the Arctic undisturbed in 50 years time. Depending on growth rates, 50–80 per cent of the Arctic may reach high levels of anthropogenic disturbance by 2050. For Fennoscandia and parts of Russia, these levels may be reached within 20–30 years. The spread of disturbance will lead to the loss of traditional lands for most of the indigenous people dependent upon reindeer husbandry and caribou hunting ... . In northern Scandinavia and parts of Russia, the current growth of infrastructure is increasingly incompatible with the requirements of traditional reindeer husbandry. At the current rate of development, most traditional Saami reindeer husbandry will not be possible in 50 years, forcing major changes in lifestyle for today's herders.

(CAFF, 2001: 99)

Oil and gas infrastructure has altered tundra habitats and the distributions of large mammals in Alaska and Siberia (CAFF, 2001: 86). Oil and gas activities can cause fragmentation of large intact areas and habitats into smaller units, isolating plant and animal populations. In addition, the report pointed at a number of threats from offshore activities. Apart from oil spills, offshore oil and gas exploration and development can have other impacts:

Seismic testing of the seafloor uses air guns, which can kill larval fish within a few meters of the gun and displace larger fish. The noise may disturb marine mammals. The installation and operation of drilling platforms, from human-built islands to mobile rigs, can disturb the ocean floor and alter coastal currents and fish movements. The migratory paths of marine mammals may shift to avoid some structures, possibly affecting their availability to hunters. Pipelines for transporting oil to land or to tankers may disrupt the seafloor.

(CAFF, 2001: 199)

The annual report from the Circumpolar Biodiversity Monitoring Program (CBMP) (CAFF, 2006), argued similarly, that the still 'relatively pristine Arctic environment' is under increasingly severe pressure from regional development in the form of roads, pipelines, oil and gas seismic lines, urbanization, forestry, mining, agriculture, hydroelectric development and more (CAFF, 2006: 2). In order to prevent further damage to biodiversity and sustainable use activities, CAFF (2001) argued that 'a great deal of conservation effort will be required to minimize disturbance from increasing industrial development' in many Arctic areas. This point was further said to be especially true for the cumulative impacts of many smaller projects and activities. According to CAFF, it is also 'essential to recognize that conservation issues must be considered in a global context' in the Arctic. Sustainable development 'is one response to the combined effects of these various forms of change. The distinctive physical, ecological and social features of the Arctic are a vital part of the diversity of life on earth, and need collective efforts to conserve them' (CAFF, 2002: 2).

Framing Arctic oil and gas

As we have seen in this chapter, sustainable development is a complex concept that can be related to the Arctic and oil and gas activities in the Arctic in a number of ways. We have argued that, within the approach to sustainable development provided by the World Commission on Environment and Development (1987), it is possible to speak sensibly about oil and gas activities in the Arctic despite the fact that oil and gas are non-renewable resources. This, however, does not imply that oil and gas activities in the Arctic are unproblematic from a sustainability perspective. As we have seen, the Arctic Council Working Groups have identified and discussed a number of social, economic and environmental challenges relating to oil and gas.

From the Working Groups different approaches to sustainable development, the sustainable development agenda in the Arctic Council seems, on the one hand, overfull. The sheer number of issues and challenges, which may be contradictory, makes it difficult for anyone to develop clear sustainable development targets in the Arctic. The Arctic Council is also a consensus-based organization, where the member states with different political agendas can block proposals and more or less implement what they want. As argued by Koivurova and Vanderzwaag (2007: 191), moving 'from words to actions have been challenging in light of the Council's limited role as a discussional and catalytic forum – a soft law body – with law and policy controls remaining within individual member states'. As such, there seems to be more of a focus on monitoring and assessment in the Arctic Council than on implementation, which is essentially left to the member states.

On the other hand, the sustainable development agenda in the Arctic Council is also partial and somewhat biased against a larger discourse on carbon dioxide (CO_2) impacts. Seen in relation to the global sustainable development agenda, there are some issues that seem to be off the agenda. This is especially visible in the specific framing of climate change under the auspices of the Arctic Council. In the ACIA framing, climate change is seen as a global problem with regional effects. The effects of oil and gas production in the Arctic are not considered relevant in the larger picture, since emissions first and foremost are related to the consumption of oil and gas, which primarily takes place outside the Arctic. Even the forthcoming AMAP study, *Assessment of Oil and Gas Activities in the Arctic*, explicitly states that the assessment 'specifically does *not* include the relation between Arctic oil and gas development and the global CO_2 emissions and greenhouse warming' (AMAP, 2006c). This topic, it is argued, is covered in other assessments: ACIA, Intergovernmental Panel on Climate Change (IPCC) and national assessments.

This is only partially true. None of the activities within the Arctic Council, including ACIA, look at the effects of Arctic oil and gas on global CO_2 emissions and greenhouse warming. Instead, the context in which mitigation efforts are placed is within the commitments under 'the UNFCCC [United Nations Framework Convention on Climate Change] and other agreements', and the 'countries' share in total global greenhouse gas emissions' (Arctic Council, 2004d: 5). Thus, the link between oil and gas production in the Arctic and climate change is lifted out of the Arctic and placed at the global and national level. In a sense, the framing of oil and gas *production* in relation to climate change in the Arctic is off the agenda, and placed at the global and national levels.

From the perspective of global sustainability, however, it is not obvious that this should be the case. The question of expanding the oil and gas activities or not in the Arctic involves sustainability questions that are regional and local in nature, and sustainability issues that go *beyond* the Arctic. Some actors question, for instance, the legitimacy of increased extraction and use of fossil fuels in light of climate change. Thus, the questions that can be, and are being raised from a sustainable development perspective are, in many cases, global in nature and cannot be isolated from the debate on oil and gas activities in the Arctic. In order to put conservation issues higher on the agenda of the Arctic Council, this is exactly what CAFF attempts to do by arguing that conservation issues must be considered in a global context.

There are, however, no neat and easy answers to how the global level should be included in the equation. We have argued that there are two balancing acts that are necessary, and these are also essentially linked. The first relates to the balancing of the social, economic and environmental dimensions. What is most important: to secure and protect the environment or to secure a reasonable standard of living? The second necessary balancing act is the balancing of global, regional and local levels. How should local and regional gains and losses be balanced against global concerns, and vice versa? What is most important: to secure the livelihoods and traditional ways of living for indigenous people, or to secure global energy supply and global energy security? And finally, do we need to choose between these concerns, or are there ways to balance and take all these concerns into account? There are different opinions on this issue. The overall question is *how* a sustainable development path can be secured and implemented in the Arctic. In relation to oil and gas in the Arctic, as argued by Fjellheim and Henriksen (2006: 22), there is a striking paradox:

> Most of the Arctic region has potential Oil and Gas reservoirs just waiting to be exploited. The paradox of facing new industrial challenges as the Arctic is warming due to the products of the very same industries now entering the Arctic region is obvious and has not only global implications, but also very concrete local/regional effects.

Or is it so obvious? The rest of the book tries to grasp and understand different answers to the above questions, and the possible paradox of oil and gas in the Arctic.

Notes

1 There are, however, also non-anthropocentric perspectives to be found (see WCED, 1987: 13, 57; Benton, 1999; Wetlesen, 1999).
2 The problem of climate change is addressed throughout *Our Common Future* (WCED, 1987: 2, 5, 8, 14, 22, 32, 33, 37, 58–59, 172–176). Leader of the World Commission,

Gro Harlem Brundtland (1991: 35) has also stated the following: 'The most global – and the potentially most serious – of all the issues facing us today is how we should deal with the threats to the world's atmosphere'.
3 The strategy was developed by the eight Arctic states and assisted by the Inuit Circumpolar Conference, Nordic Saami Council, USSR Association of Small Peoples of the North, Federal Republic of Germany, Poland, United Kingdom, United Nations Economic Commission for Europe, United Nations Environment Programme and International Arctic Science Committee (Arctic Environmental Protection Strategy, 1991: 1).
4 See Arctic Council (http://www.arctic-council.org/) for an overview.
5 As noted in the AHDR (2004: 212), the idea of an Arctic political body was taken up by a number of Canadian non-governmental organizations in the report *To Establish an International Arctic Council. A Framework Report* (1991): 'The idea of some kind of circumpolar political body had been suggested some twenty years earlier and was taken up again at the end of the 1980s in a study by Canadian non-governmental organizations. It proposed an umbrella-type political forum for governments, indigenous organizations and different interest groups, and was paralleled by an official Canadian initiative to create an Arctic Council'. See: http://www.carc.org/pubs/v19no2/2.htm (Accessed 26 February 2007).
6 There are currently six indigenous people organizations granted Permanent Participants status in the Arctic Council. The Aleut International Association (AIA) established in 1998, representing the Russian and Alaskan Aleut people. The Arctic Athabaskan Council (AAC), established in 2000, representing the interests of United States and Canadian Athabaskan member First Nation governments. The Gwich'in Council International (GCI), established in 1999, representing the regions of the Gwich'in Nation in the Northwest Territories, Yukon and Alaska. The Inuit Circumpolar Conference (ICC), founded in 1977, representing approximately 150,000 Inuit of Alaska, Canada, Greenland and Chukotka (Russia). The Saami Council, established in 1956, representing the Sami of Norway, Sweden, Finland and Russian Federation. And the Russian Association of Indigenous People of the North (RAIPON), founded in 1990, representing 31 indigenous peoples of the Russian North with a total population of over 200,000. These organizations, no doubt, play an important role in the Arctic Council. As already noted, The Inuit Circumpolar Conference also played an important role in the transition from the AEPS strategy to the establishment of the Arctic Council.
7 Although the difference between a task force and a working group is not clear in official terms, a task force or group has been regarded as a less formal organization than is the case with working groups (Vanderzwaag, Huebert and Ferrara, 2003: 152).
8 Arctic Climate Impact Assessment (ACIA) was a joint project by the Arctic Council and the International Arctic Science Committee (IASC), with the aim to evaluate and synthesize knowledge on climate variability, climate change and increased ultraviolet radiation and its consequences. In the following, ultraviolet radiation is not discussed.
9 As stated in the report: '... many longer-term impacts could be reduced significantly by reducing global emissions over the course of this century. This assessment did not analyze strategies for achieving such reductions, which are the subject of other bodies' (ACIA, 2004: 9).
10 Arctic haze is explained as 'a persistent winter diffuse layer in the Arctic atmosphere whose origin is thought to be related to long-range transport of continental pollutants' (AMAP, 2006b).

References

Adams, W. M. (1990) *Green Development. Environment and Sustainability in the Third World*, London: Routledge.

Archer, C. and Scrivener, D. (2000) 'International co-operation in the Arctic environment', in M. Nuttall and T. V. Callaghan (eds) *The Arctic Environment, People, Policy*, Amsterdam: Harwood Academic Publishers.

Arctic Athabaskan Council (2007). *Improving the Efficiency and Effectiveness of the Arctic Council: A Discussion Paper*, March 2007, Arctic Athabaskan Council.

Arctic Climate Impact Assessment (ACIA; 2004) *Impacts of a Warmer Arctic: Arctic Climate Impact Assessment*, Cambridge: Cambridge University Press.

ACIA (2005) *Arctic Climate Impact Assessment*, Cambridge: Cambridge University Press.

Arctic Council (1996) *Declaration on the Establishment of the Arctic Council*. Online. Available HTTP: <http://www.arctic-council.org/Archives/Founding%20Docs/Arctic%20Council_Declaration%20on%20the%20Establishment%20of%20the%20Arctic%20Council.htm> (Accessed 6 November 2007).

Arctic Council (1998a) *Iqaluit Declaration. The First Ministerial Meeting Canada*, 1998. Available HTTP: <http://www.arctic-council.org/Meetings/Ministeral/1998/Default.htm> (Accessed 8 November 2007).

Arctic Council (1998b) *Terms of Reference for a Sustainable Development Program*. Available HTTP: <http://www.arctic-council.org/Meetings/Ohers/CBW/ACTermsofReference2.pdf> (Accessed 8 November 2007).

Arctic Council (2000) *Framework Document (Chapeau) for the Sustainable Development Program*. Online. Available HTTP: <http://www.arctic-council.org/Archives/Founding%20Docs/Arctic%20Council%C2%A0_%C2%A0Framework%20Document%20(Chapeau)%20for%20the%20Sustainable%20Development%20Program.htm> (Accessed 8 November 2007).

Arctic Council (2004a) *Reykjavik Declaration*. Online. Available HTTP:<http://www.dfaitmaeci.gc.ca/circumpolar/pdf/Reykjavik_Declaration-en.pdf> (Accessed 8 November 2007).

Arctic Council (2004b) *Sustainable Development Action Plan (SDAP)*, Version 13.10.2004. Online. Available HTTP: <http://www.arctic-council.org/Meetings/SAO/2004%20Se/SDAP100304eng.pdf> (Accessed 8 November 2007).

Arctic Council (2004c) *Programme of the Russian Federation Chairmanship of the Arctic Council in 2004–2006*. Online. Available HTTP: <http://www.rusembassy.fi/RArcticCounsilLenta.htm> (Accessed 6 November 2007).

Arctic Council (2004d) *Arctic Climate Impact Assessment. Policy Document*. Online. Available HTTP: <http://www.acia.uaf.edu/PDFs/ACIA_Policy_Document.pdf> (Accessed 8 November 2007).

Arctic Environmental Protection Strategy (AEPS; 1991) *Declaration on the Protection of Arctic Environment*, June 14. Online. Available HTTP: <http://www.arctic-council.org/Archives/AEPS%20Docs/artic_environment.pdf> (Accessed 6 November 2007).

Arctic Human Development Report (AHDR; 2004), Akureyri: Stefansson Arctic Institute.

Arctic Monitoring and Assessment Program (AMAP; 1997) *Arctic Pollution Issues: A State of the Arctic Environment Report*, Oslo: AMAP. Online. Available HTTP: <http://www.amap.no/> (Accessed 11 November 2007).

AMAP (2002) *Arctic Pollution 2002*, Oslo: AMAP. Online. Available HTTP: <http://www.amap.no/> (Accessed 11 November 2007).

AMAP (2006a) *Arctic Assessment 2006. Acidifying Pollutants, Arctic Haze, and Acidification in the Arctic*, Oslo: AMAP. Online. Available HTTP: <http://www.amap.no/> (Accessed 11 November 2007).

AMAP (2006b) Arctic Pollution 2006. *Acidification and Arctic Haze*, Oslo: AMAP. Online. Available HTTP: <http://www.amap.no/> (Accessed 11 November 2007).

AMAP (2006c) *Assessment of Oil and Gas Activities in the Arctic. Process and Outline Content. Online.* Available HTTP: <http://arcticportal.org/en/amap> (Accessed 11 November 2007).
Benton, T. (1999) 'Sustainable development and accumulation of capital: reconciling the irreconcilable?', in Dobson, A. (ed.) *Fairness and Futurity. Essays on Environmental Sustainability and Social Justice*, Oxford: Oxford University Press.
Birch, C. et al. (1979) *Faith, Science, and the Future, Preparatory Readings for a World Conference Organised by the World Council of Churches at the Massachusetts Institute of Technology, Cambridge, Mass., USA.* Church and Society. World Council of Churches, Geneva, Switzerland.
Brundtland, G. H. (1991) 'Sustainable development: the challenge ahead', in Stokke, O. (ed.) *Sustainable Development*, London: Frank Cass.
Conservation of Arctic Flora and Fauna (CAFF; 2001) *Arctic Flora and Fauna. Conservation and Status*, Helsinki: Edita. Online. Available HTTP: <http://arcticportal.org/arctic-council/working-groups/caff-document-library/arctic-flora-and-fauna> (Accessed 11 November 2007).
CAFF (2002) *Arctic Flora and Fauna: Recommendations for Conservation*, 14 pp. Online. Available HTTP: <http://arcticportal.org/arctic-council/working-groups/caff-document-library/arctic-flora-and-fauna> (Accessed 17 June 2007).
CAFF (2006) *Circumpolar Biodiversity Monitoring Program 2006 Annual Report. CAFF CBMP Annual Report*, CAFF International Secretariat, Akureyri: Iceland. Online. Available HTTP: <http://archive.arcticportal.org/270/01/annual-report.pdf> (Accessed 11 November 2007).
Convention on Biological Diversity (CBD; 2004) 'Strategic Plan: future evaluation of progress', *COP 7 Decision VII/30*, Kuala Lumpur, 9-20 February 2004. Online. Available HTTP:<http://www.cbd.int/convention/cop-7-dec.shtml?m=COP-07&id=7767&lg=0> (Accessed 6 November 2007).
Dixon, J. A. and Fallon, L. A. (1989) *The Concept of Sustainability: Origins, Extensions, and Usefulness for Policy*, Washington: The World Bank, Environment Department, Division, Working Paper No. 1989-1.
Dresner, S. (2002) *The Principles of Sustainability*, London: Earthscan.
Dryzek, J. S. (1997) *The Politics of the Earth. Environmental Discourses*, Oxford: Oxford University Press.
Fjellheim, R. and Henriksen, J. B. (2006). 'Oil and gas exploitation on Arctic indigenous peoples' territories. Human rights, international law and corporate social responsibility', *Gáldu Cála. Journal of Indigenous Peoples Rights*, 4: 8–23.
Grubb, M., Koch, M., Thomson, K., Munson, A. and Sullivan, F. (1993) *The Earth Summit Agreements. A Guide and Assessment*, London: Earthscan.
Haavisto, P. (2001) *Review of the Arctic Council Structures. Consultant's Study*, Helsinki, 29.06.2001. Online. Available HTTP:<http://arctic-council.org/Meetings/SAO/2001%20Ro/re_AC_final.pdf> (Accessed 6 November 2007).
Heininen, L. (2004) 'Circumpolar international relations and geopolitics', in *AHDR (Arctic Human Development Report)* (2004), Akureyri: Stefansson Arctic Institute.
International Union for Conservation of Nature/World-Wide Fund for Nature/United Nations Environment Programme (IUCN/WWF/UNEP; 1980) *World Conservation Strategy. Living Resource Conservation for Sustainable Development*. Gland, Switzerland: International Union for the Conservation of Nature and Natural Resources.
Jacob, M. L. (1996) *Sustainable Development: A Reconstructive Critique of the United Nations Debate*, Gothenburg: University of Gothenburg.

Jacobs, M. (1999) 'Sustainable development as a contested concept', in Dobson, A. (ed.) *Fairness and Futurity. Essays on Environmental Sustainability and Social Justice*, Oxford: Oxford University Press.

Kirkby, J., O'Keefe, P. and Timberlake, L. (eds) (1995) *The Earthscan Reader in Sustainable Development*, London: Earthscan.

Koivurova, T. and Vanderzwaag, D. L. (2007) 'The Arctic Council at 10 years: retrospect and prospects', *UBC Law Review*, 40 (1): 121–194.

Lafferty, W. M. (1996) 'The politics of sustainable development: global norms for national implementation', *Environmental Politics*, 5 (2): 185–208.

Lafferty, W.M. and Langhelle, O. (eds) (1999) *Towards Sustainable Development. On the Goals of Development – and the Conditions of Sustainability*, Basingstoke: Macmillan Press Ltd.

Lafferty, W. M. and Meadowcroft, J. (eds) (2000) *Implementing Sustainable Development. Strategies and Initiatives in High Consumption Societies*, Oxford: Oxford University Press.

Langhelle, O. (1999) 'Sustainable development: exploring the ethics of our common future', *International Political Science Review*, 20 (2): 129–149.

Langhelle, O. (2000) 'Sustainable development and social justice – expanding the Rawlsian framework of global justice', *Environmental Values*, 9 (2000). 295–323.

Malnes, R. (1990) *The Environment and Duties to Future Generations*, Lysaker: Fridtjof Nansen Institute.

McManus, P. (1996). 'Contested terrains: politics, stories and discourses of sustainability', *Environmental Politics*, 5 (1): 48–73.

Meadowcroft, J. (1997) 'Planning, democracy and the challenge of sustainable development', *International Political Science Review*, 18: 167–190.

Mugaas, P. (1997) 'Fra Stockholm via Rio til New York', in Lafferty, W. M., Langhelle, O., Mugaas, P. and Ruge, M. H. (eds) *Rio + 5. Norges oppfølging av FN-konferansen om miljø og utvikling*, Otta: Tano Aschehoug.

Nuttall, M. (2000) 'Indigenous peoples, self-determination, and the Arctic environment', in Nuttall, M. and Callaghan,. T. V. (eds) *The Arctic Environment, People, Policy*, Amsterdam: Harwood Academic Publishers.

The Nuuk Declaration, (1993). Online. Available HTTP: <http://www.arctic-council.org/Archives/AEPS%20Docs/Arctic%20Council%C2%A0_%C2%A0The%20Nuuk%20Declaration.htm> (Accessed 6 November 2007).

O'Riordan, T. (1993). 'The politics of sustainability', in Turner, K. R. (ed.) *Sustainable Environmental Economics and Management: Principles and Practice*, London: Belhaven Press.

Protection of the Arctic Marine Environment (PAME; 2002) *Arctic Offshore Oil & Gas Guidelines*, Online. Available HTTP: <http://old.pame.is/sidur/uploads/ArcticGuidelines.pdf> (Accessed 6 November 2007).

PAME (2004) *Guidelines for Transfer of Refined Oil and Oil Products in Arctic Waters (TROOP)*. Online. Available HTTP: <http://old.pame.is/sidur/uploads/ArcticGuidelines.pdf> (Accessed 6 November 2007).

Redclift, M. (1993). 'Sustainable development: needs, values, rights', *Environmental Values*, 2 (1): 3–20.

Reid, D. (1995) *Sustainable Development. An Introductory Guide*, London: Earthscan.

Reimer, C. (1993) 'Moving toward co-operation: Inuit circumpolar policies and the Arctic environmental protection strategy' *CARC - Northern Perspectives*, 21 (4). Online. Available HTTP: <http://www.carc.org/pubs/v21no4/moving.htm> (Accessed 6 November 2007).

Scrivener, D. (1989) *Gorbachev's Murmansk Speech: The Soviet Initiative and Western Response*, Oslo: Norwegian Atlantic Committee.

Scrivener, D. (1996) *Environmental Cooperation in the Arctic: From Strategy to Council*. Oslo: Norwegian Atlantic Committee.

Scrivener, D. (1999) 'Arctic environmental cooperation in transition', *Polar Record* 35 (192): 51–58.

Sustainable Development Working Group (SDWG; 2006). *Work Plan for 2006–2008*. Online. Available HTTP:<http://portal.sdwg.org/media.php?mid=468> (Accessed 6 November 2007).

United Nations (1989) *General Assembly. GA/res. 44/228*. Online. Available HTTP: <http://www.un.org/documents/ga/res/44/ares44-228.htm> (Accessed 6 November 2007).

United Nations (1993) *Report of the United Nations Conference on Environment and Development. Rio de Janeiro, 3-14 June 1992. Volume 1. Resolutions Adopted by the Conference*, New York: United Nations.

United Nations (2002) *Johannesburg Declaration on Sustainable Development*. Online. Available HTTP: <http://www.un.org/esa/sustdev/documents/WSSD_POI_PD/English/POI_PD.htm> (Accessed 6 November 2007).

Vanderzwaag, D., Huebert, R. and Ferrara, S. (2003) 'The Arctic Environmental Protection Strategy, Arctic Council and multilateral environmental initiatives: tinkering while the Arctic marine environment totters', *Denver Journal of International Law and Policy*, 30 (2): 131–171.

World Commission on Environment and Development (WCED; 1987) *Our Common Future*. Oxford: Oxford University Press.

Wetlesen, J. (1999) 'A global ethic of sustainability?', in Lafferty, W. M. and Langhelle, O. (eds) *Towards Sustainable Development. On the Goals of Development – and the Conditions of Sustainability*, Basingstoke: Macmillan Press Ltd.

Worster, D. (1993) 'The shaky ground of sustainability', in Sachs, W. (ed.) *Global Ecology: A New Arena of Political Conflict*, London: Zed Books.

WWF Arctic Bulletin (1996) 'Different views: environmental protection under the Arctic Council', *WWF Arctic Bulletin*, No. 4, pp. 6–8.

Young, O. R. and Einarsson, N. (2004a) 'Introduction', in *Arctic Human Development Report*, Akureyri: Stefansson Arctic Institute.

Young, O. R. and Einarsson, N. (2004b) 'A human development agenda for the Arctic: major findings and emerging issues', in *Arctic Human Development Report*, Akureyri: Stefansson Arctic Institute.

3 Climate change and consequences for the Arctic

Torleiv Bilstad

Introduction

The United Nations-sponsored Intergovernmental Panel on Climate Change (IPCC, 2007) Working Group I (WGI) declared in February 2007 that the world is warming and that humans are mostly to blame. Another IPCC working group, WGII, reported in April 2007 that humans are also behind many of the physical and biological changes that are occurring. Owing to increasing levels of greenhouse gases, mostly carbon dioxide (CO_2) emissions, the world is experiencing receding glaciers, early-blooming trees, bleached corals, acidifying oceans, killer heat waves, fires, pests and diseases, butterflies retreating up mountain sides, oceans warming and cod reproducing further North towards the Arctic Ocean compared to only a few years ago.

The ice sheet losses have contributed to observed sea-level rises between 1993 and 2003 of 3.1 mm/year, compared to an average rate of 1.8 mm/year between 1961 and 2003. All these warming effects are summarized as a combination of solar radiation, land-use changes, and the prevalence of greenhouse gases and aerosols in the atmosphere. The net contribution on the energy balance in the Earth–atmosphere system is described in terms of watts per square metre, W/m^2. This change is called radioactive forcing and is an indicator of the climate change impact. The WGI document includes excellent illustrations of the various warming and cooling influences. Greenhouse gases far outweigh any other influence, with CO_2 contributing more than three times that of methane – 1.66 and 0.48 W/m^2, respectively.

The closing section of the summary of the report 'Climate Change 2007: The Physical Science Basis' (IPCC, 2007) concludes that there is a high level of scientific understanding about the mechanisms and effects of greenhouse gases. Projections of climate change are given for various emission scenarios based on observations and a large number of simulations available from a broad range of models.

The April 2007 IPCC report sees a bleak future if humans persist in their ways. The negative climate impacts fall hardest on those least capable of adapting to change: the poor developing countries, as well as the globe's flora and fauna. Drying will be concentrated at low latitudes, causing damage to crop production.

Disease and death from heat waves, floods and drought will increase. The report also predicts that 30 per cent of species will be at risk of extinction by 2050. Even if greenhouse gas emissions are immediately reduced, changes are inevitable. Humans will have to adapt to survive.

It is not only about future impacts. The World Health Organization (WHO) estimates that climate change is already causing at least 150,000 excess deaths per year. One major killer is malaria: warmer average temperatures allow the mosquito that transmits malaria to spread into the highlands. Ongoing effects of global warming vary greatly from one region to the next, with perhaps the most striking example being shifting precipitation. There has been a 20 per cent decline in snowmelt in the United States Pacific Northwest in the last 50 years as the region has warmed. An increase in dryness has started and will accelerate in the Caribbean Region as well as all around the Mediterranean. An increasing wetness at high latitudes will also occur in the Northern hemisphere. Here an increase of 1–3°C is beneficial for the major cereals, whereas in the low latitudes even a 1°C increase results in an almost immediate decrease in yield. A more than 3°C warming is detrimental to growth also at high latitudes. According to climate models, a 3°C increase is possible in this century if CO_2 emissions continue at present rates.

More local warming impacts are the melting of glaciers around the globe. Several snow-covered peaks have already disappeared in the last 50 years. As more glaciers melt in places like the Himalayas and the Andes, spring and summer melt will become history. Water storage in the form of snow and ice will disappear with grave consequences for irrigation and domestic potable water supply. Sea-level rise from melting glaciers will flood low-lying coastal areas all around the globe. Heat waves experienced in Europe in 2003, where perhaps 30,000 people died, will in the next few years become ever more extreme, frequent and deadly for people, flora and fauna. Widespread mortality and bleaching of corals already living near their upper limits of temperatures, have already become an uncomfortable certainty.

No one region of the globe seems exempt. Although the WGII report emphasizes that not everyone will bear the consequences equally, the global net damage costs resulting from a climate change will become significant. Some people will be exposed to severe climate change, some will be more sensitive and some will be less able to adapt to changes. People living in low-latitude countries where drying will predominate, have economies based largely on agriculture and are already today sensitive to droughts. Adaptation to climate change is simply a matter of survival. It is not enough to stem the greenhouse gas emissions, as the current level is already enough to sustain a temperature increase for another hundred years. The UN's Millennium Development Goals, ensuring environmental sustainability, will not now be met as a result of climate change.

A changing Arctic

The edges of Greenland are losing ice at an alarming rate, faster and faster compared to the ever-updated estimates. The IPCC (2007) findings suggest an underestimation of CO_2 importance and that the future loss of Arctic sea ice may be more rapid and extensive then predicted. Likewise, sea-level rise may be responding even more quickly than climate models indicate.

Climate change and global warming are considered major environmental and economic threats to the global community. Commitments to prevent and minimize its causes and to mitigate adverse effects are a priority on the political and scientific agendas. Strategies are implemented to achieve reductions in CO_2, which is considered to be the major human-induced greenhouse gas (GHG). The atmospheric CO_2 concentration of 385 parts per million (ppm) is the highest it has been in 10 million years (Corell, 2006). The Kyoto Protocol is a driving force in negotiations for ambitious reduction targets for GHG. Even if these commitments were fully implemented, however, significant changes in climatic conditions will occur throughout the twenty-first century.

The Arctic is rich in hydrocarbons (HCs) and production of oil and gas will improve the global energy security. However, burning of HCs, irrespective of where they are mined, advances global warming. Models suggest that, over the next century, our planet will get warmer to the tune of 1.4–5.8°C on average, depending on the levels of GHG emissions. Changes in climate influence snow and ice as well as the way of life in the Arctic. Human adaptation, therefore, becomes a necessity.

The Arctic is generally characterized as hostile and cold. However, maritime and continental influences contribute in shaping significant climate variables with unpredictable weather for the region. At high latitudes, the solar radiation is low or completely absent in the winter, contrasting with high levels during long summer days. The low elevation of the sun above the horizon during the Arctic winter makes even small topographic features significant in shaping local differences in microclimate due to shading. The high albedo owing to snow and ice reduces the absorption of solar energy. As a consequence, the heat gained during long summer days is small and highly dependent on surface properties. The Arctic is highly vulnerable to effects of global warming as solid snow and ice, including tundra, undergoes a phase shift to liquid water at temperatures above 0°C.

In Russia, Canada and Alaska, subsistence activities of the native populations are directly or indirectly threatened by the oil and gas industry. The direct effects are connected to the pollution of fragile land and water, as well as construction and operation of infrastructure for oil and gas developments that deflect animals from traditional migration routes and make access for hunters difficult. The indirect effects are related to emissions and pollution from the burning of fossils for energy production anywhere, with CO_2 interfering with natural climate variations. The transportation of chemicals by air and sea from industrial nations to the Arctic also affects the nutrition chain through bioaccumulation, which is very destructive to societies relying on a subsistence lifestyle. Climate change also threatens traditional knowledge of ice and wildlife among native peoples. Biodiversity and sustainability of the environment are at risk.

Science behind the greenhouse phenomenon

The text 'Global Warming, the Complete Briefing' (Houghton, 2004) covers relevant research reviewed in this paragraph. Jean-Baptiste Fourier first recognized the warming effect of the greenhouse gases, CO_2 and water vapour, in 1827. He pointed to the similarity between what happens in the atmosphere and under the glass of a greenhouse, i.e. the 'greenhouse effect'. Around 1860, John Tyndall measured the absorption of infrared radiation by CO_2 and water vapour. He also suggested that a cause of the Ice Ages might have been a decrease in the greenhouse gas CO_2 in the atmosphere. Svante Arrhenius in 1896 calculated the effect of an increasing concentration of greenhouse gases. A doubling of the CO_2 concentration will increase the average global temperature by 5–6°C. This estimate is still valid a century later. G. S. Callendar in 1940 was the first to calculate the warming due to increasing CO_2 from burning of fossil fuels.

The first expression of concern about an impending climate change was in 1957 when Revelle and Suess at the Scripps Institute of Oceanography published a paper pointing to a build-up of CO_2 in the atmosphere (Houghton, 2004). Routine measurements of in 1896 from the observatory on Mauna Kea in Hawaii commenced. The rapid increasing use of fossil fuels since 1957 has led to the topic of global warming through a large-scale geophysical experiment. The natural greenhouse effect is due to methane (CH_4), water vapour, CO_2 and nitrous oxide (N_2O). The amount of water vapour in the atmosphere depends mostly on air temperature at the ocean surface with associated evaporation from the oceans. Carbon dioxide is different, however: the level has increased by 30 per cent since the Industrial Revolution owing to the burning of fossil fuels, and the removal of forests and plants. In the absence of such controlling factors, the rate of increase in atmospheric CO_2 will accelerate and double in the next hundred years (Houghton, 2004).

The global carbon balance

The carbon content of the Earth's atmosphere has increased from 360 Gt mainly as CO_2 at the last glacial maximum to 560 Gt during pre-industrial times to 730 Gt today (Zimov, Scuur and Chapiin, 2006). These changes reflect redistributions among the main global carbon reservoirs. The largest such reservoir is the ocean, which contains 40,000 Gt, including 2500 Gt organic carbons, followed by soils at 1500 Gt and vegetation at 650 Gt. There is also a large geological reservoir from which 6.5 Gt carbons are released annually by burning fossil fuels.

Permafrost is an additional large carbon reservoir that is rarely incorporated into analyses of changes in global carbon reservoirs. During the glacial age, frozen loess, windblown dust or yedoma in Siberia, was deposited to an average depth of 25 m covering more than 1 million km² in northern Siberia and Alaska. These frozen soils have an average carbon content of 10–30 times that found in deep non-permafrost mineral soils. The carbon reservoir from yedoma is 500 Gt, plus other permafrost soils at 450 Gt. An additional reservoir is included in peat lands of Western Siberia, which is estimated at 70 Gt of carbon. Considering that the carbon content in the atmosphere is 730 Gt today, the potential release from permafrost could become devastating.

How fast will it thaw? Organic matter in yedoma decomposes quickly when thawed, up to 40 g carbon/m^3/day (Zimov, Scuur and Chapiin, 2006). Field observations suggest that these rates are sustained and that carbon in yedoma will be released within a century; a striking contrast to the preservation of carbon for tens of thousands of years when frozen in permafrost. Permafrost is defined as bedrock, organic or earth material that has temperatures below freezing persisting over more than two consecutive years. The impact of permafrost thaw on both climate and socio-economic conditions is already serious. The southern border for permafrost has been migrating northward by hundreds of miles over the last couple of decades. Figure 3.1 (Arctic Climate Impact Assessment; ACIA, 2004) shows a rapid decrease in the ice roads necessary for travel on the North Slope of Alaska.

Tundra lakes are net emitters of methane when thawing. Methane is 23 times as effective as a greenhouse gas compared to carbon dioxide. Fluctuations of CO_2 and CH_4 in the atmosphere over the past 400,000 years are depicted in Figure 3.2 (Petit, Jouzel, Raynaud et al., 1999). In Siberia, previously frozen tundra lakes are emitting and bubbling CH_4 as the greenhouse gas is released due to natural degradation of carbon-organics under anoxic conditions and increasing temperature.

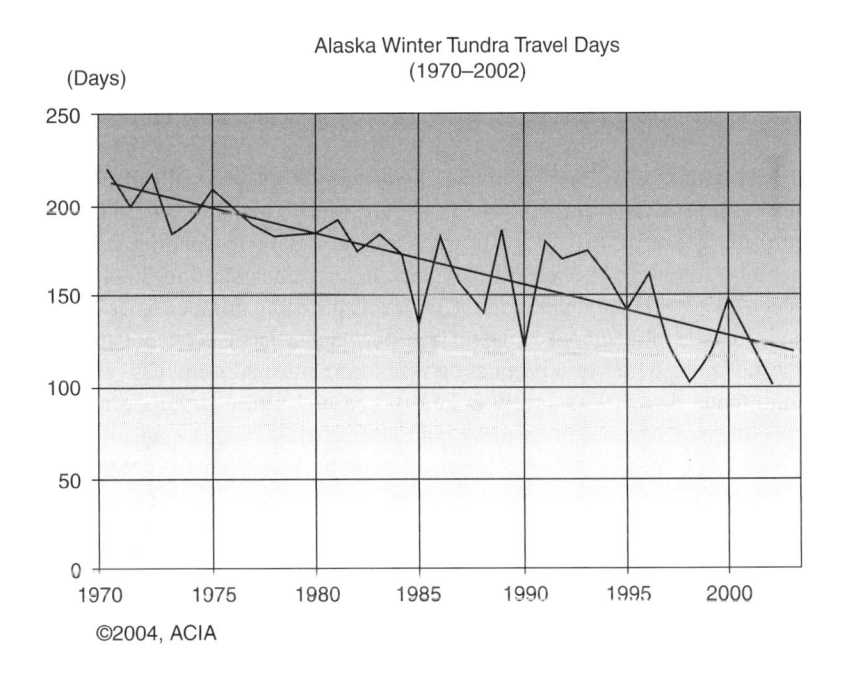

Figure 3.1 Maximum oil exploration travel activities on the tundra (ACIA, 2004).

50 *Torleiv Bilstad*

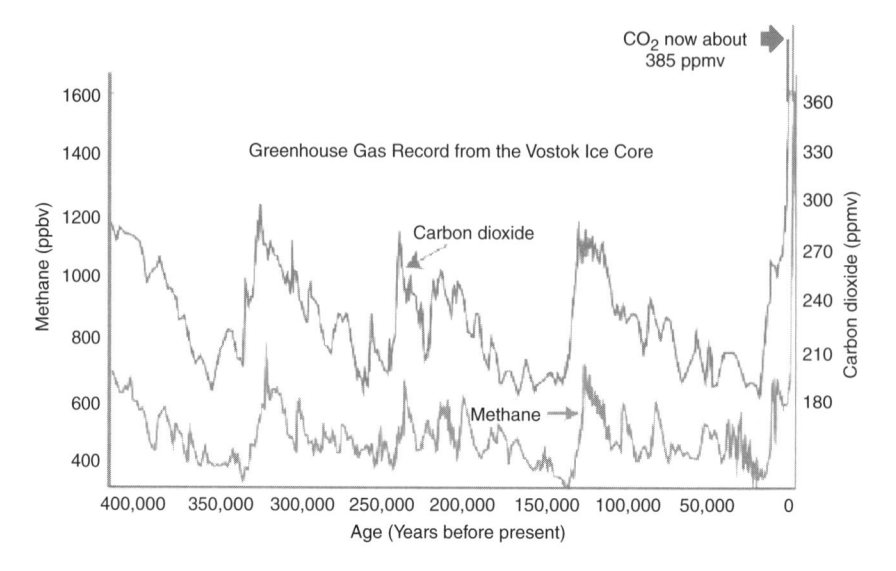

Figure 3.2 Methane concentration doubling over the last 150 years (Petit, Jouzel, Raynaud et al., 1999).

Preservation of carbon in permafrost has lasted for tens of thousands of years, but the lakes are rapidly disappearing now, as the melting process continues.

Ice is important. We are, however, losing it and, therefore, also losing important archives informing us about past climates. The relationship between greenhouse gas levels and temperatures, evident in data from ice cores, illuminates climates in the geological past, and may be a useful guide to the future (Kennedy and Hanson, 2006). Fifty million years ago CO_2 levels may have topped 1000 ppm with sea levels 50 m higher than today. CO_2 gradually decreased as marine organisms fixed carbon through photosynthesis and then buried it by sinking into the ocean basins. This CO_2 reduction and a corresponding decrease in temperature allowed ice sheets to develop in Antarctica starting 30–40 million years ago. By 3–4 million years ago, CO_2 levels probably dropped to or below the pre-industrial level of about 290 ppm and permanent ice sheets appeared in the Northern Hemisphere. As subsequent glaciations came and went, CO_2 concentration and temperature were tightly linked. When both decreased, ice sheets grew and sea levels were lowered by more than 100 m below today's level. When both went up, there were relatively stable periods with high sea levels. A central feature of this baseline is that at no time in at least the past 10 million years have CO_2 levels in the atmosphere exceeded 385 ppm, the 2007 level.

Polar warming

Polar warming is happening now: the Arctic ice is melting. By looking back, we may predict the Earth's climate future. The Pliocene epoch 3–4 million years ago experienced 3°C warmer air than today, with a similar CO_2 content in the atmosphere (Fedorov, Dekens, McCarthy et al., 2006). Sea levels, however, were 25 m higher. Humans, therefore, may already have put the world on a path back to that epoch. The accelerating decay of ice sheets in Greenland and Antarctica, coupled to concurrent rise in CO_2 concentrations, suggest that we should expect other changes as well. The significant warming observed in the Arctic regions from 1920 to 1940 and the subsequent cooling from 1940 to 1960 was natural climate variability. The present warming, also largest in the Arctic, is primarily caused by human activities through increasing release of greenhouse gases to the atmosphere.

Air temperatures over the continent of Antarctica have risen three times faster than the rest of the world over the past 30 years (Turner, Lachlan-Cope, Colwell et al., 2006). By analysing historical data from high-altitude weather balloons, it was concluded that temperatures in the lower 9 km of the atmosphere above Antarctica have risen by more than 2°C since the early 1970s. This is the largest regional warming on Earth at this level. Scientists are keen to understand the change in temperatures over the continent as the region holds enough water in its ice to raise sea levels by 57 m. Temperature rises on parts of the surface of Antarctica have been seen for some time. The western side of the Antarctic Peninsula is known to have the largest annual warming seen anywhere in the world with increases of over 2.5°C in the last 50 years. Such new findings about Antarctic warming are particularly important because, until now, researchers had only partial – and often conflicting – temperature readings from a few surface stations on the icy continent. The finding opens many new research questions for polar climate scientists. The rapid surface warming of the Antarctic Peninsula and the enhanced global warming signal over the whole continent show the complexity of climate change. If greenhouse gases are having an even greater impact in Antarctica than across the rest of the world, the question is why?

The findings from the award-winning Climate and Environment Changes in the Arctic (CECA) project (Nansen Centre, 2005) indicate large potential climate-change consequences of both a positive and negative nature affecting fisheries, oil and gas activities, transport through the Northern Sea Route, ocean circulation, including the North Atlantic Current, and climate in Europe and the Arctic. The ice cover in the Arctic regions has decreased by 3 per cent per decade since 1978. The thicker multi-year ice has been reduced by 7 per cent per decade, which indicates that the Arctic sea ice cover may be in for major transformation (Figure 3.3). A doubling of the atmospheric CO_2 is expected within a hundred years. Simulations by climate models show that the ice could disappear during the summer months. During the winter months the reduction may be about 20 per cent, with the Barents Sea being open all year around.

The increase of greenhouse gases will influence the low-pressure systems between Greenland and Iceland. They will increase in strength and lead to a warmer, wetter and wilder climate in Northern Europe. Bindschadler (2006) shows accelerated sea-level rise and increasing frequency of glacial earthquakes as a result of rapid climate change in the polar regions. Owing to a large temperature

52 *Torleiv Bilstad*

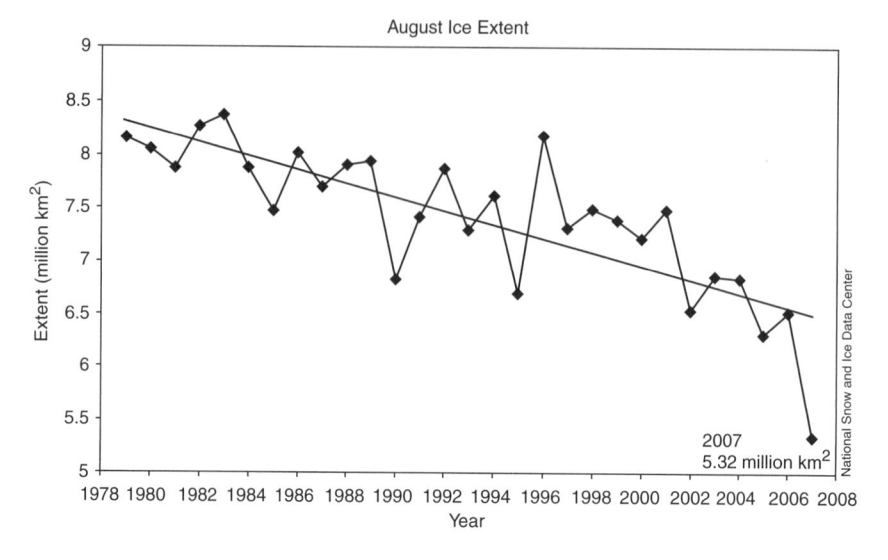

Figure 3.3 Arctic ice cover (Norsk Romsenter/National Snow and Ice Center, 2007).

contrast between the warm ocean water and the cold ice, melting occurs at a very rapid rate. The melting reduces the friction that slows the advances of these glaciers, allowing them to accelerate.

Ekstrom, Nettles and Tsai (2006) also recorded a significant and unexpected increase in the number of glacial earthquakes. Glaciers are thought of as 'inert' and slow moving, but in fact they can also move rather quickly. Some of Greenland's glaciers are as large as Manhattan and may move 10 m in less than a minute, a jolt that is sufficient to generate moderate seismic waves. This increase in the glaciers' rates of flow offer additional potential evidence of global warming.

How fast will the ice sheets of Greenland and Antarctica disappear? How fast and how far will there be a rise in sea level? At the moment, ice loss from Greenland and West Antarctica combined is contributing less than half of the ongoing 2 mm/year rise in sea level. The rest comes from melting mountain glaciers around the globe and the simple thermal expansion of sea water (Kerr, 2006a). Climate model simulations are integrated into ice-sheet models and paleo-climate data to show that the northern latitudes, particularly the Arctic, were significantly warmer during the last inter-glaciations with sea level several metres higher than at present. Models estimate that the Greenland Ice Sheet contributed between 2.2 and 3.4 m of sea-level rise in the penultimate deglaciations. Models predict that surface temperatures will be as high by year 2100 as they were 130,000 years ago; between the last two Ice Ages. The mass of ice decreased by 152 ± 80 km^3 from 2002 to 2005, mostly from the West Antarctic Ice Sheet. The loss of mass from Greenland doubled in the last 10 years (Kerr, 2006a). Ice loss around the margins of the island is proceeding faster than at the centre.

Climate change and consequences for the Arctic 53

Large and sudden sliding of glaciers triggers glacial earthquakes. Around Greenland the annual number of glacial earthquakes has doubled since 2002, as verified by satellite and airborne radar and laser altimeter measurements.

The Greenland ice sheet is 1,700,000 km² in area and up to 3 km thick and would, if melted completely, raise global sea level by 7 m. The velocity of several large glaciers draining the ice sheet to the sea has recently doubled to reach a level of more than 12 km/year. This implies that current estimates of global sea-level rise of about 0.5 (± 0.4) m before 2100 may be too low. Figure 3.4 shows the Northern Hemisphere surface air temperature and CO_2 as measured from ice-core samples.

A compilation of temperature records (Osborn and Briffa, 2006) shows that the geographic extent of recent warmth is greater than the notable Medieval Warm Period of the years 890–1170. The last 100 years are more striking than either the Medieval Warm Period or Little Ice Age. It is a period of widespread warmth affecting nearly all the records analysed from the same time. These observations add more convincing evidence of the impacts of increasing greenhouse gas emissions. As such, the findings support evidence pointing to unprecedented recent climate warming linked with human activity. The Earth in 2006 was absorbing 0.85 ± 0.15 W/m² more energy from the sun than it is re-emitting to space. This imbalance is confirmed by precise measurements of increasing ocean-accumulated heat over the past 10 years. Theoretically this should lead to an additional global warming of about 0.6°C without further change in atmospheric composition (Turner, Lachlan-Cope, Colwell et al., 2006).

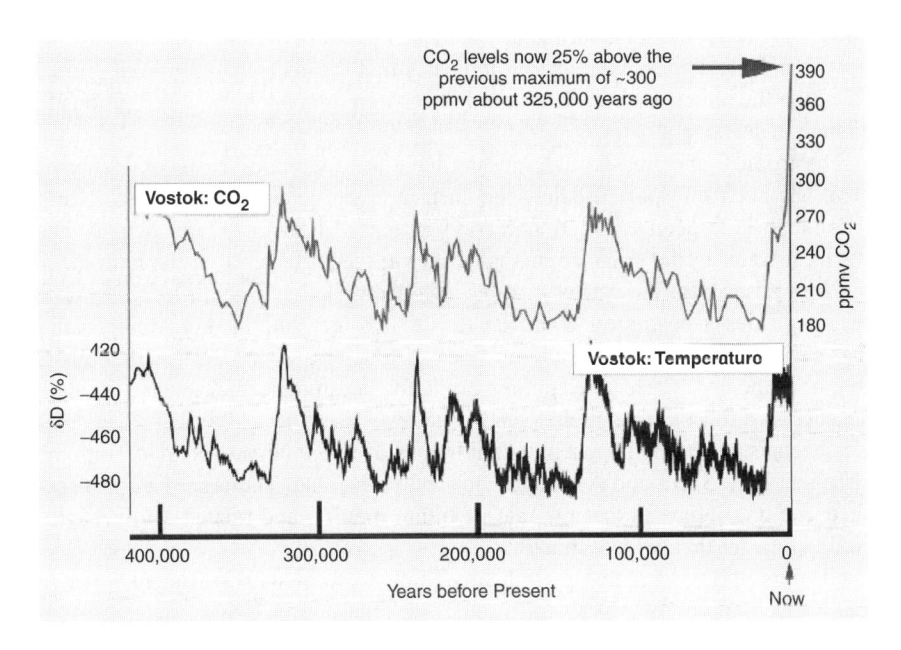

Figure 3.4 Concentrations of CO_2 and temperature from 400,000 years Vostok Ice Core data set (Corell, 2006).

Open ocean – warm ocean water from the tropics

Oceans contain 97 per cent of the Earth's water and, therefore, are fundamental in the global hydrological cycle. Water evaporation from the oceans averages 86 per cent of the total annual evaporation and, as such, is central in planetary energy exchanges. Oceans receive 78 per cent of global precipitation. As an example of the magnitude of this number, a 1 per cent change in the precipitation of the Atlantic equals the annual runoff to the Mississippi River (Kerr, 2006b).

The oceans control the timing and magnitude of changes in the global climate system by acting as a heat sink. Summertime Arctic Ocean ice may soon disappear. Its melting, however, does nothing to increase the volume of ocean water: the ice is already floating in the ocean. The melt water from receding mountain glaciers and ice caps, on the other hand, is certainly raising sea level. Nobody prior to Roald Amundsen in 1906 had tackled the Northwest Passage, and returned alive. It took Amundsen three years to make the journey. In 2006, the transit is easily done in a couple of weeks in open Arctic waters, which in Amundsen's time were ice-clogged. Something has happened in a hundred years. The heat-carrying ocean conveyor that warms the far northern Atlantic is overheating. If the greenhouse disturbs this conveyor, it means trouble (Kerr, 2006b).

A major warming of the Antarctic winter troposphere, larger than any previously identified regional troposphere warming on Earth, came to light by quality-controlled Antarctic radiosonde observations. The data show that regional mid-troposphere temperatures have increased at a statistically significant rate of 0.5–0.7°C per decade over the past 30 years (Turner, Lachlan-Cope, Colwell et al., 2006). There is also evidence that warmer subsurface waters are reaching the Earth's polar latitudes. Moreover, it indicates that the ocean plays a more critical role than the atmosphere in determining near-term glaciological contributions to change in sea level.

The tropical and mid-latitude oceans have been warming in recent decades. Heat has been transported below the surface, even below 1000 m in the North Atlantic, where deep convection carried increased heat to greater depths. The warmest water in the polar oceans is neither at the surface nor at the bottom. In the Amundsen Sea, the warmest water is concentrated at a depth of 600 m. The recent glacier acceleration could be due to warmer intermediate-depth water that can access the deep grounding lines of glaciers, where the ice first floats free from the bed. These glaciers flow out to the ocean in deep channels – floating ice shelves a few hundred metres thick. Sea-level rise from melting of these polar ice sheets is one of the largest potential threats of future climate change. Both the Greenland Ice Sheet and Antarctica may be vulnerable. The record of past ice-sheet melting indicates that the rate of future melting and related sea-level rise could be faster than widely thought.

Glaciers and ice streams periodically lurch forward with sufficient force to generate emissions of elastic waves that are recorded on seismometers worldwide. Such glacial earthquakes on Greenland show a strong seasonality as well as doubling of their rate of occurrence over the past five years. These temporal patterns suggest a link to the hydrological cycle and are indicative of a dynamic glacial response to changing climate conditions (Ekstrom, Nettles and Tsai et al., 2006).

Concluding remarks

At no time in at least the past 10 million years has the atmospheric CO_2 concentration exceeded the present value of 385 ppm. In Alaska and western Canada, the average winter temperatures have increased by as much as 3–4°C over the past 60 years. The global average increase over the past 100 years has been 0.6 (± 0.2)°C.

Glaciers are disappearing and we are losing unique archives of the Earth's climate history. The twentieth-century warming is apparent in the Arctic ice cores as well as in the low-latitude, high-altitude ice cores. The loss of low-latitude glaciers threatens the water resources in many parts of the world. These 'water towers of the world' affect nearly half of the world's population in Asia, Africa and South America. Glaciers are our most visible evidence of global warming. They integrate many climate variables in the Earth's system.

Changes in absorption of heat by the ocean as the ice disappears means a switch from 85 per cent reflected (albedo) to 85 per cent absorbed. Population growth is the main driver for climate change. The never-ending search for energy results in burning of fossils and forests at an accelerating rate adding greenhouse gases to the atmosphere.

Increasing effort is required in order to develop new energy technologies, currently far from the market. Governments must play a greater role in stimulating this process, by investing in research and development. The world needs a new global 'Apollo' project. This would stimulate education and enrolment at universities in science and technology.

References

Arctic Climate Impact Assessment (ACIA 2004) *Impacts of a Warmer Arctic: Arctic Climate Impact Assessment*, Cambridge: Cambridge University Press.
Bindschadler, R. (2006) 'Hitting the ice sheets where it hurts', *Science*, 311 (March 24): 1720.
Corell, R. (2006) *Presentation at Transatlantic Symposium*, June 15. Environmental and Energy Study Institute. Online. Available HTTP: <www.ccsi.org/bricfings/2006/Energy & Climate/6.15.06_ClimateSymposium>.
Ekstrom, G., Nettles, M. and Tsai, V.C. (2006) 'Seasonal and increasing frequency of Greenland glacial earthquakes', *Science*, 311 (March 24): 1756.
Fedorov, A. V., Dekens, P. S., McCarthy, M., Ravelo, A.C., deMenocal, P.B., Barreiro, M., Pacanowski, R. C. and Philander, S. G. (2006) 'The Pliocene paradox', *Science*, 312 (June 9): 1485.
Houghton, J. (2004) *Global Warming: The Complete Briefing*, 3rd edition, Cambridge: Cambridge University Press.
Intergovernmental Panel on Climate Change (IPCC, 2007). Online. Available HTTP: <http://www.ipcc.ch/> (Accessed 19 November 2007).
Kennedy, D. and Hanson, B. (2006) 'Ice and history', *Science*, 311 (March 24): 1673.

Kerr, R. A. (2006a) 'An entrepreneur does climate science', *Science*, 311 (February 24): 1088.
Kerr, R. A. (2006b) 'A worrying trend of less ice, higher seas', *Science*, 311 (March 24): 1698.
Nansen Environmental and Remote Sensing Centre (2005) *Annual Report*.
Norsk Romsenter/National Snow and Ice Data Center (2007).
Osborn, T. and Briffa, K. (2006) 'The spatial extent of 20th-century warmth in the context of the past 1200 years', *Science*, 311 (February 10): 841.
Petit, J. R., Jouzel, J., Raynaud, D., et al. (1999). 'Climate and atmospheric history of the past 420,000 years from the Vostok ice core', Antarctica, *Nature* 399: 429–436.
Turner, J., Lachlan-Cope, T. A., Colwell, S, Marshall, G. J. and Connolley, W. M. (2006) 'Significant warming of the Antarctic winter troposphere', *Science*, 311 (March 31): 1914.
Zimov, S. A., Scuur, A. G. and Chapiin, III F. S. (2006) 'Permafrost and the global carbon budget', *Science*, 312 (June 16): 1612.

4 Corporate social responsibility

The economic and institutional responsibility of business in society

Bjørn-Tore Blindheim

Introduction

The oil and gas industry is expanding its activities in the Arctic. Corporate activity in the region has the potential to affect Arctic societies both positively and negatively. According to the *Arctic Human Development Report* (2004: 10), Arctic societies today face an:

> unprecedented combination of rapid and stressful changes involving environmental processes (e.g. the impacts of climate change), cultural developments (e.g. the erosion of indigenous languages), economic changes (e.g. the emergence of narrowly based mixed economies), industrial developments (e.g. the growing role of multinational corporations engaged in the extraction of natural resources), and political changes (e.g. the devolution of political authority).[1]

A fundamental question is what responsibility the oil and gas companies have to address such challenges, which are all, in some way or another, tied together as challenges of sustainable development.

Since the 1950s, the role of business in, and the responsibility of companies towards, society has increasingly been addressed through the concept of corporate social responsibility (CSR). The idea of social responsibility has its modern roots in the work of Berle and Means (1932). They documented a separation of ownership from control in large United States (US) corporations, thus resulting in a:

> small (managerial) group, sitting at the head of enormous organizations, with the power to build, and destroy, communities, to generate great productivity and wealth, but also to control the distribution of that wealth, without regard for those who elected them (the stockholders) or those who depended on them (the larger public).
>
> (Mizruchi, 2004)

In line with Berle and Means' concern that increased corporate and managerial power could harm public interests, early definitions of CSR were tied more to society's interests than to those of the firm. As such, CSR built upon moral ideas about the primacy of human interests over corporate ones and the desire to modify many of the negative consequences of corporate power – environmental degradation and poisoning, unhealthy products, inhumane workplaces, and more (Logsdon and Wood, 2002). In 1953, Bowen defined CSR as the obligations of businessmen to pursue those policies, to make those decisions, or to follow those lines of action that are desirable in terms of the objectives and values of society. Later, Davis (1973) defined CSR as the firm's consideration of, and response to, issues beyond its narrow economic, technical and legal requirements to accomplish social benefits, along with the traditional economic gains that the firm seeks. Crane and Matten (2004) argue that probably the most established and accepted conceptualization of CSR is the 'four-part model of corporate social responsibility' initially proposed by Carroll (1979), who suggested CSR as a multi-layered concept that can be differentiated into the four interrelated aspects of economic, legal, ethical and philanthropic responsibilities. Carroll and Buckholtz (2000) define CSR as 'the economic, legal, ethical, and philanthropic expectations placed on organizations by society at a given point in time'. The concept later evolved into different approaches, covering other related terms such as social responsiveness (Frederick, 1987), corporate social performance (Wood, 1991), the stakeholder approach (Freeman, 1984), corporate citizenship (Crane and Matten, 2004), the 'triple bottom line' approach (Elkington, 1994, 1997) and corporate sustainability (Marrewijk, 2003). In other words, CSR may be understood as an umbrella term covering economic, social and environmental issues (Welford, 2003), wherein the relationship between business and society is studied.

Today, the idea that CSR may promote complex societal challenges and the common good seems well established (Pogutz, 2007). Internationally influential organizations, particularly the United Nations (UN), the European Union and the World Bank, depict a positive relationship between CSR and sustainable development, and point towards how CSR may promote a sustainable development path. For example, the Brundtland Report (1987) stated that: 'Industry's response to pollution and resource degradation has not been and should not be limited to compliance with regulations. It should accept a broad sense of social responsibility and ensure an awareness of environmental considerations at all levels' (World Commission on Environment and Development; WCED, 1987: 222). In a similar vein, the 2002 Report of the World Summit on Sustainable Development (WSSD) held in Johannesburg, South Africa, stated that the business sector – pursuing its legitimate activities – has a duty to contribute to the evolution of equitable and sustainable communities and societies. On this background, the report calls for enhanced CSR and accountability. In summary, at both the international and national levels, CSR policy is developed in the context of an acceptance of CSR as an important contributor to the wider goal of sustainable development (Buckland, Albareda, Lozano et al., 2006).

As I will argue, the contemporary discourse and conceptualization of CSR lies largely within what may be called the 'social activist' (Brummer, 1991) or ethical (Garriga and Melé, 2004) approach to CSR, implying an expanded and radical role of business in society, compared to the more classical economic or functionalist conceptualization of the role of business in society. My argument is that such an approach to CSR – postulating a positive impact on broader societal interests and sustainable development, based upon companies' and managers' voluntary efforts to address social misery – may in fact contribute to the opposite, which is to undermine a sustainable development path. As such, this chapter offers a critical look at CSR and its ability to further societal interests, and I agree with Blowfield (2005) about the limits to what can be expected of business and its contribution to the common good. On this background, I suggest a conceptualization of CSR that both takes seriously the basic aim of every business organization (making profit), and acknowledge the need for 'a political order where economic rationality is circumscribed by democratic institutions and procedures' (Scherer and Palazzo, 2007: 1097) in order to promote corporate responsible behaviour.

Towards a 'radicalized' role for business in society?

Since the time of Berle and Means (1932), Bowen (1953), and Davis (1973), the field of CSR has grown significantly and today contains a great proliferation of approaches. Several typologies and classifications have been suggested to bring some order into the business in society literature (e.g. Frederick, 1987, 1998; Carroll, 1999; Garriga and Melé, 2004). In order to discuss the contemporary discourse and agenda of CSR, and in order to frame my own conceptualization of CSR, I chose to build on the work of Brummer (1991). He suggested that the spectrum of approaches to CSR could be ordered in the classical, stakeholder, social demanding and social activist approaches to CSR.

Corporate social responsibility – from 'business as usual' to managers as social and moral leaders

The classical approach to CSR, which can also be framed as arguments against CSR, comes in two different variants. Building on classical Parsonian pluralism (Parsons, 1951), it could be argued that other institutions in society – like political institutions and civil society institutions – exist to perform the types of functions required by social responsibility (Jones, 1999). The functional theory argument largely defines CSR along the same economic dimension as identified by Friedman (1962). His property rights argument against CSR – above what is profitable – has its roots in classical capitalism. This perspective maintains that managers have no right to act other than to enhance shareholder value. To do otherwise constitutes a violation of the management's legal, moral and fiduciary responsibilities. In sum, the social obligations of business are confined to satisfying legal and economic criteria.

Contrary to the classical perspective, the stakeholder perspective suggests that responsibilities of a business extend beyond shareholders to include the company's stakeholders. In general, stakeholder theory is focused on those interests and actors who affect, or in turn are affected by, the corporation (Freeman, 1984). Stakeholders can be defined as persons or groups with legitimate interests in procedural or substantive aspects of corporate activity (Donaldson and Preston, 1995: 67). It is their interests in the corporation that identify the stakeholders, whether or not the corporation has any corresponding functional interest in them. Freeman's stakeholder theory asserts that managers must satisfy a variety of different individuals or groups in or outside the corporation. This could be a 'primary' stakeholder like the providers of capital, customers, employees and suppliers, but also more 'secondary' stakeholders like governments, local community organizations, indigenous people and non-governmental organizations (NGOs). Stakeholder theory implies that it can be beneficial for the firm to engage in certain CSR activities that stakeholders define as important. Otherwise, stakeholders might withdraw their support from the firm.

The social demandingness approach holds that corporations are responsible to carry out those activities that society (not just stakeholders) demands and expects of them. A foundational idea is that, since business depends on society for its existence and growth, business should integrate social demands and expectations into its activities so that they operate in accordance with the prevailing social values. As such, the approach is inherently relativistic: It does not state any specific action that corporations and their managers are always responsible to perform. The actual content of CSR is dependent both upon time and place, that is, what society currently defines as their societal responsibility.

In contrast to the social demandingness approach to CSR, the social activist approach to CSR holds that universal standards or values should determine corporate and managerial decision-making and action, independent of the view of shifting coalitions of stakeholders or expectations from society at large. Brummer (1991: 190) summarizes the social activist approach to CSR in the following way:

> It (the social activist approach to CSR) holds that executives are responsible for pursuing social or moral goals from voluntary motives, even when doing so compromises the firm's profit performance (at least in the short term). Corporations or their members are required to perform acts that benefit shareholders, stakeholders, and the general public, both in the primary areas of their business decision making (where the direct effects of their actions are more likely to be noticed) and in secondary and tertiary areas as well (where the indirect effects become more prominent). Last, in considering the interests and welfare of others, corporate executives are to respond to the formers' ideal or rational interests rather than merely their expressed or current interests.

The degrees of responsibility assigned to businesses and companies in the different approaches to CSR are summarized in Figure 4.1:

Limited responsibility | Classical   Stakeholder   Social demandingness   Social activist | Expanded responsibility

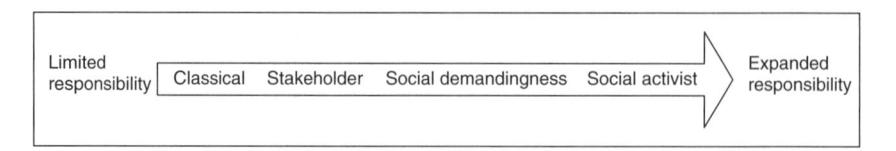

Figure 4.1 The degrees of responsibility assigned to business in different approaches to CSR.

The arrow pointing towards the right indicates increasing degrees of responsibility towards society, from limited to expanded responsibility. The classical approach to CSR falls closest to the left end of the arrow ('the only responsibility of business is to make profit'), while we find the social activist approach at the other end of the spectrum ('corporate managers as moral and social leaders'). As I will argue below, much of the contemporary discourse, agenda and conceptualization of CSR seems to fall within the social activist approach to CSR, depicting not only an expanded role for business in society, but also an alternative model of societal governance that may have the potential to undermine the institutional conditions for sustainable development.

The contemporary discourse and agenda of CSR: towards a radicalized role of business in society and an alternative model of societal governance?

A common feature between the classical, stakeholder and social demandingness approaches to CSR is that the limit of business' responsibility towards society is largely set with the companies' primary, and to some degree, secondary sphere of influence or area of decision-making. Today, there is clear evidence that the CSR agenda is widening to also encompass business responsibility for the more indirect effects of their operations, or to their tertiary sphere of influence, more in line with the social activist approach to CSR. While early definitions of the concept emphasized that social responsibility was about minimizing the negative impact of corporate activities on society (Blowfield, 2005), several organizations today see CSR as a positive contribution to development and a possible answer to complex societal challenges. The scope of CSR has been broadened to include not only aspects of corporate conduct that impinge on social, environmental and human rights issues, but also the role of business in relation to poverty reduction in the developing world and to questions of development, in general (Prieto-Carrón, Lund-Thomsen, Chan et al., 2006). This development agenda is, for example, a very important part of the best known and largest CSR initiative, the 'United Nations Global Compact' (Global Compact, 2005: 8), which explicitly points towards business responsibility beyond business' direct sphere of influence:

> By developing and implementing policies in the four areas of the Global Compact – human rights, labour, the environment and anti-corruption – companies are, by definition, contributing to the process of sustainable development. In addition, by forging partnerships with other stakeholders, businesses have the opportunity to scale up action within and even beyond their direct sphere of influence. The full integration of the ten principles, particularly in low-income countries ... can make companies a driving force for development.

Similarly, the World Bank defines CSR as the 'commitment of business to contribute to sustainable economic development, working with employees, their families, the local community and society at large to improve their quality of life, in ways that are both good for business and good for development' (Ward, 2004). Clearly, working with employee's families, local community and the society at large goes beyond how the relationship between business and society has been depicted in the past, and what has been considered as the primary responsibility and function of business. The role of the manager is not only to be a business leader, but – as emphasized by the social activist approach to CSR – to assume the role of social and moral leader. As stated in a report about the role of business in society from the World Business Council for Sustainable Development (WBCSD, 2006: 11): 'the challenges for business are to understand the roots and nature of poverty ...'. The expanded CSR agenda is not only a reality in words and definitions, but also in corporate practice. For example, oil companies now help to build schools and hospitals, launch micro-credit schemes for local people and assist youth employment programmes in developing countries (Frynas, 2005). 'Social investments' constitute part of the total activity plan for social responsibility activities for the Norwegian oil and gas company Statoil, although they define development projects and social investments as outside the company's area of responsibility (Statoil, 2005). According to the company's sustainable development report (Statoil, 2005: 61), the projects supported by Statoil aim to build local capacity, promote human rights and transparency, and improve local conditions relating to health, safety and the environment. In 2005, Statoil's social investments totalled about US $8 million and included projects in 11 countries. One of the projects that received support in 2005 was the Akassa development project in the Niger Delta. Initiated in 1997, this project covers activities in the areas of abolition of poverty, local capacity building, environment, infrastructure and institutional capacity development. Under these headings, Statoil has contributed to establishing – among other things – micro-credit loans, health stations and pharmacies, nursery schools, educational units, and the building of bridges and schools.

In a survey of global leaders in business, civil society and the media, Nelson, Hodges, Deri et al. (2005) found evidence that development issues today are firmly on the CSR agenda, both in the minds of business leaders and in company practice. All of the business leaders in the study reported that their companies are actively involved in supporting the local communities in which they invest and market their products and services. On the question of what kinds of development project were seen as most promising or important for business to perform, the respondents from the private sector identified the companies' capacity to create jobs and build local businesses as the essential foundation for long-term development, and the area where business could contribute the most or make the most impact. Issues like ensuring environmental sustainability, training the country's local workforce, tackling bribery and corruption, investing in infrastructure, promoting gender equality and empowering women, were also considered as important areas where business should play a role. Further down the list, but also recognized as important areas where business could and should make a contribution, came issues like combating human immunodeficiency virus/acquired immunodeficiency syndrome (HIV/AIDS) and other diseases, investing in higher education and new technology, and fostering universal primary education. Interestingly, all respondents reported heightened expectations from society in general of the role that business can and should play in development challenges.

Although there were concerns that the expectations placed upon business often put too much faith in the ability of business to solve development challenges, respondents from all groups agreed that development issues will and should play an increasingly important role in corporate strategy in the future. Nelson, Hodges, Deri et al. (2005: 21) note that: 'As globalization continues, the private sector will continue to expand into emerging markets around the world, and corporations will consequently assume greater responsibility for the well-being of those to whom they market their products and services'.

Together, the development from understanding CSR as being about minimizing the negative impact of corporate activities on society, to understanding it as a positive contribution to societal development where corporate managers take on the role as social and moral leaders, not only fits the social activist approach to CSR, but also indicates that a new and alternative model of societal governance is emerging. Buckland, Albareda, Lozano et al. (2006: 7) argue that the rise of CSR must be understood in the context of the new globalized economy and a crisis in the welfare state. This has led people and governments to look for 'new ways of developing and funding collective action to deal with social demands that cannot be met by the state alone such as poverty, unemployment, lack of economic development and social exclusion of key groups'. An important point is that the expanded CSR agenda is seen not only as the 'solution' to development challenges in the developing world, but also as a framework within which new ways of collaboration and partnership between business, governments and civil society are used as a mechanism for developing new models of governance to address the major social problems faced by post-industrial societies (Buckland, Albareda, Lozano et al., 2006). In a similar vein, Salin-Andersson (2006) argues that CSR must be understood as a mobilization of corporate actors to assist the development aid to states. This trend, she argues, is one in which corporations are seen as complementing and sometimes replacing states as the primary structures and shapers of the world, implying a more active role for business in society than before. Matten and Crane (2005) argue that companies today are involved in the administration of citizen's civil, social and political rights, an area and responsibility normally considered to lie within the sphere of politics and government. Moon, Crane and Matten (2005) have suggested two ways in which companies share in governing: first by contributing to societal governance issues outside the firm, often in partnerships with governmental or non-governmental organizations; and, second, by administering rights within the normal operations of the companies.

While the new global 'soft-law' regulatory regime of CSR may be understood as a criticism of corporations, the trend of corporations as 'development agencies' does not reflect such criticism, but rather builds on the view of corporations as strong and legitimate players in building a 'global welfare state' (Salin-Andersson, 2006). For example, Dunfee and Hess (2000) argue that 'private firms are uniquely positioned to provide significant relief to the misery that pervades the developing world' and that private firms have a 'competitive advantage' over nation-states and NGOs in the provision of aid. Then the question becomes: what is the potential outcome of such a 'radicalized' role of business in society? My argument – outlined in the next section – is that it may contribute to undermining and not promoting societal interest and sustainable development. It may do so by undermining what can be called the institutional conditions for sustainable development.

The potential impact of CSR on sustainable development

The question I address in this section is how CSR may impact on society in general and on sustainable development in particular. An analysis of the relationship between CSR and sustainable development is, however, full of complications. Although sustainable development has regional and local implications (see Chapter 2), it is first and foremost a macro-level concept and its challenges are global in character. CSR, on the other hand, operates at the micro-level and, as argued by Zadek (2001: 122), has to do 'with keeping an organization going and at best doing some good in the process and not too much harm'. Further, we do not know enough about the system to understand the critical aspects of the relationship between the decisions and activities of a business organization and its impact on the whole (Zadek, 2001). Understanding what decisions and actions really make a contribution to sustainable development in the long run is – on this background – not easy. Still, our argument is that the very foundational ideas and the expanded agenda of CSR may undermine what can be called the institutional condition for sustainable development.

The institutional condition for sustainable development

In this book, we mainly build on the interpretation of sustainable development as suggested by the Brundtland Report (WCED, 1987) (accounted for in length in Chapter 2), which identified economic prosperity, social equity and environmental integrity as the three principles that ground the concept. The direction towards sustainability is, however, not easily identified. Being the result of an optimization process for independent but interacting targets (i.e. economic development and environmental protection), it necessarily does not have a single, clear-cut solution, but includes a range of options to choose from (Spangenberg, 2002). However, building on the Agenda 21 (UN 1993), Spangenberg (2002) suggests that sustainable development not only has economic, social and environmental dimensions, but also an institutional dimension, important in its own right, that also functions as a *condition* to address the challenges within the other dimensions of sustainable development.

In general, the institutional dimension of sustainable development (in economics referred to as 'social capital') covers the systems of rules governing the interaction of members of society (Czada, 1995). As a condition for sustainable development, the institutional dimension provides the means for societal decision-making, determining the form of economic and social activity, thus also influencing the impact of social and economic activity on the physical environment.

Building on the Brundtland Report (1987), and with respect to the institutional requirements for sustainable development, public policy and the design of laws may be characterized as perhaps the most important mechanism for obtaining a sustainable development path. The Brundtland Report is very clear about the need for increased state responsibility and legal means to address the challenges of sustainable development (WCED, 1987: 330):

> legal regimes are being rapidly outdistanced by the accelerating pace and expanding scale of impacts on the environmental base of development. Human laws must be reformulated to keep human activities in harmony with the unchanging and universal laws of nature.

Corporate social responsibility: undermining the institutional condition for sustainable development

Sustainable development rest on an institutional condition (Spangenberg, 2002): the power and capability of the states and political institutions to shape and implement national and international policy, laws and regulations for more sustainable forms of development. CSR has the potential to undermine this condition. The central reason is that CSR may imply increased power to business in society at the possible expense of political and civil society power. At the same time, the foundational features of capitalism may work against using this power to integrate broader social and environmental considerations voluntarily in business decision-making and activity.

Corporate social responsibility: a defence against governmental regulation and public criticism

Originally, and although early definitions of CSR were tied more to society's interests than to those of the firm, the concept never intended to serve broader societal interests and the common good. Rather, the concept was intended to protect business from criticism from civil society and against mandatory state regulations. The origins of business and societal thinking and the concept of CSR had two interconnected causes (Frederick, 1987). First of all, the late nineteenth and early twentieth centuries were a time of intense concern and fear about the economic and social consequences of the power of the giant corporations being formed in that period in the US. One result of this fear was liberal and radical criticism directed against business in general, and especially against the large corporations. Second, and as a response to increased corporate power, new, mandatory regulations were imposed on business in the form of antitrust laws, banking regulations, food and drug regulations and public utility guidelines to a wide range of corporations. It was:

> in this climate of increasing public alarm about business power and of expanding government control [that] we find business executives beginning to speak of their social responsibilities. Their ideas laid the foundation of what we now recognize as the ... theory of corporate social responsibility.
> (Frederick, 1987: 143)

In this way, CSR was first and foremost a possible defence against social criticism and governmental power in order to secure the continued pursuit of profit:

> Corporate social responsibility offered a welcome alternative to government interventions in private affairs, as well as shelter from the charge that a heartless, profit-minded business system cared so little for the general public that it deserved to be abolished or severely curbed. All that was needed to counter such criticism, according to the CSR doctrine, was for business to accept its social responsibilities.
> (Fredrick, 1987: 145)

As argued by Henderson (2001), the driving force behind the new 'CSR wave' during the 1990s was exactly the same as when the concept was first formulated. On this background, there are reasons to be highly sceptical about the ability of CSR to promote societal interest and sustainable development.

Corporate social responsibility: a new governance model implying increased corporate power?

Originally, CSR rested on two foundational principles, charity and stewardship, and six fundamental precepts (Frederick, 1987):

1. Power begets responsibility.
2. A voluntary assumption of responsibility is preferable to government intervention and regulation.
3. Voluntary social responsibility requires business leaders to acknowledge and accept the legitimate claims, rights, and needs of other groups in society.
4. CSR requires a respect for law and for the rules of the game that govern marketplace relations.
5. An attitude of 'enlightened self-interest' leads socially responsible business firms to take a long-run view of profits.
6. Greater economic, social and political stability – and therefore a lower level of social criticism directed towards the private enterprise

system – will result if all business adopts a socially responsible posture.

Waddock (2004) argues that the general precepts or foundational ideas behind CSR are pretty much the same today. As argued by Jones (1996), it is exactly the paternalistic stewardship principle behind social responsibility that in its full context serves to legitimate the hierarchical domination of business in society rather than to encourage democratic pluralism. In slightly other words, in these foundational ideas and precepts lies the foundation for potential increased power to business in society, a power that today manifests itself in a new model of societal governance, where business and companies stand out as legitimate interpreters of the common good and providers of societal welfare. Business today takes part in the provision and administration of citizenship rights (Matten and Crane, 2005), both in the Western world and developing countries. As such, CSR has changed the roles and responsibilities of governments, business and civil society in delivering public welfare, and in promoting social and environmental practice (Matten and Moon, 2005; Midttun, 2005; Roome, 2005). This implies that the role of business is not only to conduct its activity within the economic sphere of society, but also to address and be part of the solution to broader societal challenges and sustainable development. Furthermore, corporate managers not only have a role to play as business leaders, but also must take on a role of moral, social and political leaders. In fact, as shown by Jenkins (2005), policymakers are now advocating CSR as an alternative route to the traditional public delivery of development. However, there is a serious 'pitfall' to this new governance model (Salin-Andersson, 2006):

> Within this framework there are no forces to limit the size and power of corporations and this trend may lead to a shifting of boundaries among states, society and corporations. Just as the trend builds on the strength and power of corporations, it may reinforce this power and add to a transfer of responsibilities and resources from states and civil society to corporations.

A possible transfer of power, responsibilities and resources from states to business may imply that the political systems' ability to address broader societal goals and the challenges of sustainable development is reduced. Frynas (2005) argues that, in developing countries – and in regions or local communities with a weak institutional framework – this may take place in the following way: government failure to deliver basic services like schools, health and education may lead to demands and expectations for a role of CSR in development, that is, in delivering such services. In spite of the fact that CSR normally will not be capable of playing such a development role, this nevertheless will ease the pressure on the government to take the responsibility for welfare services and securing the basic rights of its citizens.

The development of a 'CSR governance model' means that the distinction between the roles and responsibilities of the specialized institutions in society – heavily emphasized within the functional theory version of the classical approach to CSR – is blurred. Expressed in a slightly different way, CSR challenges the view that social welfare is best enhanced when the specialized institutions of society stick to their respective core objectives and activities. Levitt (1958: 44) very strongly argued against such a development on the basis of functional theory. He feared that companies – assuming an ever-widening social responsibility and power – would reshape not only the economic, but also the 'institutional, social, cultural and political topography of society'. And here is the danger as expressed by Levitt, that while the corporation will transform itself in the process, at bottom its outlook will remain narrowly economic and materialistic:

> What we have, then, is the frightening spectacle of a powerful economic functional group whose future and perception are shaped in a high materialistic context of money and things but which imposes its narrow ideas about a broad spectrum of unrelated noneconomic subjects on the mass of man and society.
> (Levitt, 1958: 45)

As argued by Jones (1996: 22), an empirical examination of the 'contemporary institutional landscape ... reveal(s) an increasing dominance of institutions associated with economic rationality, and the progressing colonization of noneconomic institutions by economic ones'.

Corporate social responsibility: mystifying the essential capitalist forces driving business activity

The voluntary nature of CSR may be interpreted as part of a wider revisiting of the role of government, underpinned by the assumption that companies are capable of policing themselves in the absence of binding international and national laws to regulate corporate behaviour (Blowfield and Frynas, 2005). As argued by Jones (1996), the whole concept of CSR and the discourse around it mystifies the essential forces driving business. There is very little reason to believe that business on a voluntary basis will integrate broader social and environmental considerations in decision-making and action. The concept of CSR – and especially the ethical approach to it – largely ignores how the foundational features of capitalism and the basic purpose of the firm itself within the capitalist system, structure and determine the pursuit of profit above all other considerations. Jones (1996) describes capitalism as an economic system based upon private property, production for profit, wage labour and use of the market mechanism for allocating society's productive resources efficiently: 'Capitalism is all about seeking profits by avoiding or eliminating competition, maximizing organizational productivity, and socializing the costs of production'. Within this capitalist system, the basic purpose of the firm is to increase its profit:

> Any capitalist firm essentially represents a package of human, physical, (and) capital resources that have been organized for a single overriding purpose: the pursuit of profit for its owners. These organizations do not exist

to solve society's problems, or to provide enriching jobs for their members (unless there is a positive linkage between job satisfaction and labour productivity), or to satisfy customers' needs. Employees are a resource to be utilized, a means to an end; society provides critical resources (e.g. customers, legitimacy) that the organization must obtain for survival and growth, as well as a site for externalizing the costs of production; customers' needs are to be met (as well as created) not as an end in itself, but a means to secure profits.

(Jones, 1996: 15)

Although capitalism comes in various forms (Hall and Soskice, 2001), and there are many nuances to the picture depicted above, the point is that capitalism imposes great limitations on the voluntary integration of broader societal considerations in decision-making and activity. One would simply not expect capitalist organizations voluntarily to adopt behaviour that 'flies directly in the face of their basic institutional rationality' (Jones, 1996: 25). As also pointed out by Herman (1981) and Mintzberg (1983), ethic-based arguments in support of social responsibility are not sustainable in light of an understanding of the dynamics of capitalist-bureaucratic organizations. Several empirical studies support this point. Herman (1981), Schwartz (1987) and Zeitlin (1989) all found that managers very much act in accordance with the profit-maximizing tenets of classical capitalism, and not with the more multi-dimensional CSR model of the firm, suggesting that companies and managers on a voluntary basis will not adopt broader societal interests.

On the background of the original aim of CSR – that of being a line of defence against state regulation and public criticism – and the structural features of capitalism, it is no wonder that there seems to be a big gap between the 'talk' and 'realities' of CSR (Campbell, 2007), and that CSR is struggling with the objective of promoting development. Having studied the CSR initiatives of oil companies in the Niger Delta, Frynas (2005) found that most CSR initiatives did not go beyond narrowly philanthropic gestures, as, for example, donating objects such as schoolbooks, mosquito nets or lifejackets to local communities. Frynas (2005) also found that the oil and gas companies failed to involve the intended beneficiaries of CSR, the local communities, the companies did not have the proper human resources to address complicated societal challenges, and the companies failed to integrate their CSR efforts into larger development plans, making the efforts highly inefficient from a societal point of view. In addition, one company's CSR efforts – whether a local community project or codes of conduct imposed on suppliers – are seldom coordinated with other companies' CSR efforts and initiatives. In this way, the potentially positive effects and outcome of CSR are reduced compared to what might have been.

Corporate social responsibility: effectively closing the door for regulative efforts to address the challenges of sustainable development

It may also be argued that the concept of CSR, and the new governance model rising around it, co-opt and diffuse potentially countervailing forces. As argued by Jones (1996), if one were to believe business is socially responsible, what plausible reason could there be to support government regulations or any other measures that impinge on the activities of benevolent, enlightened professional management? In this way, the social activist approach to CSR suggests a 'kinder, gentler' capitalism that does not require the vigilance of countervailing forces to keep it honest because it is essentially benign, or at least can be made so through existing and voluntary organizational and managerial procedural mechanisms.

In summary, it may be a mistake to view CSR only as a 'toolbox' for managing social issues, as suggested within the social responsiveness approach to CSR. Instead, CSR must be understood within the larger context of governance. From such a perspective, it becomes clear how CSR may exceed the limits of the economic sphere of society into the domains of public policy and civil society, at the possible expense of the power and resources of public policy and civil society to address complex societal and sustainable development challenges.

The economic and institutional responsibility of business

A conceptualization of CSR must take seriously the foundational institutional rationality to which any business firm is subject, that of capital accumulation (Jones, 1996) and, the *raison d'être* of any business firm, that of enhancing profit to shareholders and owners. As argued by Blowfield (2005), the strength of CSR lies not in presenting an alternative model of business, but in capturing and presenting the moral dimensions of capitalism in ways that resonate with investors and consumers, and are actionable by managers. On this background, I suggest a conceptualization of CSR that limits the developmental, moral and political aspirations of companies in the societal arena. Instead, focus should first of all be directed to the economic responsibility of business towards society. Further, focus should be directed to the responsibility of business to adapt its economic activity to the institutional framework and systems of governance defining – both through mandatory laws and voluntary guidelines – legitimate wealth generation in society. The possible strength of such a conceptualization is that it confines the power of business in society, and opens the door for government regulations and other measures to impinge on corporate activity that might harm societal interest and sustainable development, by not taking for granted that, owing to the grounding features of capitalism, business will voluntarily do well for society.

The economic and institutional responsibility of business

Our conceptualization of CSR builds on the classical approach to CSR (Brummer, 1991) and on functional theory (Parsons, 1951) and builds on the following basic assumptions:

> Society may be understood as consisting of different but interacting spheres of activity: business, political and civil society (Waddell and Brown, 1997; Waddell, 2000), all framed by the natural environment (Waddock, 2002)

constituting the external limit of the total system (Daly, 1992). For the purpose of the functioning of the total system, the basic purpose of the different spheres of society differ; hence, the institutions belonging to the different spheres of society have different basic aims, roles, tasks, and responsibility.

The basic aim of business – working within the economic sphere of society and within a system of democratic capitalism – is to make profit. Companies thus primarily have an *economic responsibility* towards society. Other institutions – political and civil society institutions – are better suited to perform tasks (e.g. the provision and administration of basic citizenship rights) that are outside the domain of wealth generation.

The activity of companies – within the economic sphere of society – is framed by the larger society, including both political and civil society institutions (hence the notion of business *in* – and not *and* – society). The basic purpose of political institutions and governments may be characterized as, based upon democratic elections and processes of public policy, defining and establishing the 'common good'. For the purpose of establishing that common good, business and companies thus have a responsibility not to undermine the capacity of political and civil society institutions to perform the task originating from their foundational role in society, an *institutional responsibility*.

The economic responsibility of business

Jones (1996) argues that any concept of CSR must be positioned with respect to an understanding of capitalist political economy. Taking this seriously, and building on functional theory, business has first and foremost an economic responsibility towards society. According to Carroll (1979, 1991), all subsequent responsibilities of corporations are based on the economic responsibility of CSR. Companies have shareholders who demand a reasonable return on their investments, they have employees who want safe and fairly paid jobs, and they have customers who demand good-quality products at a fair price. By definition, this is the reason why businesses are set up in society, and their first responsibility is to function properly as an economic unit and to stay in business. The actual content of the economic responsibility of business will tend to vary from society to society (Matten and Moon, 2005). Within an American context, this responsibility is closely related to the company as a profit-seeking unit intended to serve the economic interests of the shareholders. As noted by Hunt (2000), the economic responsibility within a European context may also include not only a responsibility towards the owners, but also to the employees and local community.

The economic responsibilities of business reach beyond the organization as a purely profit-seeking unit and also encompass ethical considerations. At the most fundamental level, the ethical responsibilities of business embody those standards, norms or expectations that reflect a concern for what society regards as fair and just, and for avoiding or minimizing the potential negative impact of business decision-making and activities on stakeholders and society in general

(Carroll, 1979, 1991). The ethical or moral responsibility of business has always been a part of the economic life (and responsibility) of companies. As stated by Milton Friedman (1970) – perhaps the strongest defender of the company as a profit-seeking instrument – the responsibility of business and managers is 'to make as much money as possible while conforming to the basic rules of society, both those embodied in the law and those embodied in ethical custom'. In sum, our conceptualization of the economic responsibility of business is not very far from the conclusion reached by Levitt (1958: 49) that the responsibility of business is to seek material gain (profit) and, at the same time, obey the 'elementary canons of everyday face-to-face civility (honesty, good faith, and so on)'.

The institutional responsibility of business

The basic idea of the institutional responsibility of business is to recognize a need for what Scherer and Palazzo (2007: 1097) call 'a political order where economic rationality is circumscribed by democratic institutions and procedures', and for businesses' role in contributing to sustaining or building this order. As such, there are two aspects or categories of the institutional responsibility of business: First, and most fundamentally, companies have a responsibility to take into account the frameworks and norms set up by national and international political institutions to regulate the activities of business. This responsibility may be termed their public responsibility (Preston and Post, 1975). Second, companies have a responsibility to contribute to enhancing the capacity for political and civil society institutions to perform their foundational tasks in society, including the capacity of political institutions to circumscribe economic rationality – corporate decision-making and activities – by democratic institutions and procedures. This responsibility may be termed institutional capacity-building. These two categories of businesses' institutional responsibility are further outlined below.

Most fundamentally, companies' responsibility for taking seriously the national and international policy and frameworks that regulate the activities of business in society have to do with respecting the formal laws and the rules of the game that govern marketplace relations and companies' activities in society (Frederick, 1987). Laws may be understood as the codification of society's moral views; therefore, abiding by these standards is a necessary prerequisite for any further reasoning about social responsibility (Crane and Matten, 2004). Carroll (1991) suggests that the satisfaction of legal responsibilities is required of all corporations seeking to be socially responsible.

Public responsibility also implies that companies are responsible for acting in accordance with the intention and spirit of the law, and generally to act in accordance – within their legitimate sphere of influence – with the content of public policy (Preston and Post, 1975). As such, the public responsibility of business goes further than 'just' to follow the law; it also includes taking into account any voluntary normative guidelines that have been legitimized through political processes at the national and international levels and the political interpretation of the foundational values constituting the content of such guidelines. The public responsibility of business also implies that business should not improperly interfere in political processes and decision-making, a demand established, for example, in the Organisation for Economic Co-operation and Development (OECD)'s 'Guidelines for Multinational Corporations' (2000).

Then the question becomes: what principles are businesses today expected to take into consideration as part of their economic functioning in society? Although it may be argued that the concept of CSR is a fairly new one and that its different meanings are only starting to percolate internationally (Boxenbaum, 2004), signs of some generally agreed upon principles in the area of CSR are appearing that constitute the beginning of an 'ethical custom' on the global level, legitimized through political processes and decisions. These generally agreed upon principles are expressed in the increasing number of initiatives, standards and guidelines aimed at creating increased corporate accountability. Such initiatives include, for example, the 'UN Global Compact', the 'International Labour Organization (ILO) Conventions' and the 'OECD Guidelines for Multinational Enterprises'. As argued by Fredrick (1991), such normative guidelines comprise a framework for identifying the essential behaviour expected of corporations. Having analysed six different intergovernmental compacts (including, e.g. the 'OECD Guidelines for Multinational Enterprises', the 'Helsinki Final Act' and the 'ILO Tripartite Declaration of Principles Concerning Multinational Enterprises and Social Policy'), Frederick (1991) found that these guidelines expressed general principles and attempts to influence corporate practice in the areas of employment relations, consumer protection, environmental pollution, political participation and basic human rights. Waddock (2004) argues that generally agreed upon principles seem to exist in the areas of human rights, labour standards, environment and anti-corruption initiatives. What is important about these generally agreed upon principles is that they have gained legitimacy through political processes and government decisions. Together, these principles may constitute what Campbell (2007: 951) calls a 'minimum behavioural standard, below which corporate behaviour becomes socially irresponsible', or what Waddock (2004) calls foundational values, defining the 'floor' of responsible corporate action. The minimum behavioural standard originating from such values and principles is not to suggest an alternative model of business, drawing it away from its core economic purpose in society, but to recognize a need for a political framing of the activity of business in order to make it responsible.

Within the system of democratic capitalism, there has never been such a thing as a 'free enterprise system' and the political system has many legitimate roles to play in economic life (Novak, 1982). However, while business may have a legitimate role to play in the political and civil society spheres of society, it also has a responsibility not to undermine the power and resources of political and civil society institutions. That brings us to the second category of companies' institutional responsibility: institutional capacity building.

In this chapter, we have argued that CSR – through 'mystifying' the essential driving force of capitalism, and through effectively closing the door for public policy and mandatory regulations – may contribute to undermining the institutional conditions for a sustainable development and for addressing complex societal challenges. On this background, from a societal, and not a corporate-centred perspective, CSR might better be conceptualized in a more 'minimalist fashion' than that envisaged within the social activist approach to CSR, by postulating that managers, in addition to being corporate leaders, should also take on the role of social, moral and political leaders to grapple with, for example, 'understanding the roots of poverty'. However, in some regions of the world, political institutions do not have the power, resources and ability to provide citizens with basic welfare services or, more generally, to secure basic citizenship rights. In slightly different words, many institutions have an obvious lack of capacity to perform their basic tasks properly (Webb, 2007). On this background, there is a need to define a role for business that takes seriously the inadequate capacity of political institutions to perform their role and mission in society, and the fact that business is already deeply engaged in developmental activities, while at the same time ensuring that role does not undermine the state and public institutions' prime responsibility for securing the common good. From a functionalist viewpoint, it is within a framework of government leadership that the private sector can most effectively play a constructive role in enhancing sustainable development. I suggest that business, through institutional capacity building, might play a limited role in strengthening both this leadership, and the political and civil society institutions' capacity to conduct their foundational roles in society.

The term 'capacity building' is increasingly used by NGOs, governments and business (Webb, 2007), and its definitions and approaches are divergent and wide-ranging (Backer, 2000). The United Nations Development Program (UNDP), the 'official' capacity development agency within the UN system, defines capacity building as 'the creation of an enabling environment with appropriate policy and legal frameworks, institutional development, including community participation, human resources and strengthening of managerial systems'. Milen (2001) defines capacity building as 'an ability of individuals, organizations, or systems to perform appropriate functions effectively, efficiently, and sustainably'. This involves 'the continuing process of strengthening the abilities to perform core functions, solve problems, define and achieve objectives and understand and deal with development needs'. Within such an understanding, capacity building may include human resource development, organizational development and institutional and legal framework development. More generally, capacity building may also include working with governments to improve infrastructure; sharing international business practices and standards in such areas as health, safety and the environment, ethical and corporate governance, human rights and labour; supporting local business development, and transferring technology. The special case of institutional capacity building aims at improving the legal and regulatory business environment through such initiatives as, for example, building up the capacity of local authorities and institutional structures, promoting enhanced dialogue and consciousness raising on justice and human rights, and encouraging cooperation to build appropriate policy frameworks and new institutional structures to address environmental and sustainability issues on a sector (Sørensen and Pettersen, 2006).

So far, the role of business in institutional capacity building is a vastly underexplored area (Webb, 2007), and is also largely ignored by the CSR literature (Frynas, 2005). However, theorists are beginning to suggest that businesses may play a limited but constructive role in contributing to better governance by strengthening the ability of political institutions to perform the tasks given to them through their foundational roles in society. Business responsibility for institutional capacity building was also acknowledged by the Brundtland Report (WCED, 1987). Although the report indicates that voluntary business initiatives are promising, it stated that what is really needed is for business to work more closely together with governments 'in helping to shape and implement policy, laws, and regulations for more sustainable forms of development' (WCED, 1987: 329).

A central idea in capacity building is that the receiving institutions themselves should be in the 'driver's seat', setting the agenda, defining the important problems (Milen, 2001), and deciding if business participation in capacity building projects is desirable and, eventually, how and on what terms. The advantage of institutional capacity building, at least in theory, is that businesses, rather than assuming a direct responsibility for providing and administering welfare services, such as housing, education and health, might instead, where appropriate, play a limited role in strengthening the ability of political institutions to 'perform core functions, solve problems, define and achieve objectives and understand and deal with development needs'. As such, capacity building – as an aspect of CSR – may contribute to strengthening the institutional framework that constitutes a condition for long-term societal development, rather than to undermining it. As such, suggesting capacity building (and institutional responsibility) as an aspect of CSR answers the current call for a more politically enlarged conceptualization of CSR (Margolis and Walsh, 2003; Dubbink, 2004; Matten and Crane, 2005), where companies' efforts in the societal arena are circumscribed by political processes and institutions (Scherer and Palazzo, 2007).

Some aspects of Statoil's CSR activities in Venezuela may be termed as an example of institutional capacity building. Through funding, Statoil is supporting the efforts of the Venezuelan society to modernize its legal structures, particularly in those areas related to implementing human rights (Sørensen and Pettersen, 2006: 43). Some aspects of the Akassa development project in the Niger Delta may also be termed institutional capacity building.

Although institutional capacity building may work to strengthen rather than to undermine the political and governance framework necessary to both regulate business and address complex societal challenges, it also has several potential 'downsides'. Institutional capacity building may imply an increased influence of business in political discourses and agenda setting and a weakening of representative democracy (Martens, 2007), just to mentioning two important challenges. The dilemma, of course, is that the capacity of those political institutions setting the agenda and defining the terms and conditions for capacity-building projects may be the lowest precisely where the need for capacity building is strongest.

Towards a political conceptualization of corporate social responsibility

In sum then, the social responsibility of business may be summarized as follows (Table 4.1). Business has first of all an economic responsibility towards society, that is, to making profit and enhancing value for shareholders, employees, the local community and the society at large, while at the same time obeying what Levitt (1958) called 'the canons of everyday life' (honesty, good faith and so on). This responsibility derives from the basic aim of business in society within the system of democratic capitalism. Second, business has an institutional responsibility towards society, implying both a responsibility to confirm to the institutional frameworks of society (both mandatory laws and voluntary agreements legitimized through political processes and decisions) and to play a limited role in strengthening the very same institutional frameworks.

Suggesting a functional approach to CSR – arguing that other institutions exist in society to perform the tasks envisaged as a corporate responsibility within the social activist approach to CSR – is not to suggest very strict boundaries between the different spheres of society. Business, governments and civil society institutions constantly interact and influence each other. Companies' historically grown institutional frameworks (Matten and Moon, 2005) – national history, culture, business systems, systems of societal governance – are likely to heavily influence managers' perceptions of CSR, of legitimate values and of considerations in business and managerial decision-making, thereby constituting different agendas of CSR from society to society. At the same time, from a neo-institutional viewpoint (DiMaggio and Powell, 1991), new international standards and CSR guidelines – often developed through multilateralism involving governments,

Table 4.1 The social responsibility of business

Institutional responsibility	Capacity building	Supporting a strengthened institutional framework
	Public responsibility	Compliance with politically sanctioned standards and guidelines
		Compliance with national and international laws and regulations
Economic responsibility	Making profit	Obeying the 'the canons of everyday life'
		Enhancing value for shareholders, employees, the local community and the society at large

the business sector and civil society institutions – diffuse and may make managerial understanding of CSR and corporate practice more similar across societies. The point is that the different spheres of society are interconnected in many ways and are likely to influence each other, resulting, for example, in companies and managers taking a broader view of profit than is suggested within a strictly neoclassical view of economics. However, what is important from a functional viewpoint is that political power and resources are necessary to address many contemporary, complex, societal challenges such as poverty, environmental degradation, social inequity and lack of sustainable development in general. A steadily evolving CSR agenda may undermine the capacity of political institutions to perform their foundational tasks in society and the possibility to achieve sustainable political solutions to societal challenges.

On this background, our conceptualization of CSR is based on a primacy of politics and democracy to philosophy (Habermas, 1996; Rorty, 1991). It does not start with philosophical principles, but with a recognition of a changing interplay between governments, civil society actors and business, and the consequences of that dynamic (Scherer and Palazzo, 2007). Such a conceptualization of CSR is different from a social activist approach in several important respects. Most importantly, the social activist approach holds that universal standards exist for determining responsible corporate decisions and actions independent of the view of other, including political interests. These standards typically have an ethical, religious or metaphysical basis (Brummer, 1991). This implies that, in considering the welfare of others, companies are to respond to their constituencies' ideals rather than to expressed or current interests. Consequently, a CSR policy based on social activism is decoupled not only from the positions and interests of its current stakeholders, but also from processes of public policy. As such, the ideal CSR agenda – from a social activism position – is given based upon philosophical principles and moral reasoning outside the framework of public policy. This means that CSR, rather than being complementary to political solutions, constitutes itself as an alternative and competing framework for solving social ills and challenges of sustainable development. Because corporate managers, in the role of moral leaders, voluntarily will address societal challenges because that is the right thing to do, there is no need for broader political solutions (implying increased risks for mandatory regulations).

Arctic implications

First, within the framework of capitalist democracy, business has an economic responsibility towards society. Part of this economic responsibility is, however, also to adjust economic activity to the existing ethical customs of society. Second, companies have an institutional responsibility towards society. Most fundamentally, political and civil society institutions may be understood as a circle encompassing the economic activity of business, giving direction to legitimate economic life and value creation. The institutional responsibility of business is most importantly to adapt its activity to the constraints and opportunities set by political institutions, and, as such, not undermine these institutions' capacity to perform their foundational mission in society. Further, business may – in societies where the institutional capacity to secure its inhabitants basic citizenship rights is low – contribute to building such institutional capacity, that is, to play a limited role in strengthening the circle encompassing and regulating economic life. What then, might be the possible Arctic implications of such a conceptualization of CSR?[2]

The foundational question that can be addressed under the heading of corporations' institutional responsibility is the legitimacy of *the presence itself* of the oil and gas industry in the Arctic. As pointed out by Fjellheim and Henriksen (2006):

> The universal dilemma is that we are witnessing an increase in the Oil and Gas exploration in a warmer and more accessible Arctic in the very same moment as we conclude that it is exactly the human use of fossil fuel that is the single largest reason why the Arctic is getting warmer in the first place.

However, building on our conceptualization of CSR above, the question of the presence of the industry in the Arctic region is, above all, a political question. As long as there is a political request for the oil industry to increase exploration in the Arctic, and that is the case in all the Arctic countries, then we might conclude that it is not the companies' responsibility to address the foundational question of presence.

On the other hand, building on our conceptualization of CSR, what we might question is the corporate and industry use of power and resources, through lobbying and other instruments, to influence political institutions and society in general to give them permission to expand their activity in the Arctic. Although companies are legitimate actors in the political process and discourse leading to decisions on permission and, although it is hard to draw a line between appropriate and inappropriate use of corporate power in the political arena (OECD, 2000), some argue that the industry has gone too far in influencing the political processes. Professor Thomas Chr. Wyller wrote in a Norwegian newspaper[3] that we are experiencing a new phase in Norwegian oil politics in the political struggle of expanding oil and gas activities to the Northern areas. He said that this issue has dramatic dimensions where there are deep conflicts between business and political interests, environmental considerations, fisheries, workplaces and economic and security political aspects. He continued by saying that the issue is an incendiary question and concerns our society, our globe and our government. His article discusses how Statoil, through big commercials in the papers, tries to influence the Norwegian population. He says: 'this is not information, it is agitation, and it tries to influence the value choice during a political process of opening the areas for exploration or not'. The state still holds the majority of shares in Statoil, and Wyller, a professor emeritus in political science, questions the democratic legitimacy of the way Statoil uses its power to influence the decisions of its main owner – the Norwegian State.

Corporate social responsibility 79

Building on our conceptualization of CSR – and regardless of whether or not the corporations and industry have crossed the line for appropriate participation in the political processes leading to permissions to expand oil exploration in the Arctic – the industry has an institutional responsibility not to interfere inappropriately in the current and future political processes about industry's presence in the Arctic or in the political processes aimed at regulating their operations in the region. This responsibility becomes even more important owing to the tendency of oil and gas corporations to increase in size (e.g. through merging, as with the former Statoil and Hydro – StatoilHydro since October 1, 2007) and, therefore, in power and resources. As argued by Davis (1960, 1967), business is a social institution that must use its power responsibly, and with increased power comes increased responsibility ('the social power equation').

Turning then to the first layer of responsibility in our conceptualization of CSR, business – on the basis of its foundational task within a system of democratic capitalism – has an economic responsibility towards society. Most importantly, this responsibility is to contribute to economic development in society. Further, this responsibility has to be conducted within 'the everyday canons of everyday life' (Levitt, 1958) or 'the ethical customs of society' (Friedman, 1970), implying considerations beyond mere profit seeking for the purpose of minimizing the potential negative impact of corporate activity on society and the natural environment. Considering the substantial impact that oil and gas activity has on society and the natural environment, this is a huge responsibility and an enormous task. However, the exact nature of the economic responsibility of business cannot be determined without paying attention to the political and institutional framing of business activities. To do this, I use the Arctic Council's *Arctic Oil & Gas Guidelines* (2002) as a point of departure.

The 2002 Guidelines were adopted by the second Ministerial Meeting of the Arctic Council, October 9–10, 2002, in Inari, Finland, with the following statement: 'We ... endorse the updated Offshore version of these Guidelines and encourage the concerned stakeholders to apply them' (Arctic Council, 2002). Although national authorities are the main 'concerned stakeholders' for the Guidelines, as stated when describing the general purpose of the Guidelines, they 'may also be of help to the industry when planning for oil and gas activities ...' (Arctic Council, 2002).

The 2002 version of the Guidelines was completed by the Protection of the Arctic Marine Environment (PAME) working group under the Arctic Council. The new version of the Guidelines – building on the principles of sustainable development (Arctic Council, 2002) – however, represented a combined effort from the other working groups under the Arctic Council, with assistance and commentary by representatives of Arctic, regional and other governments, NGOs, industry, indigenous people and the scientific community (Arctic Council, 2002). As such, the Guidelines were updated with participation from important stakeholders within the framework of a regional political body. Thus, building upon the principle of public responsibility, they provide a solid foundation for defining the issues and processes of CSR for the oil and gas industry in the Arctic.

The Guidelines are intended, first of all, to address the environmental aspects and issues of oil and gas activities in the Arctic. However, they also suggest that the potential socio-economic effects of offshore oil and gas activities are important to consider and integrate into the planning and conduct of exploration and development (Arctic Council, 2002). The general goals for environmental protection during oil and gas activities in the Arctic are stated as follows.

> Offshore oil and gas activities in the Arctic should be planned and conducted so as to avoid (Arctic Council, 2002):
> - adverse effects on climate and weather patterns;
> - significant adverse effects on air and water quality;
> - significant changes in the atmospheric, terrestrial (including aquatic), glacial or marine environments in the Arctic;
> - detrimental changes in the distribution, abundance or productivity of species or populations of species;
> - further jeopardy to endangered or threatened species or populations of such species;
> - degradation of, or substantial risk to, areas of biological, cultural, scientific, historic, aesthetic or wilderness significance, and;
> - adverse effects in livelihoods, societies, cultures and traditional lifestyles for northern and indigenous people.

In sum, the Guidelines particularly emphasize three interdependent areas of concerns, here understood as important issues of CSR in relation to the oil and gas industry's expanding activities in the region: (1) conservation of living resources for sustainable use, (2) indigenous people, livelihood opportunities and cultural values and (3) air and water emission and climate change. Concerning the first point, conservation of living resources, the Guidelines state that:

> Necessary measures should be taken to ensure that Arctic flora and fauna and the ecosystems on which they depend are protected during all phases of offshore oil and gas activity. Special attention ... is required for species ... which are resources for human use, particularly by indigenous people ...
> (Arctic Council, 2002)

The conservation of living resources is seen as vital not only for conservation in itself, but for the interests of human safety and the well-being of indigenous people in the Arctic. This brings us to the second point, directly addressing the indigenous population, livelihood opportunities and cultural values. The Guidelines state that:

> In planning and executing offshore oil and gas operations, necessary measures should be taken, in consultation with neighbouring communities, to recognize and accommodate the cultural heritage, values, practices, rights and resource use of indigenous residents. Arctic states, in cooperation with

the oil and gas industry, should address the economic, social, health, and educational needs based on equal partnerships with indigenous people.

(Arctic Council, 2002)

Third, and although cautiously, the Guidelines identify the oil and gas corporations' responsibility for reducing their emissions of CO_2 as a third major area of CSR in the Arctic: 'Offshore oil and gas activities entail considerable inputs of gases to the air ... . Air emissions may have effects on the climate' (Arctic Council, 2002).

Concluding remarks

From a corporate perspective, CSR may be highly important. Corporations may sustain their own legitimacy through integrating the demands and interests of society in general, and stakeholders in particular, in corporate decision-making and activity. However, from a societal and sustainable development perspective, CSR may be counterproductive to, or may undermine, the institutional conditions required for a sustainable development path. It does this by legitimizing increased corporate power in society at the possible expense of the civil and political sphere of society, and by applying voluntary solutions rather than mandatory decisions to the challenges of sustainable development. On this background, I suggest a conceptualization of CSR within the structural features of capitalism and the aims of every business organization, 'stripping' business of ambitions on the broader societal arena and thus confining – and not expanding – the power of business in society.

Notes

1 The challenges of Arctic societies are outlined in more detail in Chapter 2.
2 As Chapter 13 in its entirety discusses managerial and corporate implications of CSR for the oil and gas industry in the Arctic, we keep the discussion here short and in general terms.
3 *Stavanger Aftenblad*, March 24, 2006.

References

Arctic Council (2002) *Arctic Offshore Oil & Gas Guidelines*, Protection of the Arctic Marine Environment Working Group. Online. Available HTTP: http://old.pame.is/sidur/uploads/ArcticGuidelines.pdf (Accessed 15 October 2007).
Arctic Human Development Report (2004) Akureyri: Stefansson Arctic Institute.
Backer, T. (2000) *Strengthening Nonprofits: Capacity Building and Philanthropy*, Human Interaction Research Institute, Working paper, March 2000, Encino, CA: Human Interaction Research Institute.
Berle, A. A. and Means, G. C. (1932) *The Modern Corporation and Private Property*, New York: Harcourt, Brace & World.

Blowfield, M. (2005) 'Corporate Social Responsibility: reinventing the meaning of development?', *International Affairs*, 81 (3): 515–524.

Blowfield, M. and Frynas, G. (2005) 'Setting new agendas: critical perspectives on Corporate Social Responsibility in the developing world', *International Affairs*, 81 (3): 499–513.

Bowen, H. (1953) *Social Responsibilities of the Businessman*, New York: Harper.

Boxenbaum, E. (2004) *Institutional Innovation Processes: a Hybridization of Institutional Logics*, Paper presented at the annual meeting of the Academy of Management, New Orleans.

Brummer, J. J. (1991) *Corporate Responsibility and Legitimacy. An Interdisciplinary Analysis*, New York, Westport, London: Greenwood Press.

Buckland, H., Albareda, L., Lozano, J. M., Tencati, A., Perrini, F. and A. Midttun (2006) *The Changing Role of Government in Corporate Responsibility*, ESADE, SDA Bocconi and NSM.

Campbell, J. L. (2007) 'Why would corporations behave in socially responsible Ways? An institutional theory of corporate social responsibility', *Academy of Management Review*, 32 (3): 946–967.

Carroll, A. B. (1979) 'A three-dimensional conceptual model of corporate social performance', *Academy of Management Review*, 4: 497–505.

Carroll, A. B. (1991) 'The pyramid of corporate social responsibility: toward the moral management of organizational stakeholders', *Business Horizons*, 34 (4): 39–48.

Carroll, A. B. (1999) 'Corporate Social Responsibility. evolution of definitional construct', *Business and Society*, 38 (3): 268–295.

Carroll, A. B. and Buckholtz, A. K. (2000) *Business and Society: Ethics and Stakeholder Management*, 4th edition, Cincinnati: South-Western College.

Crane, A. and Matten, D. (2004) *Business Ethics: A European Perspective*, Oxford: Oxford University Press.

Czada, R. (1995) 'Institutionelle Theorien der Politik', in Norhlen, D. and Schultze, H-O. (eds) *Lexikon der Politikk*, Vol. 1, Munich: Droemer-Knaur.

Daly, H. (1992) *Steady-state Economics*, London: EarthScan Publications.

Davis, K. (1960) 'Can business afford to ignore corporate social responsibilities?', *California Management Review*, 2: 70–76.

Davis, K. (1967) 'Understanding the social responsibility puzzle', *Business Horizons*, 10 (4): 45–51.

Davis, K. (1973) 'The case for and against business assumptions of social responsibilities', *Academy of Management Journal*, 16: 312–322.

DiMaggio, P. J. and Powell, W. W. (1991) 'The iron cage revisited: institutional isomorphism and collective rationality in organization fields', in Powell, W. W and DiMaggio, P. J. (eds) *The New Institutionalism in Organizational Analysis*, Chicago: The University of Chicago Press.

Donaldson, T. and Preston, L. E. (1995) 'The stakeholder theory of the corporation: concepts, evidence, and implications', *Academy of Management Review*, 20 (1): 65–91.

Dubbink, W. (2004) 'The fragile structure of free-market society – the radical implications of corporate social responsibility', *Business Ethics Quarterly*, 14 (1): 23–46.

Dunfee, T. W. and Hess, D. (2000) 'The legitimacy of direct corporate humanitarian investment', *Business Ethics Quarterly*, 10 (1): 95–109.

Elkington, J. (1994) 'Towards the sustainable corporation: win-win-win business strategies for sustainable development', *California Management Review*, 36 (2): 90–100.

Elkington, J. (1997) *Cannibals with Forks: The Triple Bottom Line of the 21st Century Business*, Oxford: Capstone.

Fjellheim, R. S. and Henriksen, J. B. (2006) 'Oil and gas exploration on Arctic indigenous peoples' territories. Human rights, international law and corporate social responsibility', *Journal of Indigenous Peoples Rights*, 4: 8–23.

Frederick, W. C. (1991) 'The moral authority of transnational corporate codes', *Journal of Business Ethics*, 10: 165–177.

Frederick, W. C. (1998) 'Moving to CSR4', *Business and Society*, 37 (1): 40–60.

Frederick, W. S. (1987) 'Theories of corporate social performance', in Sethi, S. P. and Falbe, C. M. (eds) *Business and Society: Dimensions of Conflict and Cooperation*, New York: Lexington Books.

Freeman, R. E. (1984) *Strategic Management: A Stakeholder Approach*, Boston: Pitman.

Friedman, M. (1962) *Capitalism and Freedom*. Chicago: University of Chicago Press.

Friedman, M. (1970) 'The social responsibility of business is to increase its profit', *The New York Times Magazine*, September 13.

Friedman, M. and Friedman, R. (1962) *Capitalism and Freedom*, Chicago: University of Chicago Press.

Frynas, G. (2005) 'The false development promise of Corporate Social Responsibility: evidence from multinational oil companies', *International Affairs*, 81 (3): 581–598.

Garriga, E. and Melè, D. (2004) 'Corporate social responsibility theories: mapping the territory', *Journal of Business Ethics*, 53: 51–71.

Global Compact (2005) *The United Nations Global Compact: Advancing Corporate Citizenship*, The Global Compact Office, June. Online. Available HTTP: http://www.unglobalcompact.org/docs/about_the_gc/2.0.2.pdf (Accessed 15. October 2007).

Habermas, J. (1996) *Between Facts and Norms*, Cambridge, MA: MIT Press.

Hall, P. A. and Soskice, D. (2001) *Varieties of Capitalism. The Institutional Foundations of Comparative Advantages*, Oxford: Oxford University Press.

Henderson, D. (2001) 'The case against corporate social responsibility', *Policy*, 17 (2): 28–32.

Herman, E. S. (1981) *Corporate Control, Corporate Power*, Cambridge: Cambridge University Press.

Hunt, B. (2000) 'The new battleground for capitalism', *Financial Times* (Mastering management), October.

Jenkins, R. (2005) 'Globalization, corporate social responsibility and poverty', *International Affairs*, 81 (3): 525–540.

Jones, M. T. (1996) 'Missing the forest for the trees: a critique of the social responsibility concept and discourse', *Business & Society*, 35 (1): 7–41.

Jones, M. T. (1999) 'The institutional determinants of social responsibility', *Journal of Business Ethics*, 20 (2): 163–179.

Levitt, T. (1958) 'The dangers of social responsibility', *Harvard Business Review*, September-October.

Logsdon, J. M. and Wood, D. J. (2002) 'Business citizenship: from domestic to global level of analysis', *Business Ethics Quarterly*, 12 (2): 155–187.

Margolis, J. D. and Walsh, J. P. (2003) 'Misery loves companies: rethinking social initiatives by business', *Administrative Science Quarterly*, 48 (2): 268–305.

Martens, J. (2007) 'Multistakeholder partnerships – future models of multilateralism?' *Dialogue on Globalization*, No. 29. Friedrich Ebert Stiftung. Online. Available HTTP: http://globalpolicy.igc.org/eu/en/publ/martens_multistakeholder partnerships online version.pdf (Accessed 15 October 2007).

Marrewijk, M. (2003) 'Concepts and definitions of CSR and corporate sustainability: between agency and communion', *Journal of Business Ethics*, 44: 95–105.

Matten, D. and Crane, A. (2005) 'Corporate citizenship: Toward an extended theoretical conceptualization', *Academy of Management Review*, 30: 166–179.

Matten, D., Crane, A. and Chapple, W. (2003) 'Behind the mask: revealing the true face of corporate citizenship', *Journal of Business Ethics*, 45 (1–2): 108–120.

Matten, D. and Moon, J. (2005) 'A conceptual framework for understanding CSR', in Habisch, A., Jonker, J., Wegner, M. and Schmidpeter, R. (eds) *Corporate Social Responsibility Across Europe*, Berlin: Springer.

Midttun, A. (2005) 'Realigning business, government and civil society: embedded relational governance beyond the (neo) liberal and welfare state models', *Corporate Governance: The International Journal of Business in Society*, 5 (3): 159–174.

Milen, A. (2001) *What Do We Know About Capacity Building: An Overview of Existing Knowledge and Good Practice*, Geneva: Department of Health Service Provision, World Health Organization. Online. Available HTTP: http://whqlibdoc.who.int/hq/2001/a76996.pdf/ (Accessed 15. October 2007).

Mintzberg, H. (1983) 'The case for corporate social responsibility', *Journal of Business Strategy*, 4 (2): 3–15.

Mizruchi, M. S. (2004) 'Berle and Means revisited: the governance and power of large U.S. corporations', *Theory and Society*, 33: 579–617.

Moon, J., Crane, A. and Matten, D. (2005) 'Can corporations be citizens? Corporate citizenship as a metaphor for business participation in society', *Business Ethics Quarterly*, 15 (3): 429–453.

Nelson, J., Hodges, A., Deri, C., Schneider, M. and Ruder, A. (2005) 'Business and international development: opportunities, responsibilities and expectations. A survey of global opinion leaders in business, civil society and the media'. *Edelmann, John F. Kennedy School of Government, The Prince of Wales International Business Leaders Forum*. Online. Available HTTP: http://www.ksg.harvard.edu/m-rcbg/CSRI/prog_bid.html (Accessed 15 October 2007).

Novak, M. (1982) *The Spirit of Democratic Capitalism*. New York: An American Enterprise Institute/Simon & Schuster Publication.

Organisation for Economic Co-operation and Development (OECD; 2000) '*Guidelines for Multinational Corporations*'. Online. Available HTTP: http://www.oecd.org/document/24/0,3343,en_2649_201185_1875736_1_1_1_1,00.html (Accessed 15. October 2007).

Palazzo, G. and Scherer, A. G. (2006) 'Corporate legitimacy as deliberation: a communicative framework', *Journal of Business Ethics*, 66: 71–88.

Parsons, T. (1951) *The Social System*, Glencoe, IL: The Free Press.

Pogutz, S. (2007) *Sustainable Development, Corporate Sustainability, and Corporate Social Responsibility: the Need for an Integrative Framework*, Paper presented at the International Conference of the Greening of Industry Network, June 15–17, 2007.

Preston, L. E. and Post, J. E. (1975) *Private Management and Public Policy: the Principle of Public Responsibility*, Englewood, NJ: Prentice-Hall.

Prieto-Carròn, M., Lund-Thomsen, P., Chan, A., Muro, A N. A. and Bhushan, C. (2006) 'Critical perspectives on CSR and development: what we know, what we don't know, and what we need to know', *International Affairs*, 82 (5): 977–987.

Roome, N. (2005) 'Some implications of national agendas for CSR', in Habisch, A., Jonker, J., Wegner M. and Schmidpeter, R. (eds.) *Corporate Social Responsibility Across Europe*, Berlin: Springer.

Rorty, R. (1991) 'The priority of democracy to philosophy', in Rorty, R. (ed.) *Objectivity, Relativism, and Truth: Philosophical Papers*, Vol. I, Cambridge: Cambridge University Press, 175–196.

Salin-Andersson, K. (2006) 'Corporate social responsibility: a trend and a movement, but of what and for what?', *Corporate Governance*, 6 (5): 595–608.

Scherer, A. G. and Palazzo, G. (2007) 'Toward a political conception of corporate responsibility: business and society seen from a Habermasian perspective', *Academy of Management Review*, 32 (4):1096–1120.

Schwartz, M. (1987) *The Structure of Power in America*. New York: Homer & Meier.

Spangenberg, J. H. (2002) 'Institutional sustainability indicators: an analysis of the institutions in Agenda 21 and a draft set of indicators for monitoring their effectiveness', *Sustainable Development*, 10: 103–115.

Statoil (2005) 'Global challenges – local solutions', *Statoil and Sustainable Development 2005*. Online. Available HTTP: http://www.statoilhydro.com/en/EnvironmentSociety/dataandreports/SustainabilityReports/Pages/SustainabilityReport2005.aspx (Accessed 15 October 2007).

Sørensen, M. B. and Pettersen, S. M. (2006) *Partnering for Development – Making it Happen*, United Nations Development Program (UNDP). Online. Available HTTP:http://www.unglobalcompact.org/docs/issues_doc/7.3/UNDP_PartneringforDevelopment.pdf (Accessed 15 October 2007).

United Nations (ed.) (1993) *Earth Summit: Agenda 21, the United Nations Programme of Action from Rio*. New York, United Nations.

Waddell, S. (2000) *Business–Government–Civil Society Collaborations: A Brief Review of Key Conceptual Foundations*, For the Interaction Institute for Social Change, Working Paper, Cambridge, MA: Interaction Institute for Social Change.

Waddell, S. and Brown, L. D. (1997) 'Fostering intersectoral partnering: a guide to promoting cooperation among government, business, and civil society actors', *IDR Reports*, 13 (3).

Waddock, S. (2002) *Leading Corporate Citizens: Vision, Values, Values Added*, New York: McGraw-Hill.

Waddock, S. (2004) 'Creating corporate accountability: foundational principles to make corporate citizenship real', *Journal of Business Ethics*, 50: 313–327.

Ward, H. (2004) *Public Sector Role in Strengthening Corporate Social Responsibility: Taking Stock*, The World Bank, International Finance Corporation, The World Bank Group. Online. Available HTTP: http://www.ifc.org/ifcext/economics.nsf/AttachmentsByTitle/CSR-CSR_interior.pdf/$FILE/CSR-CSR_interior.pdf (Accessed 15 October 2007).

Webb, T. (2007) *Business and Better Governance: Companies and Institutional Capacity Building in Developing Countries*, Paper presented at the European Academy of Business in Society's 6th Annual Colloquium: The Emerging Global Governance Paradigm: The Role of Business and its Implications for Companies, Stakeholders and Society, September 20 and 21.

Welford, R. (2003) *Corporate Social Responsibility in Europe and Asia: Critical Elements and Best Practices*, Corporate Environmental Governance Programme. Project Report 5. The Centre of Urban Planning and Environmental Management, The University of Hong Kong.

Wood, D. J. (1991) 'Corporate social performance revisited', *Academy of Management Review*, 16 (4): 691–718.

World Business Council for Sustainable Development (WBCSD; 2006) *From Challenge to Opportunity. The Role of Business in Tomorrow's Society*, A paper from the Tomorrow's Leaders Group of the WBCSD. Online. Available HTTP: http://www.wbcsd.org/Plugins/DocSearch/details.asp?DocTypeId=25&ObjectId=MTgyMTM&URLBack=%2Ftemplates%2FTemplateWBCSD5%2Flayout%2Easp%3Ftypc%3Dp%26MenuId%3DMTE0NQ%26doOpen%3D1%26ClickMenu%3DLeftMenu (Accessed 15 October 2007).

World Commission on Environment and Development (WCED; 1987) *Our Common Future*, Oxford: Oxford University Press.

World Summit on Sustainable Development (WSSD; 2002) *The Johannesburg Declaration on Sustainable Development*, WSSD, Johannesburg: United Nations.

Zadek, S. (2001) *The Civil Corporation – The New Economy of Corporate Citizenship*, London: Earthscan.

Zeitlin, M. (1989) *The Large Corporation and Contemporary Classes*, New Brunswick, NJ: Rutgers University Press.

5 Framework and methodology

Regulation and discourse analysis as a research strategy

Ove Heitmann Hansen, Oluf Langhelle and Robert Anderson

Introduction

The purpose of this chapter is to give an overview of how the project 'Arctic Oil and Gas – Sustainability at Risk' was developed and carried out, and to describe the framework and methodological approach that underpins the country studies of Alaska, Canada, Norway and Russia (Chapters 7 to 10). Our approach to collecting, understanding and presenting the data is based on a framework flowing from regulation theory and discourse analysis as our research methodology.

This book is a comparative case study of the different stories brought forward by stakeholders about oil and gas activities in the Arctic. The Arctic Human Development Report (AHDR) contained a reflection on what was described as the 'general perspectives that stakeholders bring to consideration of human development in the Arctic'. This was further said to provide a 'broader context for the AHDR accounts of conditions prevailing in the Arctic'. This was elaborated the following way:

> It is tempting to assume that we can characterize a region like the Arctic in objective terms that will somehow capture the perspectives of all those who are active in the area as well as those who are interested in the region even though they are not players in the Arctic in any direct sense. But efforts of this sort are doomed to failure. There are many visions of the Arctic, and the appeal of individual visions varies as a function of the vantage points and interests of individual actors. Thomas Berger's familiar phrase "Northern Frontier, Northern Homeland," for instance, captures the distinction between those who see the circumpolar Arctic as a storehouse of natural resources of interest to industrialized societies to the south and those who reside in the Arctic and see themselves as the current representatives of peoples who have lived in the region.
>
> Important as it is, this dichotomy does not do justice to the range of Arctic visions that have framed northern issues and shaped the interests of individuals and stakeholder groups during the course of modern history. Many of the visions have given rise to distinct mindsets. Because these mindsets tend to spawn dramatically different and sometimes conflicting

88 *Ove Heitmann Hansen, Oluf Langhelle and Robert Anderson*

approaches to Arctic issues of public importance, they provide an important backdrop for the Arctic Human Development Report.

(Young and Einarsson, 2004: 22)

There are strong reasons to believe that oil and gas activities in the Arctic are framed and structured very differently in different regions and countries, depending on the legal and institutional framework, level of development, social welfare, state of the environment and community viability and the historical path of the discourse. This is partly what lies behind the choice of a comparative case-study approach. We have used regulation theory and Hajer's discourse analytical approach to describe, and compare, the 'interaction' between contextual factors and political processes concerning oil and gas activities in different localities in the Arctic.

This chapter has two major sections, the first dealing with the theoretical framework and the second dealing with the research methodology. In the first section, we introduce regulation theory and the concepts of Regimes of Accumulation and Modes of Social Regulation and how the two of them come together in particular times and places to produce Modes of Development. Then, we introduce a framework based on these concepts (Figure 5.1) that allows us to explore the interactions among communities, states, supra-national bodies,

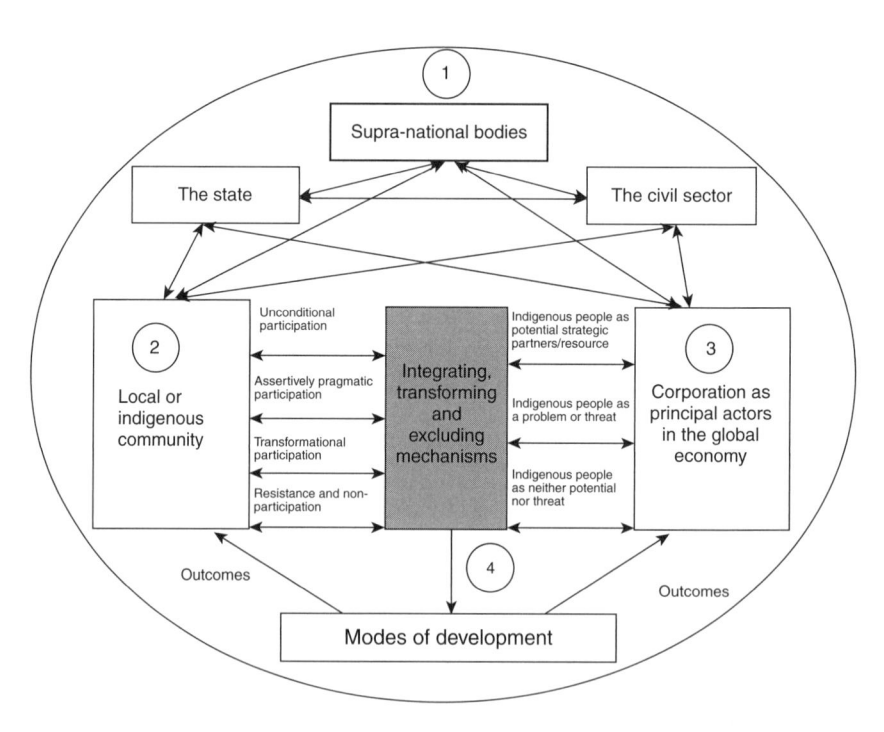

Figure 5.1 Major actors associated with the oil and gas industry in the Arctic (Anderson, MacAulay, Kayseas et al., 2007).

organizations of the civil sector and corporations in the context of developing of oil and gas resources in the Arctic. The interactions we analyze are the discourses in which these are engaged.

In the methodology section, we outline discourse analyses as a research strategy. We explain how we use the concepts of storylines and discourse coalitions to study the stakeholders' relations to, and interests in, oil and gas activities in the Arctic. This allows us to bring the relationships identified in the framework to life and explore the interactions that are giving rise to modes of social regulation and hence modes of development, sustainable or not. At the end of the chapter, we discuss some limitations of the study.

A regulation theory framework

Modernization theory has dominated much of the development paradigm and practice in the middle decades of the twentieth century. There are a number of notions that are part of this theory. First, progress is seen as passing through various stages. It implies that, in order to progress and develop, traditional societies have to move towards modernity (Crewe and Harrison, 1998). 'Modernization' and 'development' came to be used as synonymous terms. Second, monetary income and, therefore, economic growth are regarded as key elements in measuring the quality of life. Third, humans are, or should be, motivated by self-interest and rational economic behaviour (Burkey, 1993; Crewe and Harrison, 1998). From this point of view, development is measured in Western economic terms, with the expectation that underdeveloped nations over time will assume the qualities of industrialized nations (Burkey, 1993). One of the underlying assumptions of modernization is that traditional culture and social structures are barriers to progress. This can be seen in the following quotation:

> Pre-existing social relations ... family, kinship and community, constitute obstacles to business enterprises and achievement ... Successful capitalism involves some rupturing of existing social relations and possibly the diminution of affective relations to leave more space to impersonal, calculating forms of social interaction believed to characterise the market economy.
> (Moore, 1997: 289)

Dependency theory emerged as a critique of the failure of the modernization agenda to deliver the anticipated development outcomes, but even more fundamentally to draw attention to what was seen as a new form of colonization. In this analysis, the corporation and the developed states (particularly the United States), the World Bank, the International Monetary Fund (IMF) and the General Agreement on Tariffs and Trade (GATT) are cast as the villain or villains (Hancock, 1989; Klitgaard, 1990). Rather than leading the underdeveloped to a developed state, the actions of the developed world are seen as the original (through conquest and colonialism) and continuing (through economic exploitation) causes of underdevelopment. According to the dependency critiques, participation by the underdeveloped in the global capitalist economy as it is currently constructed can only exacerbate their circumstances, not improve them. The evidence since the Second World War certainly offers some support for this view. The gap between the rich and the poor within and among states has widened, not closed, in spite of six decades of development efforts of various types.

Even as modified in recent years, the modernization and dependency perspectives present incompatible views of the relationship between a developing people/region and the developed world. In particular circumstances, one or the other of these approaches can often adequately explain what happened. However, when applied in any particular circumstance to offer insight into what might happen, the two produce conflicting answers, thus providing contradictory guidance to groups searching for a path to development, as they perceive it.

In the closing three decades of the twentieth century, the conflict between the modernization and dependency perspectives led many to conclude that both are incomplete (as opposed to wrong), with each describing a possible but not inevitable outcome of interaction between local regions seeking what they regard as a better form of life, and the global economy. In this vein, Corbridge says that there has been a powerful trend towards 'theories of capitalist development which emphasise contingency ... a new emphasis on human agency and the provisional and highly skilled task of reproducing social relations' (Corbridge, 1989: 633).

As Tucker states, this allows 'for the possibility of incorporating the experience of other peoples, other perspectives and other cultures into the development discourse' (Tucker, 1999: 16). Development need not be as defined by the 'developed world' and the interaction between a particular people and the global economy need not be as envisaged by the modernization or dependency perspectives; it can be something else entirely. Why not that which is being sought by indigenous people – development as they define it?

Regulation theory is one of the new approaches to development that emphasizes contingency and human agency. Hirst and Zeitlin say that it executes:

> A slalom between the orthodoxies of neo-classical equilibrium theory and classical Marxism to produce a rigorous but nondeterministic account of the phases of capitalist development that leaves considerable scope for historical variation and national diversity.
>
> (Hirst and Zeitlin, 1992: 84)

Expanding on this notion of variation and diversity, Elam says that, on one hand, national and regional units are constantly in a state of flux as they adjust to the influences of the global economy. All must accommodate themselves at least to some extent to its hegemony. At the same time, these broader global influences 'are seen as having essentially local origins' (Elam, 1994: 66). This translates into a counter-hegemonic potential in terms of the activities actually undertaken by people as they negotiate their way locally through the global economy. It is not simply a case of conform or fail.

In the vocabulary of regulation theory, the 'regime of accumulation' is 'a historically specific production apparatus ... through which surplus is generated, appropriated, and redeployed' (Scott, 1988: 8), at the level of the international economy as a whole (Hirst and Zeitlin, 1992: 85), that is, the global economy.

A mode of social regulation (MSR) is 'a complex of institutions and norms which secure, at least for a certain period, the adjustment of individual agents and social groups to the over arching principle of the accumulation regime' (Hirst and Zeitlin, 1992: 85). The 'mode of development' is the combination of the currently ascendant regime of accumulation and the various modes of social regulation (Hirst and Zeitlin, 1992: 84–85).

While regulation theory does not prescribe the exact nature of a particular mode of social regulation, it is generally agreed that: 'a regime of accumulation does not create or require a particular mode of social regulation, each regime, in short, may be regulated in a multiplicity of ways (Scott, 1988: 9). Because modes of social regulation are based on such things as 'habits and customs, social norms, enforceable laws and state forms' (Peck and Tickell, 1992: 349) unique modes 'can exist at virtually any territorial level – local, regional, national, global' (Storper and Walker, 1989: 215).

Another aspect of regulation theory – its historicity – adds further strength to the argument that modes of social regulation, and therefore modes of development differing considerably one from another, can and do emerge at every geographic scale. Corbridge (1989) says regulation theory indicates that the global economic system has gone through four stages in the twentieth century. In stage one, the system was in equilibrium. Stage two was a period of crisis or disequilibrium resulting from a shift from the extensive to the Fordist regime of accumulation. Equilibrium returned in stage three when suitable modes of social regulation emerged. The fourth (current) stage is also one of crisis caused by a failure of the monopolistic mode of social regulation (in all its variants) to accommodate a 'selective move from mass production (the Fordist regime accumulation) to various forms of flexible production' (Norcliffe, 1994: 2).

Forces resulting in the shift to the new flexible regime of accumulation include:

1 technical limits to rigid fixed capital production techniques;
2 working class resistance to Taylorist and Fordist forms of work organization (Jessop, 1989);
3 a change in consumption patterns 'toward a greater variety of use values ... [that] cannot be easily satisfied through mass production' (Amin, 1994: 12);
4 the increasing mobility of capital and the resulting ability of transnational corporations (TNCs) to move among spatially bounded regulatory jurisdictions in the pursuit of greater profits (Leyshon, 1992);
5 in the face of this internationalization of capital, the inability of national Keynesian policies (all variants of the monopolistic mode of social regulation) to avert crisis (Komninos, 1989).

Everywhere and at every geographic scale – community, sub-national region, national, supra-national region and globally – indigenous or not, people are struggling to develop modes of social regulation that will allow them to interact with this new flexible regime of accumulation on their own terms. While it is a 'work in process', the nature of the flexible regime of accumulation is becoming clearer and multiple overlapping modes of social regulation are emerging. Dunning says:

> We are moving out of an age of hierarchical capitalism and into an age of alliance capitalism. This is placing a premium on the virtues needed for fruitful and sustainable coalitions and partnerships (be they within or among institutions), such as trust, reciprocity, and due diligence.
>
> (Dunning, 2003: 24)

As a result of the emergence of the flexible regime of accumulation, there has been a shift in which companies consider to be stakeholders and how they behave towards these groups. Nowhere is this truer than in the relationship between companies and the communities where their customers and employees live and where they conduct their activities. In spite of globalization, everything a company does, it does somewhere, every employee and every customer lives somewhere, and inputs of raw material and capital goods come from somewhere; and all these 'somewheres' are located in communities of some sort. Because of this, as companies forge networks of suppliers, sub-contractors and marketing channel partners and seek to control them through 'collective social and institutional order in place of hierarchical control' (Storper and Walker, 1989: 152), they are much more likely to see communities as valued members of networks as opposed to something external to them. This general predisposition applies no less to indigenous communities who increasingly control key resources or represent an important component of the labour force in certain areas.

In response to the change in the regime of accumulation, the nature of regulation is changing. The locus of regulation is shifting from the nation-state in two directions – to the supra-national and the local. For example, Amin (1994: 1–40) say the crisis in the global economy has resulted in 'new opportunities for the location of economic activities' and that 'the geography of post-Fordist production is said to be at once local and global'. Scott (1988:108) says that new industrial spaces result from 'a very specific articulation of local social conditions with wider coordinates of capitalist development in general'. Finally, Dicken (1992: 307) emphasizes that successful participation in the global economic system 'is created and sustained through a highly localised process' and that 'economic structures, values, cultures, institutions and histories contribute profoundly to that success'.

Local modes of social regulation can be, in Gramscian terms, both hegemonic and counter-hegemonic to the extent to which they consent to the capitalist global economy, attempt to transform it or dissent from it. Hegemonic and non-hegemonic responses are associated with three different analytic or intuitive starting points that motivate policies and programmes in local modes of social regulation. The first of these starting points is an analysis that claims that peripheral communities have been excluded from capitalism and that the objective is to remedy this through inclusion. The second is an analysis that claims that capitalism is at least in part culturally alien and that it is necessary to transform the 'alien' aspects of it as part of the process of participating in it. The third is an analysis that claims that capitalism is exploitative and beyond redemption and that the need is to exclude or resist it. These analytical/intuitive starting points are not simply abstract concepts. They, and the beliefs and discourses about the economy associated with them, are present in varying combinations and varying strengths among the members of indigenous and all other communities.

People in a particular community do not adopt their perspective on the global economy in isolation or in abstract. It emerges in response to their direct experience with actors in the global economic system. The four groups of actors with whom they are most familiar (and, therefore, constitute the face of the global economy) are:

1 the exogenous businesses (henceforth called corporations for simplicity sake) with which they interact as customers, employees or suppliers;
2 the 'state' at local, sub-national, national and international levels;
3 myriad groups of the civil sector, including non-government organizations (NGOs) of all types, and special interest groups, such as Amnesty International, the World Council of Indigenous People, the Sierra Club and so on;
4 global and transnational cooperation, such as the World Trade Organization, the World Bank, the European Economic Union, North American Free Trade Agreement (NAFTA).

Corporations are most closely associated with the regime of accumulation; indeed for many they are the face of the regime of accumulation. That it is not to say that corporations are not influenced by and do not influence modes of social regulation; of course they are, and they do. The state at all its levels is most closely tied to the modes of social regulation. Indeed, the sum of the actions of the state at all levels constitutes the bulk of the modes of social regulation at any particular time and place; the bulk but not the entirety. The organizations of the civil sector also play an important role directly and through their influence on the state and on corporations. Increasingly, transnational institutions are taking on a powerful role in the economy that is more than the expression of the collective voice of member states. Transnational organizations and agreements are becoming a regulatory force, with considerable impact on states, corporations and communities.

The institutional landscape is also changing in the Arctic. As argued by Young (2005: 9), the Arctic has experienced 'a dramatic shift from the status of sensitive theatre of operations for the deployment of strategic weapons systems to that of focal point for a range of initiatives involving transnational cooperation'. These initiatives take a variety of forms and involve different actors:

> Some, like the establishment of the Arctic Environmental Protection Strategy (1991) and the creation of the Arctic Council (1996), involve straightforward

intergovernmental agreements. Others, such as the construction of the Northern Forum (1991), feature leagues of subnational actors drawn together in pursuit of common interests that differ from those of national governments. Still others, like the founding of the Inuit Circumpolar Conference (1977), the development of the International Arctic Science Committee (1990), and the launching of the University of the Arctic (2001), take the form of nongovernmental arrangements designed to address specific Arctic issues and to convey a sense of the significance of these concerns to the world at large.

(Young, 2005: 9)

For the most part, these cooperative arrangements are 'neither regulatory in nature nor endowed with the authority to make binding decisions on matters of significance' (Young, 2005: 10). What is important, however, is that Arctic cooperative arrangements have:

> become vehicles for articulating regional interests and for protecting regional actors from the side effects of global processes. In most cases these efforts deal with the impacts of large-scale environmental changes and with the effects of economic and social globalization.

(Young, 2005: 12)

According to Young, these cooperative arrangements remain fragile in the sense that they have not yet 'hardened into a well-established assemblage of social practices featuring a common discourse that can turn expediential and transient calculations of political interest into a mode of operation that becomes a matter of second nature to its participants'. And to complicate it even further, global environmental change and globalization 'threaten to overwhelm efforts to carve out coherent agendas at the regional level and to pursue them without undue concern for the linkages between regional activities and planetary processes' (Young, 2005: 14).

In any particular case, a complex set of factors influence the interaction of a group of people with the forces of the global economy and the outcomes they experience. These factors include:

1   The impact of the 'state' at all levels and the 'civil sector' on the multiple overlapping modes of social regulation and, therefore, on participants in the global economy, and the influence of these participants on the 'state' and the 'civil sector'.
2   The community-in-question's approach to economic development including its history, current circumstances, objectives and approach to participation in the global economy, including strategies for participation, transformation and exclusion and actual outcomes.
3   Corporate (as the usual representative of the regime of accumulation encountered by communities) responses to the community-in-question. Particularly motivating forces include the community's control over the critical natural, human and financial resources. Community members' attractiveness as a

market and corporate strategies and objectives may also affect outcomes. Outcomes, of course, feed back to the ongoing process of development.
4   The expected mode of development and the actual mode that emerges in a particular circumstance.

The mix of integrating, transforming and excluding mechanisms adopted by a particular community in its approach to the global economy and, therefore, the mode of development that emerges, is heavily influenced by the particular faces of the state, civil sector corporations and international organizations that a community sees now and has seen in the past. This 'face to face' meeting, while heavily influenced by local circumstance, occurs within the context of the dominant global regime of accumulation and multiple, overlapping and often conflicting modes of social regulation.

In this ongoing story, we see all the players and forces in action, each influencing one another and the mode of social regulation, and hence the mode of development that is emerging. Following Young, however, our focus is on the social practices that attempt to create common discourses to turn expediential and transient calculations of political interest into actions.

A discourse analytical approach

One important force in action among the actors is how different development alternatives are framed and justified. A discourse[1] can be described as the way we speak about things such as the environmental or social consequences of oil and gas activities in the Arctic. Dryzek (1997) defines discourse as 'a shared way of looking at the world' or as a 'shared way of apprehending the world'. The adherents of a particular discourse will use 'a particular kind of language when talking about events, which in turn rests on some common definitions, judgements, assumptions and contentions'. These are embedded in language, and it makes it possible for those who subscribe to them to 'interpret bits of information and put them together into coherent stories or accounts' (Dryzek, 1997: 8). This means that different people use different bits of information and knowledge to promote their interests, but at the same time borrow arguments to support their views. Actors in the debate of an Arctic oil and gas activity will capture ideas and concepts that are useful to bring their messages to decision-makers to obtain or increase support.

A discourse is also a comprehensive and significant system that regulates what can be said and done, and what cannot be said or done within a social field or subject area. It serves as a powerful force in determining our realities or 'truths', the meaning we give to things (Foucault, 1980). Language is a particular way of looking at the world and interpreting our experience, and it is the language in use and practice as discourses we are looking for. Language reflects the social relations of power, domination and ideology. Ideology is the ability of a powerful and dominant social group to impose its interpretations and particular meanings of social reality on other groups in society through language. The point is that it is not sufficient to 'only' read texts. Institutional settings and constellations of actors must be considered as well. The focus is a communicative practice as the everyday-world site where societal and interactive forces merge. This is the discursive practice of stakeholders acting in pursuit of their goals and aspirations, and not just a matter of individuals encoding and decoding messages. To interact is to engage in an ongoing process of negotiation, both to infer what others intend to convey and to monitor how one's own contributions are received.

In the following, we will use Hajer's seminal approach to discourse analysis to identify and analyze discourses about what sustainable development entails, and what it might imply for oil and gas activities in the Arctic. Hajer (1995: 44) defines a discourse as 'a specific ensemble of ideas, concepts, and categorizations that are produced, reproduced, and transformed in a particular set of practices and through which meaning is given to physical and social realities', and we will use this definition in our discourse approach. We are interested in different discourses about expanding the oil and gas activities in the Arctic, understood as different mindsets and ways of framing oil and gas activities, with different and sometimes conflicting policy implications. Discourse analysis includes analysis of speeches and written texts, but also institutional settings influencing decision-making processes and policy outcomes. When we as researchers apply discourse analysis, however, we are ourselves part of the discourse and never outside of it. In a very fundamental sense, no social activities take place outside of discourse.

In a broad sense, the 'argumentative approach' as Hajer calls it, uses discourse analysis to analyze political struggles over meaning. Politics is here seen as a struggle for discursive hegemony in which actors try to secure support for their definition of reality (Hajer, 1995: 59). Discourse, in this perspective, fulfils a key role in processes of political, and hence also social, change. Political conflict is seen as hidden in the question of what definition is given to problems, which aspects of social reality that are included, and which are left undiscussed (Hajer, 1995: 43).

Hajer (1995: 61) puts forward two middle-range concepts, which are supposed to make it possible to show how discursive orders are maintained and transformed within politics: story lines and discourse-coalition. Story lines are defined by Hajer as narratives on social reality through which elements from many different domains are combined and that provide actors with a set of symbolic references that suggest a common understanding. Story lines are essential political devices that allow the overcoming of fragmentation and the achievement of discursive closure. A discourse-coalition consists of a set of story lines, the actors who utter these story lines, and the practices in which this discursive activity is based.

Story lines and discourse coalitions

Story lines refer to a condensed form of narrative in which metaphors are employed and used by people as 'shorthand' in discussions. Such story lines are narratives that allow stakeholders to draw upon various discursive categories to give meaning to specific physical or social phenomena. A story line about environmental effects from oil and gas activities may link particular industrial emissions to dying fish, birds, less biodiversity, long-time effects on other industries and climate change. Thus, story lines are made-up of discursive practices from many disciplines, such as ecology, chemistry, engineering and economics. A story line is therefore a 'middle-range concept', which attempts to capture the 'inter-discursive' characteristics of issues like environmental or social effects from oil and gas activities. Story lines are also simplified accounts of issues, in the sense that they do not contain all the uncertainties and contingencies of the discrete discursive practices that they combine.

Hajer explicitly argues that also sustainable development should be analyzed as a story line: the story line 'that has made it possible to create the first global discourse coalition in environmental politics' (Hajer, 1996: 14). These story lines do not only construct problems, they also play a key role 'in the creation of social and moral order in a given domain':

> Story lines are devices through which actors (stakeholders) are positioned, and though which specific ideas of blame and responsibility, and of urgency and responsible behaviour, are attributed. Through story lines stakeholders can be positioned as victims of pollution, as problem-solvers, as perpetrators, as top scientists or as scaremongers.
>
> (Hajer, 1995: 64–65)

Story lines are our way of analyzing the interaction between the four groups of actors in Figure 5.1, and help us to identify who the stakeholders are and what the substance of their messages is. Capturing story lines is also the way we identify groups or coalitions of people who have the same messages with common argumentation or using part of an argumentation to promote their view.

> The point of the story-line approach is that by uttering a specific element one effectively reinvokes the story-line as a whole. It thus essentially works as a metaphor. First of all story lines have the functional role of facilitating the reduction of the discursive complexity of a problem and creating possibilities for problem closure. Secondly, as they are accepted and more and more stakeholders start to use the story-line, they get a ritual character and give a certain permanence to the debate.
>
> (Hajer, 1995: 62–63)

The discourse-coalition refers to a group of actors that, in the context of an identifiable set of practices, shares the usage of a particular set of story lines over a particular period of time. This understanding of Hajer's discourse coalition and story-line concept is important, and when both this and the ritual use of story lines, discursive 'hegemony' can be thought of as existing. Thus, according to Hajer, stakeholders who use a part or whole of a story-line invoke this common understanding and discursive order.

Hajer's story-line and discourse-coalition concepts are essentially linked because discourse-coalitions form around the telling of shared story lines. However, Hajer's account is somewhat ambivalent on how constrained actors are within this discursive 'hegemony' of the discourse coalition associated with a common telling of a story line (or story lines). Thus, although Hajer states that: 'An essential assumption in a discourse-coalition approach is that the political power of a text is not derived from its consistency (although that may enhance its credibility) but comes from its multi-interpretability' (Hajer, 1995: 61), he more than often suggests that story lines may imply discourse closure.

If the political function that Hajer assigns to story lines – that of achieving discourse closure – is stressed, then Hajer's story-line and discourse coalition concepts are somewhat constraining in their presentation of how free stakeholders are to interpret and reinterpret the policy discourse within which they operate. Finally, a story line can be characterized as an intermediate level of discourse in society. Story lines occur at a level higher than individual discourses (such as atmospheric chemistry or financial accounting) but at a more detailed societal level than Hajer's presentation of Foucault. Thus, in summary, a story line can be thought of as:

i a particular kind of narrative account associated with a policy issue,
ii used by a number of stakeholders who make-up a discourse coalition,
iii associated with a degree of discursive hegemony or discursive closure of a particular policy issue.

Based on Hajer's view, discourses could possibly offer an interpretative framework and intellectual context for story lines. The function of story lines can be used in the same way as Star and Griesemer (1989: 393) describe 'boundary objects'. Like 'boundary objects', story lines are both flexible enough to adapt to local settings and constraints of several stakeholders employing them, yet robust enough to maintain a common identity across sites. They are on a regular basis structured in common use, and become strongly structured in individual site use. A story line can have different meanings in it to make it recognizable, and the creation and management of story lines is a key process in developing and maintaining coherence across social dimensions and geographical borders.

Hajer's notion of discourse-coalitions is linked to actors. It can, however, also be linked to the concept of stakeholders. According to Wang and Dewhurst (1992), the term 'stakeholder' was first adopted at Standford Research Institute in 1963. Since then the concept has developed and has become a prominent concept in corporate and academic communities. What is the exact definition of a 'stakeholder', and what is a stakeholder in the context of oil and gas activity in the Arctic? Using R. E. Freeman (1984), we can consider a stakeholder as 'any individual or group who can effect, or is affected by the achievement of the organisation's objectives'.

Different stakeholders can adopt the same story line (or story lines) as narratives on social reality in the above sense (Hajer, 1995: 62). A discourse coalition in Hajer's sense differs from ordinary political coalitions in the sense that it is based in linguistics, not necessarily interests. Story lines are seen as capable of changing the stakeholders' interpretation of what their interests are.

Stakeholders also use story lines to bring together previously unrelated elements of discourse, and thus allow new understanding and create new meanings. Story lines allow stakeholders to give meaning to specific problems and reduce discursive complexity.

This does not mean that the basis for the story line is always coherent. It is possible that, although stakeholders share a specific set of story lines, they can still have their own interests and their own interpretations regarding the significance of the story lines. It is the constructivist aspect of story lines that make them essential political devices in re-ordering understandings, positions, subjects and structures, and in the creation of social and moral orders. It is this aspect of story lines that makes the identification of story lines essential for understanding or interpreting sustainable development in the context of the Arctic.

We have used discourse as a shared meaning of a phenomenon. This phenomenon may be small or large, and the understanding of it may be shared by a small or large group of people on a local, national, international or global level. The stakeholders adhering to the discourse participate in various degrees to its production, reproduction and transformation through written and oral statements. In describing the discourses, we present data concerning several elements in the debate around sustainable development in the Arctic, with particular focus on identifying the main characteristics of the story.

The *Arctic Human Development Report* (Young and Einarsson, 2004) identified and discussed a number of different 'mindsets' or ways 'northern issues' have been framed which is close to our use of story lines: the idea of the Arctic as a homeland for a diverse group of indigenous peoples; as a land of discovery from a European perspective; as a magnet for cultural emissaries, in the beginning, most notably from Christian missionaries and later in the form of general impacts of Western culture; the Arctic as a storehouse of both renewable and non-renewable resources; as a theatre for military operations; as an environmental linchpin, both in relation to waterborne and airborne pollutants, accumulation of persistent organic pollutants and climate change; the scientific Arctic; the destination for adventure travellers and the Arctic of the imagination. Our study focuses especially on the mindsets that relate to Arctic oil and gas.

Design and methodology

Following Yin (2003), our study is best described as an embedded multiple-case design. An embedded multiple-case is an empirical form for descriptive studies, where the goal is to describe a phenomenon containing more than one sub-unit of analysis. The cases are the four regions or countries in general. The embedded units of analysis are different in each region/country. In the case of Alaska, the chosen area is the North Slope. In Canada, it is the areas linked to the Mackenzie Pipeline plan. The Norwegian study is less embedded in the sense that the controversies described in the Norwegian chapter are more general. The Integrated Management Plan, however, is an embedded unit of analysis.

In Russia, there are three embedded units: the Yamal Peninsula, and the Murmansk and Arkhangelsk regions. The research objects of the comparisons, however, are the discourses that dominate the framing of oil and gas in the Arctic. Discourse analysis, therefore, is what unites the case studies.

It is important to discover the various positions one can take in debates about the issues and what sorts of stakeholders that are usually on each side. On certain issues about oil and gas activities in the Arctic you know what the 'sides' are, how they are talked about, and what sort of stakeholders tend to be on specific sides. Some of these sorts of issues are known by nearly everyone in society, others are known only by specific groups of stakeholders. Certain debates can form different discourse coalitions. Discourses may not have discrete boundaries because stakeholders always create new discourses, change old ones, and contest and push the boundaries of discourses. Therefore, there are 'smaller' story lines inside bigger ones, and each story line triggers or is associated with others, in different ways in different settings and differently for different stakeholder groups.

We do not attempt to seek the one 'true' characterization of a particular issue but to explore how different stakeholders characterize their contributions within the social arena and interpret the contribution of others. The project provides statements of how oil and gas activities are framed, how they should be managed, which managerial implications of economic, environmental and social issues there are, and how the different stakeholders consider sustainability issues within the oil and gas discourse.

We have used the perspective of stakeholders and discourse analysis to achieve a shift from method to a strategy of analysis. In the following we outline how we have mainly done our research, based on Hajer's approach to discourse analyses. We present the interview guide, informants, and collection and analysis of data.

The interview guide

The aim of the analytical guide was to secure consistency and comparability across regions and countries, with a common set of research propositions and questions to be used. We have used eight categories of research propositions and questions pertaining to oil and gas activity in the Arctic:

1 drilling in the Arctic;
2 sustainability;
3 social issues;
4 environment;
5 power and politics;
6 indigenous people and local communities;
7 Arctic institutions and organizations; and
8 narratives of the Arctic.

The aims of the questions in each category were to:

- reveal the informant's general attitudes towards oil and gas activities in the Arctic, and the arguments and justification for the position(s) taken;
- uncover the informant's general opinions about sustainable development, and the priorities and implications following from a sustainable development perspective;
- focus upon and identify the social issues more specifically; the aim was to get the informant to specify the main social issues, challenges and opportunities (like human rights, well-being, health, education, living conditions, gender equality, capacity building);
- focus upon and identify the environmental issues more specifically, and to get the informant's views on the ranking of different environmental problems, and whether or not the solving of environmental issues is possible with expanding oil and gas activities;
- focus upon and identify the main actors and decision-makers at the national level, and to get the informant's views on the ranking of different actors;
- focus more specifically on indigenous peoples and local communities in order to identify attitudes, conflicts and relations between different groups and organizations in relation to oil and gas activities;
- elicit the informant's views on the importance of different international Arctic institutions and organizations, like the Arctic Council, the Northern Forum, the International Arctic Science Committee and so on; and
- identify the most important story lines of the Arctic and the implications from the particular framing of the Arctic.

The interview guide was adjusted to reflect the different context when interviewing stakeholders in the different countries. Not all questions could be used for all the stakeholders interviewed, and some questions needed to be adjusted to suit the themes and story lines told by the stakeholders.

Informants

The stakeholders have mainly been identified on the basis of general knowledge of the different countries in question. These have further been classified into stakeholder groups. The different groups do not make up homogeneous types and even within a particular stakeholder type it is hazardous to assume that any particular stakeholder is representative of all stakeholders in the group. However, we believe that it is useful as a starting point to order the stakeholders into broadly similar groupings in order to give an overall description of relevant stakeholders involved in the oil and gas policy arena. This gave us the opportunity to analyze similarities and differences between stakeholders within the same overall groupings as well as across groupings.

The key stakeholders in the debate have been categorized into the following groups: non-governmental organizations, business, government officials, academics, and indigenous peoples and local inhabitants. Table 5.1 reports the number of

Table 5.1 Interview matrix

Area	Informants						
	NGOs	Industry/ business	Government/ officials/ bureaucracy	Politicians	Academia/ research institutions	Indigenous people/local people	Media (newspapers/ TV/radio, etc.)	Others
Alaska	1	3	2	1	3	15	0	0
Canada	7	5	6	0	4	10	3	0
Norway	6	5	14	10	5	8	12	2*
Russia†	7	2	8	4	6	5	11	4

*International Association for Impact Assessment (IAIA) 26th Annual Conference, May 23–26, 2006: Power, poverty and sustainability. Meeting between The Norwegian Oil Industry Association (OLF) and Municipalities and Business in Lofoten and Vesterålen area.
†Data for the Russian case are mostly collected from secondary sources. Hence the numbers in the table are describing these sources.

interview subjects in each category. Additional information was also collected from secondary sources, such as organizational websites and annual reports.

The political programmes in different regions are themselves a part of the discourse. Different actors need to construct story lines and reinterpret the story lines of others if they are to be persuasive. The challenge this has raised for the Arctic project is how to reveal through discourse and rhetorical analysis the story lines constructed by stakeholders in the different countries. Story lines that are interpreted and reinterpreted to be persuasive will prove particularly useful in revealing the dynamics of how story lines are made and remade through a controversy. Therefore, in our context a story line has been used to describe a discursive theme and style that commonly occurs in society, has a particular sequence and rhetorical ordering associated with it, and can be interpreted and reinterpreted by stakeholders.

Empirical material and data analysis

We have used the following sources and methods for data collection: available research documents and archive data, research reports and published papers; the Internet; key informant interviews; conference participation; media (newspapers, television, radio, magazines, etc.); and Cambridge libraries. Several researchers from each country participated in the data collection and writing process. The field work was primarily carried out during 2006 and 2007.

In the discourse analysis of the different countries, we begin with an overview of the main characteristics of the countries. The oil and gas discourse is first analyzed broadly, using a discourse-framing approach that identifies stakeholders and their conceptualizations within the overall discourse. Then we focus more specifically on debates, and the different stakeholders and discourse coalitions. Finally, the specific context of each country and the story lines are drawn together and examined within the broader context of sustainable development.

The term 'policy arena' is used as a metaphor that conceptualizes a space where stakeholders communicate and interact either directly through events such as meetings or conferences, or through text such as reports, legislation or websites, or through presenting their perspectives on oil and gas activities in media, such as newspapers, television and radio in each country. Our point of departure was the assumption that stakeholders try to shape policy discussions about oil and gas in the Arctic.

Several recurring themes have been identified in the text and talk about present oil and gas activity in the Arctic. This includes discussion around environmental issues, indigenous peoples' rights, climate change, energy demand and supply, political issues and local communities. Themes have been used as a foundation for the story lines, and as headlines in the discourse and representing different story lines. The story lines are ordered in a way to support different central propositions in the themes. We have constructed the story lines, not only from words, but also from the context where the stakeholders have brought forward their opinions about certain issues in a fixed context.

The oil and gas discourse is made up of a specialized language, and we can analyze the story lines commonly used by stakeholders in the oil and gas policy arena when discussing, for example, environmental impacts. Each story line consists of a discursive theme, presented in a sequence to communicate a way of understanding the theme. Different stakeholders may attempt to interpret and reinterpret specific story lines so that it presents their perspective on the theme. But, if they want their contribution to be seen as relevant to the existing discourse, they are to some extent constrained by other stakeholders telling the story. We have several story lines linked to the same theme. Each story line communicates a way of understanding the theme, and it is important to understand how themes and story lines are connected. The themes are the basis for several stories or narratives within this part of the discourse.

Limitations

It is important not to use discourse analysis disconnected from the theoretical and methodological foundation. In discourse analysis, theory and methodology are linked. Discourse analysis is first and foremost a way to observe the world. By using discourse analysis, we use a certain type of scientific lens to ask certain questions and search for certain processes. At the same time, when using discourse analysis, we use it in a specific thematic field. In our case, it is oil and gas activities in the Arctic (specifically), and sustainable development and corporate social responsibility (generally).

The context in which story lines are constructed, that of sustainable development and corporate social responsibility, are partly our own constructions. This theoretical entry directs the questions and the way we have monitored the different debates about oil and gas in the different parts of the Arctic. Of course, a different theoretical entry could provide us with other answers and solutions. Moreover, the case studies from Alaska, Canada, Norway and Russia do not pretend to give a full and overall presentation of all story lines in each country, and the story lines represent the author's interpretations and findings. Each story brought forward, either in interviews, documents, news articles or television-programmes, have affected the way we look at the debates about oil and gas activities in the Arctic. We have chosen the stories that we think are the most prominent and important, and tried to see which stakeholders share the usage of a particular set of story lines over a particular period of time. In this setting, it is our framework that sets the premises of what are the essential stories and stakeholders.

However, the choice of interview subjects has also changed during the course of the study, and in each interview we have obtained referrals to other key stakeholders. We have not had the resources to interview all the stakeholders brought forward in the primary interviews, and we may have missed some points or stories to the discourse. We have not covered all stories and we may have lost some important aspects of the debates, or we have highlighted stories that other people would not have considered important. The main rule has been that, if many people use a story or part of a story to conceptualize the issues, and if it solidifies into institutions and organizational practice, this particular discourse is dominant. We do believe, however, that the country studies capture the most important aspects and framings of oil and gas in the Arctic.

To some extent the above strategy also varies across countries and researchers. There are cultural, political and institutional differences between the countries and regions that necessitate this. There have been rapid changes in some of the countries: the case of Russia has been especially challenging. Compared with Alaska, Canada and Norway, Russia is still in a transformation stage. Russia is a young democracy with a less developed civil society in which there is a smaller role for public discourse, which makes it more challenging to elicit and identify conflicting story lines and opposition. Nonetheless, we have tried to capture the most important and influential story lines, and also the relationships between them.

Time has also been an important factor. When should we draw an end? The discourses about oil and gas activities in the Arctic will not stop with this project, and we have had some challenges deciding when to stop the data collection. There will always be new and updated data and sources for the story lines, and these story lines will change in the struggle over oil and gas in the Arctic. Therefore, the story lines presented in this book must be read with the time of the data collection in mind, that is, 2006 and 2007. The answers we provide to the questions of how different organizations, authorities, political parties, indigenous peoples, locals and industry conceive and frame oil and gas activities in the Arctic will not be stable. However, the challenges they address, of balancing the different concerns of sustainable development, will remain for a long time to come.

Notes

1 There are a number of different approaches to discourse analysis. For useful introductions to discourse analysis, see Alvesson and Karreman (2000), Brown and Yule (1983), Coulthard (1977), Coulthard and Montgomery (1981), Gee (1999), Phillips and Jørgensen (1999), Potter (1996, 1997), Stubbs (1983), Widdowson (1979) and Wynne (1995).

References

Alvesson, M. and Karreman, D. (2000) 'Varieties of discourse: on the study of organizations through discourse analysis', *Human Relations*, 53 (9): 1125–1149.

Amin, A. (1994) 'Post-Fordism, Fantasies and Phantoms of Transition', in Amin, A. (ed.), *Post-Fordism: A Reader*, Oxford: Blackwell Publishers, pp. 1–40.

Anderson, R. B., MacAulay, S., Kayseas, B. and Hindle, K. (2007) 'On their own terms: indigenous communities, development and the new economy', in Shockley, G. E., Frank, P. M. and Stough, R. R. (eds) *Non-market Entrepreneurship: Interdisciplinary Approaches*, Cheltenham: Edward Elgar.

Brown, G. and Yule, G. (1983) *Discourse Analysis*, Cambridge: Cambridge University Press.

Burkey, S. (1993) *People First: A Guide to Self-Reliant Participatory Rural Development*, London: Zed Books.

Corbridge, S. (1989) 'Post-Marxism and development studies: beyond the impasse', *World Development*, 18 (5): 623–639.

Coulthard, M. (1977) *Introduction to Discourse Analysis*, London: Longman.
Coulthard, M. and Montgomery, M. (1981) *Studies in Discourse Analysis*, London: Routledge.
Crewe, E. and Harrison, E. (1998) *Whose Development?: An Ethnography of Aid*, London: Zed Books.
Dicken, P. (1992) 'International production in a volatile regulatory environment', *Geoforum*, 23 (3): 303–316.
Dryzek, J. S. (1997) *The Politics of the Earth: Environmental Discourses*, Oxford: Oxford University Press.
Dunning, J. H. (2003) *Making Globalization Good: The Moral Challenges of Global Capitalism*, Oxford: Oxford University Press.
Elam, M. (1994) 'Puzzling out the post-Fordist debate', in Amin, A. (ed.) *Post-Fordism: A Reader*, Oxford: Blackwell Publishers, 43–70.
Foucault, M. (1980) *Power/Knowledge: Selected Interviews and Other Writings*, New York: Pantheon.
Freeman, R. E. (1984) *Strategic Management: A Stakeholder Approach*, Boston: Pitman.
Gee, J. P. (1999) *An Introduction to Discourse Analysis: Theory and Method*, London: Routledge.
Hajer, M. A. (1995) *The Politics of Environmental Discourse: Ecological Modernization and the Policy Process*, Oxford: Oxford University Press.
Hajer, M. A. (1996) 'Ecological modernization as cultural politics', in Lash, S., Szerzynski, B. and Wynne, B. (eds) *Risk, Environment and Modernity: Towards a New Ecology*, London: Sage.
Hancock, G. (1989) *Lords of Poverty: The Power, Prestige, and Corruption of the International Aid Business*, New York: Atlantic Monthly Press.
Hirst, P. and Zeitlin, J. (1992) 'Flexible specialization versus post-Fordism', in Storper, M. and Scott, A. (eds) *Pathways to Industrialization and Regional Development*, London: Routledge, 70–115.
Jessop, B. (1989) 'Conservative regimes and the transition to post-Fordism', in Gottdiener, M. and Komninos, N. (eds) *Capitalist Development and Crisis Theory*, New York: St Martin's Press, 261–299.
Klitgaard, R. E. (1990) *Tropical Gangsters*, New York: Basic Books.
Komninos, N. (1989) 'From national to local: the Janus face of crisis', in Gottdiener, M. and Komninos, N. (eds) *Capitalist Development and Crisis Theory*, New York: St Martin's Press, 348–364.
Leyshon, A. (1992) 'The transformation of a regulatory order', *Geoforum*, 23 (3): 347–363.
Moore, M. (1997) 'Societies, polities and capitalists in developing countries: a literature survey', *Journal of Development Studies*, 33 (3): 287–363.
Peck, J. and Adam, T. (1992) 'Local modes of social regulation', *Geoforum*, 23 (3): 347–363.
Phillips, L. and Jørgensen, M. W. (1999) *Diskursanalyse som teori og metode*, Fredriksberg: Samfundslitteratur, Roskilde Universitetsforlag.
Potter, J. (1996) *Representing Reality: Discourse, Rhetoric and Social Construction*, Thousand Oaks, CA: Sage.
Potter, J. (1997) 'Discourse analysis as a way of analysing naturally occurring talk', in Silverman, D. (ed.) *Qualitative Research: Theory, Method and Practice*, Thousand Oaks, CA: Sage, 144–160.
Scott, A. J. (1988) *New Industrial Spaces: Flexible Production Organization and Regional Development in North America and Western Europe*, London: Pion Ltd.

Star, S. L. and Griesemer, J. R. (1989) 'Institutional ecology, "translations" and boundary objects: amateurs and professionals in Berkley's Museum of Vertebrate Zoology, 1907-39', *Social Studies of Science*, 19: 387–420.

Storper, M. and Walker, R. (1989) *The Capitalist Imperative: Territory, Technology and Industrial Growth*, New York: Basil Blackwell.

Stubbs, M. (1983) *Discourse Analysis: The Sociolinguistic Analysis of Natural Language*, Chicago, IL: University of Chicago Press.

Tucker, V. (1999) 'The myth of development: a critique of a Eurocentric discourse', in Munck, R. and O'Hearn, D. (eds) *Critical Development Theory: Contributions to a New Paradigm*, London: Zed Books, 1–26.

Wang, J and Dewhurst, H. D. (1992) 'Boards of directors and stakeholder orientation' *Journal of Business Ethics*, 11(2): 115–123.

Weingart, P., Engels, A. and Pansegrau, P. (2000) 'Risks of communication: discourses on climate change in science, politics, and the mass media', *Public Understanding of Science*, 9: 261–283.

Widdowson, H. (1979). 'Rules and procedures in discourse analysis', in Myers, T. (ed.), *The Development of Conversation and Discourse*, Edinburgh: Edinburgh University Press.

Wynne, B. (1995) 'Public understanding of science', in Jasanoff, S., Markle, G. E., Petersen, J. C. and Pinch, T. (eds) *Handbook of Science and Technology Studies*, London: Sage, 361–388.

Yin, R. (2003) *Case Study Research: Design and Methods*. Newbury Park: Sage Publications.

Young, O. R. (2005) 'Global insights – governing the Arctic: from Cold War theater to mosaic of cooperation', *Global Governance: a Review of Multilateralism and International Organizations*, 11 (1): 9–15.

Young, O. and Einarsson, N. (2004) '*Introduction*', in *Arctic Human Development Report*, Akureyri: Stefansson Arctic Institute.

Part II
Legal and institutional framework
Case studies

6 Legal and institutional framework

A comparative analysis

Nigel Bankes

Introduction

This chapter provides a background comparative analysis of the legal and institutional framework of each of the four countries examined in this book: Canada, Norway, Russia and the United States of America (Alaska). In deciding what to include and what to omit, the chapter takes a very functional approach. The content focuses upon the types of information that a reader might need to gain a deeper contextual understanding of the relationships between each of the three principal actors we consider in this volume: indigenous people, oil and gas operators, and the state in its various manifestations.

The chapter is organized around three main themes or sets of questions. Part one poses a set of questions about the form of government in each of the four jurisdictions. Much of the focus of this part is on the three federal states, Canada, Russia and the United States (since it is important that the reader have an understanding of the distribution of legislative power and public property in each of those states), but the chapter also discusses the unitary form of government in Norway and provides some discussion of local and regional forms of government in each of the jurisdictions.

The second set of questions deals with the ownership of land and resources and, in particular, the division between public ownership and private ownership. The premise here is that, in addition to legislative jurisdiction or the power to make laws, the bundle of rights associated with ownership provides an important source of power or authority within society as well as a place where the owner (whether an individual or a collectivity, such as an indigenous community) may have a greater degree of protection from state authority, and a space within which to exercise autonomy. At least since the writings of John Locke, the Western liberal tradition has afforded property claims a special status and, for a variety of reasons, including utilitarian, personal autonomy and efficiency reasons (McPherson, 1978). Other traditions, such as those of indigenous people, emphasize the importance of land for cultural and spiritual reasons, and frequently view property as a collective rather than individual interest.[1] But seen from either perspective, the institution of property continues to be of central importance.

The right to property and the power to make laws have conceptually distinct origins but, in some cases, those powers may be vested in the same person or entity (e.g. the government may be the owner/proprietor as well as the law-maker), leading to a tendency to conflate the two ideas. But in other cases the two sources of authority will be vested in very different persons or entities (i.e. the more usual liberal paradigm of private owner and public legislator). In any event, the oil and gas sector will always need to know who is the owner of the resource. With whom must industry negotiate for a lease or concession? To whom must it pay royalties? A sub-set of ownership questions deals with the distinction between surface rights and sub-surface rights (i.e. mineral/oil and gas rights). It is entirely possible that these rights will be vested in different persons (e.g. government may own the sub-surface rights, and a private owner or an indigenous people may own the surface). In such a case, how does the legal system mediate the conflict between these different resource owners? What are the implication for an oil and gas operator who already has an oil and gas concession from the mineral owner, but who also needs to obtain access to the surface to drill a well?

The third set of questions deals with the legal status of indigenous people and their land and resource claims in each of the four jurisdictions. Here, the sorts of questions that are of interest are: How does the jurisdiction recognize (or not) indigenous rights to land and resources? And what are the implications of this for the oil and gas sector? Have indigenous people within the jurisdiction negotiated land claim or similar agreements that clarify the respective ownership rights of the government and the indigenous people? And what are the implications of these agreements for the oil and gas sector? How do these agreements empower indigenous people in their dealings with the oil and gas sector? Do they enable them to negotiate economic participation agreements that allow them to participate in, and benefit from, resource development opportunities within traditional territories? What forms do these agreements take? As well as property rights, do indigenous people have the power to make laws within their traditional territories or some sub-set of their traditional territories? If so, what is the source of this power and how might the exercise of that power affect the oil and gas sector?

The form of government and the distribution of legislative authority

Of the four states examined here, three are federal states and only one, Norway, is a unitary state. But the labels 'federal' or 'unitary' do not tell us much about the actual distribution of power within the state. To answer that question we must peel the onion and ask a series of additional questions. For the three federal states, the sorts of questions we might have will be the following: What are the sub-units of the federation? How does the federation allocate the power to make laws? Are these laws-making powers exclusive or concurrent? In the event of a conflict, which trumps? How does the federation allocate the ownership of public property including natural resources? Taking law-making powers and ownership together, how does the federation allocate the power to collect the available economic rent from the oil and gas sector? Beyond the sub-units of the federation, what other governments have law-making powers that may affect the conduct of oil and gas operations?

Within the unitary state of Norway, our questions will focus on issues of devolution to county-level governments, and the relationships between the sector-specific departments of the central government in Oslo and regional forms of government. The recent Finnmark Act (Norway, 2005) suggests that it may also be important to ask questions about the vesting and control of public lands in those regional-level governments. In discussing the form of government in each of the four states, it is important to recall that governmental systems are never static. They change over time not only as a result of formal amendments, but also as a result of informal practices and court decisions. The constitutions of some states may appear more stable, for example, the United States (US) and others more dynamic. For example, the Russian form of government seems to be constantly evolving in these decades following the collapse of the Soviet Union. Whereas President Yeltsin encouraged the different regions to seek power, President Putin has sought to re-centralize authority in the federal government using several different techniques explored below. But Russia is not the only institutional innovator within this group of four states. For example, Norway's Finnmark Act is a notable experiment in regional devolution while the creation of Sámi parliaments in each of the three Nordic states (with Norway leading the way) affords a way of promoting Sámi interests while falling short of a territorially based and constitutionalized form of federalism. In Canada, the distribution of authority between federal and territorial governments continues to be in flux as a result of the processes of devolution and some commentators see the institutional apparatus of modern land claim agreements between the federal government and aboriginal people as a new form of 'treaty federalism', a third level of federalism within the Canadian polity (White, 2002).

All four jurisdictions profess support for a separation of powers to a greater or lesser extent, and all formally commit to a separate judicial function and the rule of law, meaning, at a minimum, that the exercise of executive and legislative authority is constrained by the operation of law guaranteed by an independent judiciary and a largely self-regulating legal profession. But the constraints and disciplines of the rule of law evolve over decades if not centuries and the concept is at a very early stage in its development in post-Soviet Russia.

The sub-units of the federation and the allocation of legislative authority: Canada, Russia and the United States

Federal systems provide one way for the nation-state to recognize and accommodate regional and linguistic differences and claims to autonomy, but there is no 'normal way' to structure a federal system. Each federal system is grounded in a particular history (Field, 1992). There may be points of similarity between different federations, but there will also likely be significant differences both in structure and outcomes. Take Canada and the US: at one level there are important similarities between these two federal systems, since they are both basically symmetrical. Each has a relatively strong central government and, in both cases, each of the sub-units of the federation (the 50 states, in the case of the US, and Canada's ten provinces) has the same set of powers (at least in formal terms) as another sub-unit of the federation. Thus, while there may be significant wealth and income disparities between provinces and between states (and in Canada there is a strong tradition of equalization payments from 'have' to 'have-not' provinces to reduce these differences[2]), each unit has the same power to make laws in relation to specified subject matters.

By contrast, the Russian form of federation appears starkly different from these two North American models. Article 1 of Russia's 1993 Constitution stipulates that Russia shall be 'a democratic federal rule-of-law state with the republican form of government' but composed (Articles 5 and 65) of some 89 subjects of the federation consisting of republics [e.g. the Republic of Sakha (Yakutia)], territories (e.g. Krasnodar Territory), regions (e.g. Murmansk Region or the Sakhalin Region), federal cities (Moscow and St Petersburg), an autonomous region (Jewish Autonomous Region) and autonomous areas (e.g. Nenets, Khanty-Mansior Yamal-Nenets Autonomous Areas). There is some geographical overlap between some of these subjects, since an Autonomous Region is also a component part of another unit (leading to the description 'nesting dolls'; Salikov, 2004; Domrin, 2006). Although the constitution suggests that all 89 subjects of the federation are to be 'equal subjects', in fact there are tremendous formal and practical asymmetries in terms of wealth and power. They vary widely in size [e.g. Yakutia (Sakha Republic), 3.1 million km^2; and Adygeya 7800 km^2) and population (Domrin, 2006). In some cases (e.g. Yakutia), the republic represents the successor to a former Autonomous Soviet Socialist Republic. Such republics originally afforded territorial recognition to an ethnic (indigenous) people (the Sakha in the case of Yakutia), but in-migration of Russians and other nationalities (often with Moscow's active encouragement) has resulted in the indigenous people becoming 'a minority in their titular republic'.

But seen from another and more formal perspective, there are also significant differences between the forms of government of Canada and the US. Not only is one, Canada, a constitutional monarchy with a Westminster-style of government and the other, the US, a presidential form of democracy with a strict separation of executive and legislative powers, but the formal statement of the division of powers is quite different. Whereas Canada's constitution contains a list of the legislative powers allocated to each of the federal and provincial governments, and allocates any residual power to the federal government,[3] the US Constitution contains a list of federal powers leaving the residue to the states, a distribution that the Tenth Amendment simply confirms: 'The powers not delegated to the United States by the Constitution nor prohibited by it to the States, are reserved to the States respectively, or to the people'.

But while these apparently fundamental differences might suggest that the central government should be much stronger in Canada than the US, this has not really turned out to be the case because of the important interpretive role performed by the final appellate courts in each jurisdiction. Thus, the US Supreme Court has tended to support a very broad interpretation of Congressional authority, while the Canadian courts have offered a narrow interpretation of federal powers including the residual clause of the constitution (Field, 1992). The importance of provincial authority in Canada is especially evident in the oil and gas sector where the combination of provincial ownership rights and the power to make laws in relation to the industry is clearly dominant. Federal authority in this sector is generally limited to certain environmental responsibilities (largely through the fisheries power) but assumes much greater importance in areas that lie beyond the provinces (the offshore areas and the northern territories – and see further discussion below).

In the case of the Russian Federation, it is Articles 71–73 of the constitution that effect the division of legislative authority between the federal jurisdiction and the subjects of the federation. Thus Article 71 lists areas of exclusive federal jurisdiction, while Article 72 lists areas of joint or concurrent jurisdiction. Article 73 provides that residual jurisdiction lies with the subjects of the federation. The areas of exclusive jurisdiction allocated to the Russian federation are extremely broad and include civil and criminal laws, all aspects of foreign and military policy and defence. The areas of joint or concurrent jurisdiction between the Russian federation and the subjects of the federation are also very broad and include several areas that are significant in the context of this volume: protection of the rights of ethnic minorities; issues of the possession, use and management of the land, mineral resources, water and other natural resources; delimitation of state property; management of natural resources, protection of the environment and ecological safety; land, water and forestry legislation; legislation on the sub-surface and environmental protection; establishment of the general guidelines for taxation and levies within the federation; and protection of the original environment and traditional way of life of small ethic communities. In cases of inconsistency between the laws of a sub-federal unit and federal law, the federal law prevails. This general paramountcy norm of federations also applies in the US and Canada.

As head of a presidential republic, the President of the Federation also has significant law-making authority insofar as Article 90 of the Constitution affords the President the authority to issue decrees and executive orders on any matter provided that they do not contravene the Constitution or federal laws. The President also has certain veto powers over laws adopted by the Duma and Federation Council.

The existence of a federal paramountcy rule combined with a broad statement of federal powers allows an assertive and active federal government to centralize authority in its own hands. While in the US the Senate (where the states are equally represented) serves as a check on this tendency, observers of the Russian Federation emphasize how the federal government in that jurisdiction has consolidated its authority in recent years notwithstanding a similarly structured bicameral system, including as the second chamber the Federation Council composed of two members from each of the 89 subjects of the federation.[4] Various measures have served to centralize authority in the Russian Federation and specifically in the office of the President. For example, as Domrin (2006) shows, shortly after his election in 2000, President Putin issued a decree dividing the Federation into seven federal zones with each zone incorporating various sub-units of the federation. The President appoints a presidential representative for each region, with the responsibility for the implementation of federal decisions within the region as well as broad powers over the use of federal property and resources within the region. Domrin (2006) argues that the creation of these seven federal superdistricts is 'just the beginning of reforms aimed at reversing a decade of fragmentation ... and restoring federal control over Russia's subject units'. Similarly, in 2004, new legislation was approved abolishing direct gubernatorial elections and sanctioning instead nomination by the President of the Russian Federation. Measures to further strengthen federal authority and to reduce the number of sub-units of the federation through mergers have continued since Putin was elected for a second term in 2004. The same trend to centralization and diminishing the authority of the subjects of the Federation is observable in the oil and gas sector. Thus amendments to the Law on the Subsoil introduced in 2004 effectively terminated the 'two-key' approach, which required that disposition decisions were to be made jointly by federal and regional authorities. Ongoing proposals to amend or replace the law on the subsoil offer further confirmation of this trend (Goudina, 2006; Skyner, 2005).[5]

Local governments (e.g. cities, municipalities, regional governments) in each of the US and Canada are generally creatures of the sub-units of the federation. In Russia, however, local self-governments have a measure of constitutional protection insofar as Article 12 of the Constitution recognizes and guarantees local self-government. The territories of the subjects of the Russian Federation are divided into raions (and sub-raion towns, township or villages and local regions) and cities (Salikov, 2004: 11–12). Goudina (2004: 72) suggests that municipal bodies play an important role in the Russian oil and gas industry, since they will frequently control surface access, which will be important for infrastructure development.

Canada's territories

While Canada is a federal state and the principal sub-federal unit of Canada is the province, that position is complicated in northern Canada by the existence of the three territories of Yukon, Northwest Territories and Nunavut. Strictly speaking these three territories are not sub-units of the federation, since Canadian constitutional documents divide the totality of law-making powers between the provinces and the federal government, and afford no independent status to territories. The three territorial governments are, therefore, creatures of federal legislation. They are not constitutionally entrenched and they lack the autonomy, and at least some of the powers, typically afforded to the provinces; their constitutional status is analogous to that of Alaska within the US federation prior to Alaska attaining statehood in 1959.

The relative lack of power and autonomy of the Canadian territories is particularly noticeable in the oil and gas sector. Historically, the federal government has retained ownership of oil and gas and mineral resources within the territories and has effectively retained the exclusive power to make laws with respect to these matters.

Consequently, the main oil and gas leasing statute in northern Canada, the Canada Petroleum Resources Act (Canada, 1985), which applies to Nunavut, the Northwest Territories (NWT) and the Arctic offshore area, is a federal statute. More recently, however, there has been a trend for the federal government to transfer both ownership and control, and the right to make laws with respect to oil and gas, to the territorial governments. This has already been accomplished for Yukon (although confined to onshore oil and gas) and there are ongoing devolution negotiations with both the NWT and Nunavut. In the case of the NWT, these negotiations are complicated by unsettled land claims and by arguments that, if resources' ownership is to be devolved, it should be devolved to aboriginal governments rather than to non-ethnic territorial governments. While resource transfers should provide an important revenue source for territorial governments, such governments will likely be dependent on federal fiscal transfers for a long time to come.

The degree to which the three territories can be seen as meeting the desires of northern indigenous people for a greater degree of autonomy and self-government varies dramatically across the three territories, depending largely upon the ethnic breakdown of the population. Historically the territorial governments were appointed rather than elected governments and did not represent indigenous populations of the territories. This changed in the last two decades of the twentieth century and the territorial governments increasingly came to represent the interests of their residents. But the demographic breakdown between indigenous/non-indigenous residents varies considerable across the northern territories, as one moves from east to west. While Inuit represent over 80 per cent of the population of Nunavut (created as a separate territory in 1999), the NWT are equally divided between indigenous and non-indigenous people, whereas in Yukon only about 25 per cent of the population is of indigenous ancestry. Hence, while the Inuit of Nunavut are generally willing to pursue autonomy through the public government structure of the territorial government, indigenous people in the NWT and Yukon are more likely to question the legitimacy of the territorial level of government and to seek instead to empower more ethnically based forms of self-government.

Local and regional government in Alaska: the North Slope Borough

In addition to the various indigenous governance organizations that serve to implement the Alaska Native Claims Settlement Act (discussed below), and in addition to the tribal organizations that preceded that Act (Case and Voluck, 2002: 317–367), state law and the state constitution also recognize local and regional governments. In some cases, these governments can create significant opportunities for local communities, including indigenous communities, to obtain important economic benefits from oil and gas developments that may occur within their sometimes extensive boundaries. The best-known example of this is the North Slope Borough (NSB). Incorporated in 1972, the NSB covers an area approximately the same size as the state of Montana. The Borough includes within its tax base the oil and gas development infrastructure of Prudhoe Bay. Comprised of some eight villages from Point Hope in the West to Kaktovik in the east, the population of the Borough, some 7500 people, is predominantly Inupiat. The large tax base permits the NSB to offer an impressive range of services and employment opportunities within each of the Borough's villages (Northern Economics Inc., 2006).

Norway as a unitary state

Norway is the one unitary state of the four countries under consideration in this volume. Like Canada, Norway is a constitutional monarchy, which observes a separation of powers between the Storting holding legislative authority, to whom the Executive (in formal terms, the King) is responsible, and a judicial branch. Furthermore, while Norway has significant agreements with the European Union (EU), and while EU policies in some areas may be very influential and may constrain Norway's freedom to manoeuvre in some policy areas, Norway remains outside the EU.

The formal implication of a unitary state is that plenary sovereignty is vested in a single authority (the King and the Storting), and no region or regional government has the constitutionally protected (and exclusive) right to make laws in relation to particular subject areas. But that does not prevent the central government from delegating (devolving) significant powers to regional or local governments, and this may afford these jurisdictions considerable practical autonomy. Furthermore, parliamentary sovereignty may also be self-limited by the adoption of a Bill of Rights or similar instrument. While Norway does not have a formal Bill of Rights, the 1999 Human Rights Act (discussed further in Chapter 11) makes the provisions of the European Convention on Human Rights and the two International Covenants on Civil and Political Rights and Economic, Social and Cultural Rights part of domestic law and further stipulates that these conventions shall take precedence over any conflicting law.

The central government is organized along conventional sectoral lines. For example, Norwegian petroleum legislation is administered by the Ministry of Petroleum and Energy, although much of the detailed responsibility for ongoing regulation has been delegated to the Norwegian Petroleum Directorate.

Regional and local government in Norway is based upon counties and municipalities. Norway has 435 municipalities and 19 counties. Municipalities generally have jurisdiction over local matters, including schooling and social services, but also land-use planning and local environmental issues. Counties have a land-use planning jurisdiction over areas outside the municipalities. The planning authority of counties and municipalities is most likely to be engaged by the development of petroleum infrastructure on land but also through responsibilities for coastal zone planning (Jentoft and Buanes, 2005). The northernmost county is Finnmark, which includes five Sámi municipalities: Karasjok, Kautokaino, Nesseby, Porsanger and Tana. Together these municipalities form the Sámi language administrative district within which the Sámi language has official status for many purposes (Sámi Act, 1987, Chapter 3; Norway, 1987). In practice, the planning power of municipal and county authorities is constrained by the authority of sector agencies like the Ministry of Petroleum and Energy, since, in the event of a conflict, the sector authority of the central government will prevail over local or regional planning rules (Jentoft and Buanes, 2005: 156).

From the perspective of this volume, it seems important to emphasize the significance of two institutional developments in Norway. The first is the creation of a Sámi Parliament, the Sameting, in 1989, and the second is the new Finnmark Estate Act. I discuss the Sameting here and leave discussion of the Finnmark Estate Act to the final part of this chapter dealing with the legal status and recognition of claims by indigenous people to title to land and resource harvesting rights.

The Sameting

The Storting created the Sameting following the recommendations of the Sámi Rights Commission, which was in turn a response to the Alta Dam controversy (see further discussion below, Sámi Act, 1987–2003). The Sameting is a body elected by Sámi for Sámi and strengthens Sámi self-identity (Broderstad, 2001: 163). Norway is divided into 13 electoral districts for purposes of elections to the Sameting, and voter eligibility is based upon a declaration that the person considers himself/herself to be Sámi and either uses Sámi in the home or whose parents, grandparents or great grandparents used Sámi, or is the child of a person listed in the Sámi census.

The terms of reference for the Sameting are broad and s. 2(1) of the Sámi Act states that the 'business of the Sameting is any matter that in the view of the parliament particularly affects the Sámi people'. But the Sameting's role is principally an advisory one, except in relation to its own budget. Section 2(2) of the Act suggests that other public bodies should give the Sameting 'an opportunity to express an opinion' before making decisions on matters that fall within the business of the Sameting but the soft ('should') language of this section leads Oskal (2001: 225) to question whether there is a legal duty to engage in this form of consultation.

The relationship between the Sámi Parliament and the government of Norway and the Storting is a dynamic one. While Broderstad (2001) asks how the Sámi Parliament and the Norwegian government can share competence, Oskal (2001: 254) suggests that it is possible to view the creation of the Sameting as 'one attempt to, and a method of, political inclusion of indigenous people as clearly indigenous people within the framework of a nation state. The challenge consists in transferring authority to the Sámi Parliament, then integrating this authority into the higher state governmental jurisdiction'.

While the Sámi Parliament is a national body, Broderstad (2001: 166–167) also draws attention to the responsibility of counties and municipalities to accommodate Sámi interests. She argues that this is especially important in those many areas outside the few municipalities within Finnmark where Sámi are in a minority. According to Broderstad, the Norwegian government has been reluctant to be too prescriptive about how to accommodate Sámi interests because of the value the government places on local autonomy. But Broderstad urges the need for a formal acknowledgement that local governments with their responsibility for land-use planning and other local responsibilities have significant practical authority to affect Sámi interests, positively or negatively, and that it is important that they exercise their authorities in a manner that respects both the constitution and Norway's responsibilities under international law.

Conclusion

This overview of the forms of government and the allocation of legislative authority within the four states shows considerable variation. It also suggests that, in some cases at least, the organization of government within the Arctic areas of these states may depart from the usual mode of organization within that state. For example, the territorial organization of northern Canada affords the Canadian government considerably more authority than it would otherwise have. Similarly, in Alaska, local or regional governments like the North Slope Borough have proven to be important players in capturing some of the benefits of oil and gas exploration for residents. By contrast, in Norway, the legislative position in relation to oil and gas developments is really no different in the Arctic from any other area in Norway and much the same would follow for Russia.

Public and private ownership of land and natural resources

All four jurisdictions that we cover in this book allow for private ownership of land but they take different positions with respect to the private ownership of oil and gas resources.[6] One jurisdiction, Norway, is completely committed to the public ownership of oil and gas resources and hence, under Norwegian law, the state is the owner of all petroleum resources both on land and sub sea [Einarsbøl, 2006; Petroleum Act (Norway, 1996) 1996: Section1-1]. Private ownership of oil and gas resources is permitted in both Canada and the US, but in both jurisdictions, the relevant governments (federal/state/provincial) have long followed policies of retaining oil and gas and mineral rights when alienating lands to private parties. As a consequence, private ownership of oil and gas resources tends to be confined to those areas of the two countries that were settled early in the history of colonization rather than later. In northern Canada, subject to the exceptions provided by modern land claim agreements, there is no private ownership of oil and gas rights. Ownership of these resources remains vested in the Crown and the same situation prevails offshore. Alaska presents a similar picture, and the Alaska Statehood Act, 1958 (US, 1958), s. 6 (i) provides that any grant by the State of lands that it received pursuant to the terms of that constitutional compact should include a reservation of all mineral rights (including oil and gas) for subsequent disposition by the State by way of leasing.

Article 9 of the Russian constitution contemplates that land and resources may be in private, state, municipal and other forms of ownership, but the ownership

Legal and institutional framework 121

rights of private parties do not extend to minerals and oil and gas resources. Federal and sub-federal laws on the (re-)establishment of obshchinas (a form of pre-Soviet indigenous territorial organization) provide a mechanism for indigenous people to obtain a measure of control over some traditional territory through the recognition of permanent usufructuary rights and certain exclusionary rights, but in all cases title to the lands remains vested in government. Furthermore, in some cases raion/ulus (county level governments) are not always receptive to the creation of obshchinas, resulting in significant geographical variation in their recognition from region to region (Fondahl, 1995: 9–12; Fondahl et al., 2001), as discussed further below.

In sum, most of the oil and gas rights in the Arctic areas of the four states are publicly owned. The principal exceptions are the oil and gas rights owned by indigenous people (or their wholly owned corporations) as a result of modern land claim agreements.

Where oil and gas rights are publicly owned, governments typically dispose of rights to these resources through generally applicable petroleum legislation. Some such schemes [e.g. the Yukon Oil and Gas Act (Yukon, 1998) and the Canada Petroleum Resources Act (Canada, 1985)] grant oil and gas rights on the basis of a competitive bidding system with a single variable bid. Such schemes are very 'objective' and offer little or no opportunity for negotiation. By contrast, the licensing scheme under the Norwegian Petroleum Act entails the pre-qualification of companies for eligibility to participate and a careful assessment of proposals based upon the expertise, financial capacity, geological understanding and experience of the applicants. There may also be negotiations about work requirements and relinquishment provisions.

Alaskan oil and gas authorities use a variety of leasing methods and may potentially use a broad range of bidding variables, including fixed royalties and bonus bids and fixed cash bonuses with a variable royalty or a share of net profits (Alaska Statutes, AS 38.05.180). Russia's petroleum or sub-soil legislation contemplates that a party may acquire oil and gas rights either by means of a licence issued as a result of a bid or by means of a negotiated production sharing agreement.[7] Under The Law on the Subsoil adopted February 1992 and the relevant regulations, bids are measured against a range of criteria including the technical characteristics of the project proposal, use of environmental friendly technologies, and potential social and economic contribution to the local area (Goundina, 2004).

As indicated above, both the US (Alaska) and Canada recognize one important exception to the proposition that rights to oil and gas are vested in the government (whether federal, state or, in the case of Yukon, the territory) in so far as both the Alaska Native Claims Settlement Act (ANCSA, 1971) and land claim agreements in Canada recognize that the title to at least some of the lands vested in Alaska natives by ANCSA or on aboriginal organizations by virtue of Canadian land claim agreements shall include oil and gas rights. Thus, regional corporations in Alaska, and First Nations and Inuit in Yukon, NWT and Nunavut are able to develop their own petroleum leasing policies. While it is difficult to generalize, it seems fair to say that most such groups have adopted leasing policies that involve negotiated agreements with pre-selected companies rather than standard form bidding arrangements. Furthermore, at least some such organizations have drawn on a broad range of international experience using as models the forms of agreements that host states generally try and negotiate with international petroleum companies (Bankes, 2000).

An important question to pose in assessing public oil and gas disposition legislation relates to the nature of any environmental review that might be required before lands are posted for disposition. For example, does a proposal to issue rights trigger an environmental or strategic impact assessment? Or is such an assessment only conducted for subsequent activities such as drilling activities? The practice varies in each of the jurisdictions. Perhaps the most formal requirements are prescribed in state and federal laws for Alaska. Under Alaskan state law, the Department of Natural Resources (DNR) can only include lands within a lease sale where the lands have been the subject of a 'best interest finding'.[8] A 'best interest finding' is a lengthy process that includes an assessment of the potential effects of oil and gas, exploration, development, production and transportation as well as possible mitigation measures. Preliminary findings are posted for public comment which opportunity is supplemented by workshops and public hearings in affected communities. If the lands are included in the coastal zone, DNR will also assess whether the proposed leasing is consistent with the Alaska Coastal Management Program and any local management plans. A 'best interest' finding is valid for ten years, although the finding may be supplemented by additional information. An equally formal process is required under federal law before federal authorities can issue leases for the Outer Continental Shelf. Prior to conducting any sales, the Minerals Management Service (MMS) of the federal Department of the Interior must prepare a detailed Environmental Impact Statement.[9]

Norway has similar requirements. Thus s. 3.1 of the Petroleum Act (Norway, 1996) requires that an evaluation be undertaken of all the interests involved before a new area is opened up for granting production licences. The evaluation must include an assessment of the social, economic and environmental implications, and must involve consultations with local public authorities as well as others.

Canadian rules are far less prescriptive and, while the provisions of relevant land claim agreements may require more formal consultations with affected communities before dispositions occur, at no point does the relevant legislation require the preparation of an environmental impact statement (EIS)or anything like a 'best interest finding'. Instead, environmental assessments are generally only triggered by particular proposed projects or activities (e.g. an authorization to drill a well or construct a pipeline) rather than a proposal to lease lands (Bankes and Rowbotham, 1993). The Russian practice under the Law on the Protection of the Environment is to conduct an environmental impact assessment of a detailed development proposal submitted by the successful bidder for a licence (Goudina, 2004: 63–71).

In conclusion, with the exception of indigenous oil and gas ownership rights recognized as part of modern land claim agreements in Alaska and Canada, oil and gas rights are vested in the government in each of the four states. Neither Norway nor Russia recognizes an exception to this rule for indigenous people. The nature of indigenous land rights in each of the four jurisdictions is explored in more detail in the following section of the chapter.

The legal status of indigenous people and their land rights and use rights

This final section of the chapter explores the evolution of the state recognition of the land rights and resource use rights of indigenous people in each of the four jurisdictions. The principal vehicle for recognizing, protecting and demarcating an aboriginal land interest in northern Canada has proven to be negotiated settlements in the form of land claim agreements. Litigation has, at various times, also proven to be instrumental in bringing government to the negotiating table. While Alaska's Organic Act of 1884 and the Statehood Act (US, 1958) recognized an indigenous but inchoate interest in the lands and resources of the state, it was not until the federal government imposed a land freeze in1966 on state land selections that serious steps, culminating in the Alaska Native Claims Settlement Act of 1971 (ANSCA, 1971), were taken to resolve aboriginal title claims in Alaska. Similarly, it seems to have taken an activist response to the proposed Alta Dam in the late1970s to bring Sámi land rights issues to the forefront in Norway where the precise basis of Sámi land rights continues to be debated. Finally, in the case of Russia, the land and resource rights of indigenous people have acquired a measure of constitutional and statutory recognition from the post-Soviet state, although much remains to be done to provide effective and practical protection of their land and resource rights.

Canada

When Canada was originally created in 1867, it succeeded to a long tradition of treaty-making between the English Crown and the indigenous people of North America. Canada continued this policy from the 1870s through to the 1920s, including two treaties – Treaty 8 (1899) and Treaty 11 (1924) – with indigenous people in the southwest of the NWT and the southeast of Yukon. But after this, there followed a long period during which treaty-making ceased in Canada (as did effective implementation of the northern treaties). Furthermore, Canada's formal legal position during this period (from the 1920s to the 1970s) was that it could not be compelled to enter into negotiations with those indigenous people who did not have treaties (especially First Nations and Inuit in northern Canada), since it characterized the Crown's historical practice of treaty-making as a policy based upon political expedience and not on the acknowledgement of a legal duty.

That attitude ultimately changed in the 1970s. Confronted with hydro developments in northern Quebec and a decision of the Supreme Court of Canada (the *Calder* decision), which confirmed the legal basis of the concept of aboriginal title, the Crown launched a new period of treaty-making, principally with Inuit

Table 6.1 Canadian land claim agreements*

Agreement	Beneficiaries	Land, surface (km²)	Land, minerals	Lump sum payments, millions
Inuvialuit, 1984	4,000	91,000	13,000	$78 (1984$)
Gwich'in, 1992	2,500	16,264	4,299	$75 (1990$)
Sahtu Dene and Metis, 1994	2,500	41,437	1,813	$75 (1990$)
Nunavut, 1993	17,500	351,000	37,000	$580 (1989$)
Tlicho, 2005	5,400	2,000	2,000	$90 (1997$)
Yukon (1993 to date)†	6,000	41,595	25,900	$146 (1989$)

* Based on information summarized in DIAND (2006); copies of the agreements are available at: http://www.ainc-inac.gc.ca/pr/agr/index_e.html (Accessed 12 November 2006).
† In the case of Yukon, an Umbrella Final Agreement (UFA) was negotiated and finalized in 1993. Thereafter each of the 14 First Nations in Yukon were to negotiate there own Final Agreement and self-government agreement within the framework of the UFA. Since then, 11 of the 14 First Nations have finalized their agreements.

and First Nations in northern Canada. This new round of treaty-making has produced a series of northern land claim agreements (and in some cases self-government agreements) as depicted in Table 6.1.

In addition to recognizing First Nation/Inuit ownership interests to two categories of land, surface only and surface and mineral rights (including oil and gas) and, in addition to providing for capital transfer payments and resource royalties, the agreements also protect traditional resource harvesting rights and create a series of co-management boards that exercise authority throughout the traditional territories of the people concerned.[10] These boards include land use planning boards, wildlife management boards, water and land use boards and environmental impact assessment boards. But the agreements also serve to confirm and recognize the Crown's title (subject to the harvesting rights of indigenous people) to the balance of the lands. These agreements, which are all constitutionally protected, significantly change the legal environment for oil and gas and other resource companies working in the Canadian Arctic. Not only do they change the locus of much decision-making authority, but they also provide a more secure legal environment for investment for both oil and gas exploration, but also necessary infrastructure such as pipelines.

Not all northern indigenous people have finalized land claim agreements with Canada. In particular, some Yukon First Nations in the southeast of Yukon and the Dehcho First Nation in southwest NWT have yet to negotiate agreements. The latter represents a particularly significant gap in the context of oil and gas development, partly because of gas discoveries in the area but also because the DehCho territory lies on the route of the projected Mackenzie Valley Gas Pipeline.

One of the key implications of land claim agreements in Canada and ANCSA in Alaska is that the industry can no longer assume that all its exploratory acreage can be acquired from government, since in some cases these oil and gas rights will be vested in an indigenous owner. Those owners may be willing to provide access to those resources by agreement, but those agreements (as we have already suggested in the last section) may resemble typical international oil and gas concession agreements rather than the competitive bid/standard form agreement more frequently used by the governments of US/Alaska, Canada and Norway.

Norway

In his survey of the legal basis of Sámi claims in Norway, Eide (2001) identifies three distinct stages in governmental policies towards the Sámi. The first stage, associated with the period prior to the middle of the nineteenth century when the general attitude was one of 'benign neglect', represents a period in which the government did not claim ownership of lands in Finnmark. A key event during this period was the 1751 treaty between Sweden (Finland then being part of Sweden) and Denmark (Norway then being part of Denmark) formally demarcating the boundary in the northern territories which, prior to that 'had not been under the explicit jurisdiction of any nation state' (Eide, 2001: 138). That treaty also included the Lapp Codicil, which dealt with the implications of the boundary for the Sámi people. But for Eide, as for other writers, the acquisition or recognition of various sovereign claims did not result in the automatic acquisition of title to land in the new territory.

The second stage begins around 1850 and is marked by a deliberate assimilation policy. For Eide this new policy is part of reaction to the threat posed by what was then Russian-controlled Finland. In addition, failed negotiations designed to deal with Norwegian Sámi pasturing reindeer in Finland led to the closing of the border in 1850. During this period, when efforts were made to have the Sámi adopt Norwegian lifestyles, the state made explicit ownership claims to non-registered land and at the same time discriminated against Sámi who wished to acquire title to land. The legal basis for the state claim to unregistered land, while never formally articulated, came to be accepted legal dogma: 'generally speaking, the institutions of the Norwegian legal order, the executive, the parliament, and the courts, have all taken it for granted that the state could assert ownership of these lands' (Eide, 2001: 140; Semb, 2001: 184–186).[11]

The third stage begins for Eide in the late 1970s, a period when the nation state was seen to be in decline and the discourse on subjects, such as the environment and human rights, was becoming increasingly internationalized. During this period, the very public controversy over the Alta project served to focus attention on Sámi issues and led to the creation of the Sámi Rights Commission in 1984. Minde (2001: 113) notes that, prior to the Alta dispute, Nordic countries generally took the view that the Sámi were not indigenous people, since they were well integrated into society: 'The thought that the Sámi were an indigenous people had indeed been foreign to Nordic authorities, the general public and to the majority of the Sámi through the end of the 1980s'. The Commission's first report (1984) 'Concerning the Legal Position of the Sámis' led to the Sámi Act of 1987 (Norway, 1987) and the creation of the Sámi Parliament in 1989 discussed in the first part of this chapter. The Commission's investigation of Sámi rights and the status of Sámi people in Norwegian law, explicitly located much of its discussion in the context of international law, drawing first upon Article 27 of the International Covenant on Civil and Political Rights and later upon Convention 169 on Indigenous and Tribal People of the International Labour Organization (ILO; ILO, 1989; Semb, 2001). An additional outcome was that the Constitution was amended in 1988 to provide explicit recognition of Sámi rights and state duties.[12]

The Commission's 1997 report to the Storting on the natural resource base for Sámi culture, led more or less directly to the introduction of the Finnmark Act in 2003. The Act is worth discussing in some greater detail here because it shows a recent and innovative approach to dealing with questions of rights and title. The Act engendered significant discussion in Norway, particularly in the context of Norway's obligations as a party to ILO Convention 169. That aspect of the discussion is covered in more detail in the human rights chapter (Chapter 11).

The Finnmark Act

Finnmark County is the northernmost county in Norway, comprising about 47,000 km^2. It is the part of the country with the largest Sámi population and the greatest concentration of reindeer husbandry. Sámi people constitute the majority in several of the municipalities within the county and the (non-)recognition of Sámi land rights and reindeer herding rights in this area has long been a contentious issue.

Informed by the work of the Sámi Rights Commission referred to above, the Finnmark Estate Bill represented the government's attempt to resolve those issues and to resolve them in a way that fulfilled Norway's obligations under ILO 169. The Minister of Justice suggested that the Bill was founded on four main principles: (1) equal treatment of all citizens in the county; (2) safeguarding of Sámi culture; (3) ecological sustainable development; and (4) clarifying title issues so as to provide a foundation for positive industrial development.[13]

Although the final form of the bill received the unanimous support of the Sámi parliament and that of a large majority of the Finnmark County Council before finally being adopted by the Storting, the Sámi had initially expressed their strong opposition to the Bill on both procedural and substantive grounds. There is further discussion of the Bill in Chapter 11 of this volume but, in its final form, the Bill protects Sámi rights but does so in an ethnically neutral way by offering protection to all traditional use rights based on prescription and immemorial usage.[14] For present purposes, the key parts of the legislation are those sections dealing with title to land in Finnmark County.

Prior to the Act, title to the lands owned or claimed by the state in Finnmark (some 95 per cent of all the lands in Finnmark) was vested in and controlled by a state resource corporation known as Statskog SF.[15] One of the key provisions of the Act effects a transfer of title from this state enterprise to a more representative body known as the Finnmark Estate. But the Act (s. 5) also makes it clear that the Estate's title is subject to the traditional use rights of Sámi and others, including reindeer herding rights.[16] It also contemplates that these rights will be investigated and confirmed through the work of the Finnmark Commission (s. 29). The Act provides that the Finnmark Estate is to be governed by a six-person board of which three members are to be elected by the Sámi parliament and three by the Finnmark County Council. One of the persons elected by the Sámi parliament shall be a representative of the reindeer industry (s. 7).

The Board is generally responsible for the management of the Finnmark Estate (s. 9) and one of its key responsibilities is to make decisions with respect to changes in the use of uncultivated lands and in doing so to assess the significance of such changes for Sámi culture, reindeer husbandry, use of non-cultivated areas, commercial activity and social life (s. 10). In assessing Sámi interests, the Board is to follow guidelines developed by the Sámi Parliament (as approved by the Ministry) under s. 4 of the Act. The Act contains a set of complex provisions for making decisions with respect to the change of use of non-cultivated areas. Where a minority bases its decision on the Sámi Guidelines, the decision requires a special majority of affirmative votes and the minority may also refer the matter to the Sámi Parliament. Where the Sámi Parliament fails to ratify the majority decision, the matter is to be referred to the King, who has the final decision. Where a decision is only approved by three members, any three members may require that the matter be reconsidered except that on the reconsideration of the matter one Board shall not take part.

Where the decision concerns the Inner Finnmark counties, it is the last member appointed by the Finnmark County Council who shall not take part and, in the case of a decision in relation to any other part of Finnmark, it is the last member appointed by the Sámi Parliament who is not the representative of the reindeer industry. Clearly these rather complex provisions are designed to provide procedural and other assurances to Sámi that the Estate will have to take serious account of Sámi interests, especially where the decisions affect those parts of Finnmark where the Sámi are in the majority.

While the Act contains these and other provisions[17] designed specifically to protect Sámi interests, much of the Act, consistent with its stated purpose (s. 1), extends many of the rights that it confers to all residents of Finnmark. These rights include: (1) resource use rights (fishing, gathering, peat cutting, and timber harvesting for limited purposes) extended to all persons resident in municipalities (s. 22); (2) more limited resource rights extended to all residents of Finnmark (s. 23); and (3) yet more limited rights extended to all persons. Reindeer herders merit special attention and are afforded all the rights of persons resident in the municipality during the period when reindeer herding takes place there (s. 22).

The title that is to be vested in the Finnmark Estate is limited to the rights of the surface owner (and indeed subject as we have said above to traditional use rights) and does not extend to ownership of the mineral estate, which remains with the state. The Act did, however, effect a number of amendments to other statutes, including the Mining Act. Under these amendments, miners proposing to do preliminary examination (prospecting) must provide advance notice (1 week) to potentially affected parties including landowners and the Sámi Parliament. More importantly, perhaps, a new section provides that the same parties are entitled to notice of licensed prospecting applications as well as applications to proceed to patent a claim and that the Commissioner of Mines should place significant emphasis on Sámi culture, reindeer husbandry, use of non-cultivated areas, commercial activity and social life in deciding whether to grant an application and in fixing appropriate conditions for any approval.

In conclusion, in recent years, Norway has taken significant steps to strengthen Sámi land rights. Much of the impetus for this has come from Norway's accession to ILO 169. While the Finnmark Act does not itself recognize areas of Sámi land title, it does offer other forms of protection. In particular, it affords Sámi a much greater role in decision-making affecting land and resource use throughout Finnmark, it acknowledges that Sámi have land rights based on long use, and it provides a mechanism in the form of a commission and special court to establish the scope and content of those rights.

Alaska

When Alaska joined the US as a result of the Treaty of Cession of 1867 between Russia and the United States, Article III of that Treaty drew a distinction between the Russian settlers of Alaska and the indigenous tribes. Inhabitants with Russian heritage were to be given the option of remaining and thereby enjoying all the rights and benefits of citizenship, or returning to Russia. The tribes, however, were to 'be subject to such laws and regulations as the United States may, from time to time, adopt in regard to aboriginal tribes of that country'. In writing about the treatment of indigenous people in Alaska under US law, Case and Voluck (2002: 6–22) identify three distinct periods: (1) the early years (1867–1905); (2) a middle period that carries us through to Alaska Native Claims Settlement Act of 1971; and (3) the post-ANCSA period.

During the first period, Case and Voluck suggest that Alaska natives were not treated in the same manner as the tribes in the lower 48 and the prevailing assumption seems to have been that Alaska natives could not maintain claims of aboriginal title or claim the same self-governing status as tribes in the lower 48. This policy of exceptionalism changed during the second period. Congress applied the Indian Reorganization Act to Alaska natives and case law developed recognizing that the US had a duty to protect native occupied lands in Alaska from acquisition by non-natives. This idea was ultimately carried forward into the Alaska Statehood Act, which provided (s. 6) that, while the State was to receive title to significant blocks of public lands (to be selected in the following 25 years), lands held by 'Indians, Eskimos or Aleuts' or held in trust for them by the US were not available for selection and were to remain under the sole jurisdiction of the US (Statehood Act, s. 4; US, 1958).

In the same year when Alaska achieved statehood, the US Supreme Court confirmed that indigenous people in Alaska could maintain a claim of aboriginal title: *Tee-Hit Ton Indians v. United States* (348 US 272; Case and Voluck, 2002: 156). State pressure to make its land selections under the Statehood Act, combined with the uncertainty surrounding the scope of native title claims ultimately led the Secretary of the Interior, Udall, to impose a land freeze in 1966, suspending issuance of all federal patents and federal approvals of proposed state land selections. The land freeze also affected proposals to bring Prudhoe Bay production to market via a pipeline once the field had been discovered in 1968. All this pressure combined to force Congress to develop a comprehensive settlement for Alaska natives, which it did with the adoption of ANCSA in 1971.

A law of 'stunning complexity' (Case and Voluck, 2002: 157), ANCSA purportedly extinguished aboriginal claims (s. 4) (Bankes, 2004a: 153–156) and in return vested title to some 45 million acres of lands in a variety of more than two hundred village and 12 regional corporations.[18] The mineral title to these lands was vested in one of 12 regional corporations subject to a duty to share revenues with other regional corporations under s. 7 of the Act. In addition to the land component, ANCSA also provided a cash settlement of approximately $1 billion. As owners of mineral rights, including oil and gas rights, the regional corporations have been able to enter into oil and gas exploration and other agreements with the oil and gas industry. For example, the Arctic Slope Regional Corporation (ASRC) has negotiated agreements with several companies, including a current agreement with Anadarko. Similarly, a producing field, the Alpine Oil Field operated by ConocoPhillips, is located partly on ASRC lands and partly on Alaska state lands (ASRC, 2005).

The ANCSA settlement dealt principally with issues of aboriginal title. It did not deal with issues related to traditional resource harvesting or subsistence rights, and it did not deal with issues of self-government. While subsistence issues were dealt with to some degree in the 1980 Alaska National Interest Lands Conservation Act, issues of self-government remain contentious partly because of the nature of ANCSA which adopted a corporate settlement model and did not employ traditional governance structures. The position, based on the US Supreme Court's conclusion in *Alaska v. Native Village of Venetie Tribal Government* (522 US 520, 1998), seems to be that native villages continue as self-governing entities but that they lack a territorial basis. In *Venetie*, the Court held that Congress' decision to transfer lands to native owned corporations rather than to tribal governments was inconsistent with the application of tribal taxing laws to those lands. That said, tribal governments still have jurisdiction over their members and other internal affairs with respect to such matters as adoption, tribal courts and internal governance.

In conclusion, the ANCSA settlement in Alaska affords the native people of Alaska clear title to specific lands, including oil and gas rights, and vests those lands in native-controlled corporations.

Russia

Nikolai Vakhtin (1992), in his study of the *Native People of the Russian Far North*, identifies a number of distinct periods in the development of Soviet and Russian policy in relation to indigenous people or the small-numbered people of the north and east.[19] During the period of first contact (beginning around the 1550s and extending to the early nineteenth century), Russian interests emphasized trade rather than the annexation of land and resources. A second period dates from the 1820s through to the revolution of 1917, and begins with the development and application of the Code of Indigenous Administration of 1822. This progressive document (consolidated in 1892 as the Statute of the Indigenous People) extended legal protection to indigenous people but its implementation lagged behind its intent. During this second period, indigenous people generally lost land and migrated to the north and east as trading and settlement extended.

The revolution abrogated this body of law and the 1918 Constitution recognized the 'equal rights of all citizens'. A key development occurred in 1924 with the creation of the Committee of the North, which included among its functions the responsibility 'to define and reserve the territories necessary for the life and cultural development of each ethic group'. The commentators generally agree that this was a positive development. Fondahl (1999), for example, notes that, during its early period, the Committee lobbied for the creation of native councils (tuzemnye sovety) and native counties (natsional'nye rayony). But the commentators also agree that this phase was short lived and that, rather than protecting indigenous people in their territories, the Soviet goal became one of opening northern lands for settlement and to exploit natural resources. Vakhtin (1992: 11), for example, refers to a conservative and a radical tradition within the Committee. At first, the conservative (protective and scholarly) tradition prevailed, but for the radical, and later dominant, tradition an indigenous minority represented an obstacle to progress. The radical tradition regarded indigenous people as culturally inferior and soon dismantled early progressive steps through processes of collectivization, sedentarization and village consolidation and relocation (Fondahl, 1995).

However, the same period also saw the establishment of nine (soon reduced to seven) national regions, which provided titular recognition to some native people (e.g. Evenkiisky National Region). In-migration, however, changed the demographics of these national regions, which soon came to be dominated by non-aboriginal people. In any event, Fondahl (1999) suggests that these regions (re-named autonomous regions in 1977–1978) were weak and subordinate to both the provinces and territories within which they were situated, as well as to the central government.

Fondahl's description of this period coincides with what Vakhtin refers to as a period of industrialization and collectivization between 1930 and 1941. These processes tended to transfer power from local administrations to central industrial ministries, and to dispossess indigenous people of favoured resource harvesting and grazing lands. But Vakhtin reserves his harshest criticism for what he describes as the 'Dark Years' of 1941–1985 when the Soviets consolidated and institutionalized their authority and supported a policy of Russification through increased settlement, the use of Russian in education and, as elsewhere in the north, the forced relocation of communities. Vakhtin (1992: 24) also notes that industrialization led to a period of 'ecological aggression' against the traditional territories of indigenous people.

With the collapse of the Soviet Union, the national republics all opted to become independent states but the national autonomous regions retained their status as sub-national units within the new Federation (Kartashkin and Abashidze, 2004: 204). However, in-migration of Russians into these regions over the course of decades had simply reinforced a trend that Fondahl and others had noted earlier; namely, that the people whose names these regions recognized had become minorities. Thus, while the 1993 Constitution empowers the autonomous regions (by affording them equal status to the other subjects of the federation), these regions cannot 'serve as the loci of indigenous self-government ... [since] the immigrant population vastly outnumbers the indigenous population' (Fondahl, 1999: 57).

However, at least at the formal levels of constructing a constitution and drafting laws, the post-Soviet government proved more sympathetic and responsive to the concerns of the numerically small people of the North than had the earlier Soviet government. Thus, Article 69 of the 1993 constitution offers explicit protection to indigenous people and 'guarantees the rights of small indigenous people in accordance with generally accepted principles and standards of international law and international treaties of the Russian Federation'. In addition to this constitutional provision, the government, at both the executive and legislative levels, has also introduced a series of measures to provide better protection for indigenous people.[20]

An early federal step was the Presidential Ukase No. 397 of 1992, urging the subjects of the federation to establish 'territories of traditional use' (TTPs) within which indigenous people might be guaranteed the right to engage in their traditional activities (Fondahl, 1997; Osherenko, 2001). Once created, a TTP is not available for industrial activities without the consent of the indigenous people. Implementation of this and other related initiatives has been 'spatially irregular' and many provinces did nothing (Fondahl, 1995: 19; 1997). More recently, the federal government has passed a number of special laws to supplement the Ukase. These laws include the laws 'On guarantees of Indigenous Minority People of the Russian Federation Rights' (1999), 'On General Principles of Organizing Communities of Indigenous Minority People [obshchinas] of the North, Siberia and the Far East of the Russian Federation' (2000) and 'On Territories of Traditional Use of Natural Resources by Indigenous Minority People of the North, Siberia and the Far East of the Russian Federation' (2001).

Fondahl et al. (2001) and Osherenko (2001) both suggest that efforts to territorialize indigenous rights have been based upon either the recognition of TTPs or the (re-)construction of obshchinas (family clans or communes). Once formed, an obshchina, may petition the okrug or oblast for the allocation of land in order to continue such traditional activities as hunting and reindeer herding. Title remains with the state. However, the actual practice has varied and Fondahl notes that, while both the original Ukase as well as the federal law of 2000 are explicitly directed at obshchinas formed by and for indigenous people, in at least some cases the laws of the sub-units of the federation (e.g. Sakha Republic) refer more generally to those engaged in traditional activities. There is also considerable diversity in the form of tenure. While the Ukase contemplated that lands (or the use of those lands) would be granted in perpetuity, in many cases the tenure granted has proven to be much more limited (Fondahl, 1999: 59). Certainly title to these lands has not been transferred and Osherenko (2001) suggests that there remains considerable opposition in Russia to the privatization of land and resources.

While the recognition of TTPs and the allocation of land to obshchinas offer protection to small-numbered people, there are also cases in which indigenous people have lost land and resource rights through the increasing industrialization of traditional territories. Osherenko (2001) offers several well-documented examples including the decision of the Murmansk administration to lease parts of the Ponoi Riverto private fishing interests without regard to the interests of the Sámi and Nenets people of the Komi peninsula. In other cases (e.g. in the Khanti-Mansiisk region of Siberia), family and clan lands have been acquired by oil and gas interests either with no regard for the traditional owners or for purely nominal consideration.

In addition to federal laws, some of the constituent entities of the Federation, for example the Republic of Sakha (Yakutia) and the Koryak Autonomous Area, have also adopted constitutions and laws committing to protect the language and cultural rights of indigenous people (Kartashkin and Abashidze, 2004: 213).

Conclusions

This chapter has provided an account of some of the basic legal and institutional features of the four states. The first part of the chapter reviewed the main forms of government in each of the four states. While this revealed a basic division between Norway and the three federal states, it also showed that a variety of governmental forms may assume importance in mediating the relationship between governments, the oil and gas sector and indigenous people. Federalism is often acknowledged as a means of conferring a degree of autonomy on distinct regions or people, but this account shows that measures falling short of federalism (e.g. the Sámi Parliament and the case of Finnmark) or even measures within but not part of the formal arrangements of federal states (e.g. the North Slope Borough, and Canadian territories, especially when accompanied by the significant devolution of authority as in the case of Yukon) may enhance autonomy and the ability of indigenous communities to benefit from industrial developments occurring within their territories.

The second and shortest part of the chapter reviewed arrangements for public and private ownership of land, and especially oil and gas. Here we noted that public ownership of oil and gas is the norm with the exception being the recognition of indigenous rights to oil and gas resources as part of modern land claim settlements in Canada and Alaska. The various state petroleum leasing policies are generally similar with some of the key variables being the degree to which strategic environmental assessment is conducted before holding leasing sales and the degree to which there is an opportunity for negotiation as to the terms of licence arrangements between the state and industry. Indigenous owners of oil and gas rights in Alaska and Canada have considerably discretion and autonomy in the forms of agreement that they are able to negotiate with industry.

The third part of the chapter examined the extent to which each of the four legal systems recognizes the special status of indigenous people and their land and resource rights. While some of this discussion continues in Chapter 11 on human rights, some key conclusions emerge. First, the land and resource rights of indigenous people receive the most secure protection under the modern settlements in Alaska and Canada. While the ANCSA settlement may not be subject to the same degree of formal constitutional protection as is afforded to land claim agreements in Canada, it seems clear that the ANCSA settlement can also be seen as part of the constitutional compact that led to Alaska statehood. Indigenous land and resource rights are less well protected in Norway and Russia. Although both constitutions recognize the special status of indigenous people, the land and resource rights of such people are less clearly recognized in these two jurisdictions.

One clear theme that does emerge in this chapter is the importance of demographics of settlement and in-migration. In many areas of Russia and northern Norway the indigenous people are in a minority position. In such cases, local and regional forms of government are unlikely to be responsive to indigenous interests. This suggests the need to create innovative institutional arrangements if the state aims to recognize the claims to autonomy of its indigenous people. There is already evidence of such innovative practices within these states, including the North Slope Borough in Alaska, the Finnmark Act in Norway and the territory of Nunavut in Canada.

Notes

1 See, for example, the discussion of collective property rights by the Inter American Court of Human Rights (ICHR) in Mayagna (Sumo) Awas Tigni v. Nicaragua (ICHR, 2001), esp. at paras 142–153, and the separate opinion of Ramirez at paras 13–14. There is an irony in mentioning in the space of one sentence Locke's views on property and the property conception of indigenous people since, as Oskal (2001: 241) points out, Locke's views on property formed part of the intellectual apparatus that was used to deny the property rights of indigenous people, pointing out that Locke viewed indigenous people as existing in a state of nature, a pre-political society and pre-law, a version of the terra nullius theory used to support the acquisition of sovereignty by colonizing powers.

2 Now reflected in s. 36 of the Constitution Act, 1982 (Canada, 1982).
3 Constitution Act, 1867, ss. 91–95 (Canada, 1867).
4 Constitution of the Russian Federation, Article 95, one member shall be from the representative and the other from the executive body; the Duma comprises 450 deputies (Russia, 1993).
5 Both Goudina (2006) and Skyner (2005) discuss a draft law that was introduced into the Duma in 2005. However, the draft was subsequently withdrawn in early 2006. It is anticipated that a new version will be tabled in early 2007: Goudina, personal communication, October 23, 2006.
6 However, public ownership of lands is perhaps more important in the Arctic areas of all four states. For example, of Alaska's 365 million acres, non-indigenous private lands constitute only 2.7 million acres; lands held by native corporations account for another 37.4 million acres. The balance represents government lands of one form or another, state lands, national park and wildlife refuge lands and other federal lands managed by the Bureau of Land Management.
7 Law on Production Sharing Agreements, December 1995. Production sharing agreements remain relatively uncommon and seem to be used principally in relation to major projects or difficult developments for which the licensing scheme seems unsuitable because of the need to negotiate specific arrangements.
8 AS 38.05.035; and for an example, see Final Best Interest Finding, Sale 87, North Slope Area Wide, March 17, 1998. Relevant documents are available on the website of the Department of Natural Resources: http://www.dog.dnr.state.ak.us/oil/.
9 See, for example, the Final Environmental Impact Statement prepared for the Beaufort Sea Planning Area, Oil and Gas Lease Sales, 186, 195 and 202, February 2003. This and other relevant documents are available at http://www.mms.gov/alaska/cproject/beaufortsale/index.htm.
10 Both the Alaskan and Canadian settlements also provide for lump sum payments to the beneficiaries of the agreements. The precise basis for these payments is not clear. To some extent they may be viewed as compensation for prior dispositions by the Crown, or in return for extinguishing title or the promise not to pursue an aboriginal title claim to those lands that are confirmed as Crown lands. Alternatively, some or all of the payment may be viewed as a mechanism for fostering economic development of indigenous communities or to meet other priorities of the communities concerned.
11 Recent decisions of the Norwegian courts and especially the Svartskog (2001) and Saelbu (2001) cases have tended to undermine the accepted legal dogma of Eide's (2001) second stage in the (non-)recognition of Sámi rights but his observation certainly reflected accepted dogma at the time.
12 Section 110a provides that: 'It is the responsibility of the authorities of the State to create conditions enabling the Sámi people to preserve and develop its language, culture and way of life'.
13 And see s. 1 of the Act, which provides that: 'The purpose of the Act is to facilitate the management of land and natural resources in the county of Finnmark in a balanced and ecologically sustainable manner for the benefit of the residents of the county and particularly as a basis for Sámi culture, reindeer husbandry, use of non-cultivated areas, commercial activity and social life.
14 See in particular The Finnmark Act: a guide issued by the Ministry of Justice and by the Ministry of Local Government and Regional Development and distributed to all households in Finnmark (n.d.), and available online at http://www.galdu.org/govat/doc/brochure_finnmark_act.pdf. The guide contains a foreword by the two Ministers as well as a commentary on the main provisions of the Act. In their foreword, the Ministers are at pains to note that clarification of traditional rights 'will take place on the basis of national law. It is the use that counts – not the ethnic origin of the user'.

Legal and institutional framework 135

Similarly, the guide itself proclaims that the Act 'is ethnically neutral in the sense that individual legal status is not dependent on whether one is a Sámi, Norwegian or Kven [Finnish] or belongs to another population group'.

15 The state transferred these lands to Statskog in 1993. The Act applies on the shoreline 'as far out to sea as private right of ownership extends'. The Act does not apply to ocean fisheries.
16 On the nature of reindeer herding rights as customary rights (and therefore as property rights protected by the constitution) rather than rights that are conferred by statute, see Strom Bull (2001), who also emphasizes the particular challenges faced by migratory pastoralists. They require access to lands within the jurisdiction of municipalities but, as non-taxpayers and non-residents within the various municipalities to whose lands they need access, their interests will likely not have a high priority.
17 These provisions include s.3 (indicating that the Act is subject to ILO 169 and is to be construed in compliance with the provisions of international law dealing with minorities and indigenous people); and s.5 (which acknowledges that Sámi have collective and individual rights to land in Finnmark by virtue of prolonged use and further acknowledges that the Act does not interfere with those rights).
18 A thirteenth regional corporation was created to provide for Alaska natives living outside Alaska. This corporation participated in the cash settlement but did not share in the land and other benefits.
19 There exist different official lists of the numbers of distinct aboriginal people in Russia and different authors also cite different numbers. Yamskov (2001: 122) refers to an original 'official list' of small-numbered people. By 1989 the list included 26 nations and by 1993 the law added another four nations, but Yamskov admits that 'other additions are possible'. RAIPON, the Russian Association of Indigenous People of the North, http://www.raipon.org/History/People/tabid/310/Default.aspx lists over 40 indigenous people: Aleut, Alutory, Besrmyan, Veps, Dolgan, Ijorcy, Itelmen, Kamchadal, Kereki, Kety, Koryak, Kumandincy, Mansi, Nagaibaki, Nanaicy, Nganasan, Negidalcy, Nenets, Nivkhy, Oroki, Orochi, Sámi, Selkup Soioty, Tazy, Telengity, Teleuty, Tofolar, Tubolar, Tuvin-Todjin, Udege, Ulchi, Khanty, Chelkancy, Chuvancy, Chukchi, Chulymcy, Shapsugi, Shorcy, Evenk, Even, Ency, Eskimosy, Inuit, Ukagiry.
20 This next section is taken from Bankes (2004b).

References

Alaska Statutes; Title 38, chapter 5, s.180. Available at: http://www.touchngo.com/lglcntr/akstats/Statutes/Title38/Chapter05/Section180.htm.

ANCSA (1971) *Alaska Native Claim Settlement Act*, 43 United States Code (U.S.C.) § 1617.

Arctic Slope Regional Corporation (2005) *Annual Report*. Online. Available HTTP: http://www.asrc.com/_pdf/_annualreports/ASRC2005.pdf (Accessed 13 November 2006).

Bankes, N. (2000) *Oil and Gas Rights Regimes and Related Access and Benefit Issues in Yukon and Northwest Territories*, Paper prepared for a conference on Oil and Gas Law: Selected Topics, convened by the Legal Education Society of Alberta, Calgary, 2000.

Bankes, N. (2004a) 'Aboriginal title to petroleum: some comparative observations on the law of Canada, Australia and the United States'. *Yearbook of New Zealand Jurisprudence*, 7: 111–157.

Bankes, N. (2004b) 'Legal systems', in Einarsson, N. and Young, O. R. (eds) *AHDR Arctic Human Development Report*. Akureyri: Stefannson Arctic Institute.

Bankes N. and Rowbotham, P. (1993) 'The oil and gas industry: some current problems in environmental law', in Thompson, G. McConnell, M. L. and Huestis, L. B. (eds) *Environmental Law and Business in Canada*, Aurora: Canada Law Book Inc.

Broderstad, E.G. (2001) 'Political autonomy and integration of authority: the understanding of Saami self-determination', *International Journal on Minority and Group Rights*, 8 (2–3): 151–175.

Canada (1867) *Constitution of Canada*. Online Available HTTP: http://laws.justice.gc.ca/en/const/index.html (Accessed 31 March 2008).

Canada (1982) *Constitution of Canada*. Online Available HTTP: http://laws.justice.gc.ca/en/const/annex_e.html (Accessed 31 March 2008).

Canada (1985) *Canada Petroleum Resources Act*, Revised Statutes of Canada (1990), Chapter 36, 2nd Supplement.

Case, D. S. and Voluck, D. A. (2002) *Alaska Natives and American Laws*, 2nd edition, Fairbanks: University of Alaska Press.

DIAND (Department of Indian Affairs and Northern Development; 2006) *Comprehensive Claims Branch, General Briefing Note on The Comprehensive Land Claims Policy of Canada and the Status of Claims*. Online. Available HTTP: http://www.ainc-inac.gc.ca/ps/clm/gbn/gbn_e.pdf (Accessed 13 November 2006).

Domrin, A. N. (2006) 'From fragmentation to balance: the shifting model of federalism in pots-soviet Russia', *Transnational Law and Contemporary Problems*, 15(2): 515.

Eide, A. (2001) 'Legal and normative bases for Sámi claims to land in the Nordic (sic)', *International Journal on Minority and Group Rights*, 8(2–3): 127–149.

Einarsbøl, E. (2006) 'Some legal considerations concerning Sámi rights in saltwater', *Journal of Indigenous People Rights*, 2006 (1): 1–48. Online. Available HTTP: http://www.galdu.org/govat/doc/sjosamisk_english.pdf (Accessed 13 November 2006).

Fondahl, G. A. (1995) 'Legacies of territorial reorganization for indigenous land claims in Northern Russia', *Polar Geography and Geology*, 19: 1–21.

Fondahl, G. A. (1997) 'Freezing the frontier? Territories of traditional nature use in the Russian North', *Progress in Planning*, 47 (4): 307–319.

Fondahl, G. A. (1999) 'Autonomous regions and indigenous rights in transition in Northern Russia', in Petersen, H. and Poppel, B. (eds) *Dependency, Autonomy, Sustainability in the Arctic*, Aldershot: Ashgate.

Fondahl, G., Lazebnik, O., Poelzer, G. and Robbek, V. (2001) 'Native "land claims" Russian style', *The Canadian Geographer*, 45 (4): 545–561.

Field, M. (1992) 'The differing federalisms of Canada and the United States', *Law and Contemporary. Problems*, 55: 107–120.

Goudina. E. (2004) *A Sustainable Development Critique of the Russian Oil and Gas Disposition System: Learning from the Canadian Experience with Intra Generational Equity*, LLM thesis, The University of Calgary.

Goudina, E. (2006) 'A new draft law on the subsoil for Russia: advantages and disadvantages', *Resources* 92/93: 8–10. Online. Available HTTP: http://www.ucalgary.ca/~cirl/pdf/Resources9293.pdf (Accessed 13 November 2006).

ICHR (2001) *Mayagna (Sumo) Awas Tigni v. Nicaragua*. Online Available HTTP: http://www.corteidh.or.cr/docs/casos/articulos/seriec_79_ing.pdf (Accessed 31 March 2008).

ILO (1989) *Convention 169 on Indigenous and Tribal People*. Online Available HTTP: http://www.ilo.org/ilolex/cgi-lex/convde.pl?C169 (Accessed 31 March 2008).

Legal and institutional framework 137

Jentoft, S. and Buanes, A. (2005) 'Challenges and myths in Norwegian coastal state management', *Coastal Management,* 33: 151–166.

Kartashkin, V. A. and Abashidze, A. K. H. (2004) 'Autonomy in the Russian Federation: theory and practice', *International Journal on Minority and Group Rights*, 10: 203–220.

McPherson, C. B. (ed.) (1978) *Property: Mainstream and Critical Positions*, Toronto: University of Toronto Press.

Minde, H. (2001) 'Sámi land rights in Norway: a test case for indigenous people', *International Journal on Minority and Group Rights,* 8: 107–125.

Northern Economics Inc. (2006) *North Slope Borough Economy: 1965 to 2005*, Prepared for the Mineral Management Service. Online. Available HTTP: http://www.mms.gov/alaska/reports/2006rpts/2006_020.pdf (Accessed 13 November 2006).

Norway (1987) *Act Concerning the Sameting (the Sami Parliament) and other Sami Legal Matters*, English translation at: http://www.ub.uio.no/ujur/ulovdata/lov-19870612-056-eng.pdf (Accessed 30 March 2008).

Norway (1996) *Act relating to Petroleum Activities*, English translation at: http://www.npd.no/regelverk/r2002/Petroleumsloven_e.htm (Accessed 30 March 2008).

Norway (2005) *Act Relating to Legal Relations and Management of Land and Natural Resources in the County of Finnmark*, English translation at: http://www.ub.uio.no/ujur/ulovdata/lov-20050617-085-eng.pdf (Accessed 30 March 2008).

Osherenko, G. (2001) 'Indigenous rights in Russia: is title to land essential for cultural survival?', *Georgetown International Environmental Law Review,* 13 (3): 695–734.

Oskal, N. (2001) 'Political inclusion of the Samias indigenous people in Norway', *International Journal of Minority and Group Rights*, 8: 235–261.

Russia (1993) The Constitution of the Russian Federation. English translation available at: http://www.departments.bucknell.edu/russian/const/constit.html (Accessed 30 March 2008).

Saelbu (2001) Judgment of 21 June 2001 Serial No. 4B/2001: *Jon Inge Sirum (et al., a total of 200 parties) versus Essand Reindeer Pasturing District and Riast/Hylling Reindeer Pasturing District*. Online. Available HTTP: http://www.galdu.org/govat/doc/selbudommen.pdf (Accessed November 12, 2006).

Salikov, M. (2004) *The Russian Federal System: Sub-National and Local Levels*, Presented at a conference on 'Subnational Constitutions and Federalism: Design and Reform' March 22–27, 2004, Bellagio, Italy and convened by the Center for State Constitutional Studies, Rutgers. Online. Available HTTP: http://www.camlaw.rutgers.edu/statecon/subpapers/salikov.pdf (Accessed 13 November 2006).

Semb, A.J. (2001) 'How norms affect policy – the case of Sami policy in Norway', *International Journal on Minority and Group Rights*, 8: 177–222.

Skyner, L. (2005) 'The regulation of subsoil resource usage: the erosion of the "two-key" principle and its inclusion into the framework of civil law', *Review of Central and East European Law,* 31. 118–110.

Strom Bull, K. (2001) 'The right to herd in the light of the report of the Sámi Law Committee', *International Journal on Minority and Group Rights,* 8: 223–234.

Svartskog (2001) Judgment of 5 October 2001 Serial No. 5B/2001, No. 340/1999: *Landowners and right-holders in Manndalen, under cadastral Nos. 29–35 in the Municipal Area of Kåfjord*. Online. Available HTTP: http://www.galdu.org/govat/doc/svartskogdommen.pdf (Accessed 13 November 2006).

US (1958) *Alaska Statehood Act*. Public Law 85-508, 72 Stat. 339, July 7, 1958.

Vakhtin, N. (1992) *Native People of the Russian Far North,* London: Minority Rights Group International.

White, G. (2002) 'Treaty federalism in Northern Canada: aboriginal-government land claim boards', *Publius* 32(3): 89.

Yamskov, A. N. (2001) 'The Rights of small-numbered people of the Russian North in the territories of traditional nature use: ownership or use?', *Journal of Legal Pluralism*, 46: 121–134.

Yukon (1998) *Oil and Gas Act*, Online Available HTTP: http://www.gov.yk.ca/legislation/acts/oiga.pdf (Accessed 30 March 2008).

7 Expanding oil and gas activities on the North Slope of Alaska

Aslaug Mikkelsen, Sharman Haley and Olaug Øygarden

The culture is the most important part of Alaska. Its cultural groups really define Alaska. When people think about Alaska they don't think about oil and gas development, they think about the Eskimos and people in the Arctic. The face of Alaska really is the Native people.

(Inupiat woman)

When you start seeing the world from the north, it's not remote – it's where we are. It's home I guess.

(An employee at Arctic Slope Regional Corporation)

Introduction

The present oil fields on the North Slope of Alaska are mature, and the companies want to expand and open new areas. The prospects are promising, both onshore and off. The primary controversy over expanding oil and gas development concerns sustainability of the subsistence economy and culture. The line of conflict runs between those who argue that oil and gas activities have minimal impact on subsistence livelihoods, and those who see the oil and gas as a threat to the local communities.

The oil and gas revenues, activities, and the increased contact with the outside world have modernized the North Slope communities. The communities have become dependent on the oil and gas revenues to maintain the new infrastructure, modern equipment and lifestyle. In this situation, *onshore* oil and gas activities have broad support on the North Slope as long as there are sufficient environmental protections. Opposition to development on land is focused on specific areas that have particular environmental, wildlife and subsistence values. Some Native[1] people also express growing concern about the increasing encroachment of development on their lifestyle, values and traditions.

Offshore development faces sharper conflict. The federal and state authorities who have the decision authority offshore want to expand oil production to increase the domestic oil supply and bring in new revenue. But they face nearly universal opposition from Inupiat people and local and regional authorities who fear that oil spills will harm marine resources, and industrial noise will disturb and deflect migrating whales and other marine mammals.

Underneath these relatively clear lines of conflict lies a much more complex arena of issues, perceptions and themes in the discourse among stakeholders. There are groups of indigenous people that are more dependent on subsistence living than others. These groups are stronger in their voice against further oil and gas development. Others believe that embracing the job opportunities that development brings is the path to the future. They hold the opinion that those who want paid work can get it, so remaining in a traditional lifestyle is a personal choice. But in counterpoint, educational attainment among Natives is relatively low, so even with improved educational opportunities, it will take an estimated 20 years for the Native population to reach the level of educational attainment that would make them attractive workers for the industry. In the meantime, they argue, the subsistence activities will be destroyed. Taking kids to schools far away also makes knowledge transfer on whaling and other subsistence activities impossible.

Social problems, such as lack of education, lack of employment, lifestyle health problems like diabetes, mental health problems, drug and alcohol abuse and poverty, characterize the present situation in the villages. The oil and gas companies are challenged to take responsibility for the situation, for mitigating negative impacts and providing a positive future for the coming generations, but the companies are of the opinion that they are paying enough and that the governmental authorities have the obligation to provide infrastructure and help to the communities.

These issues will be explored from the various stakeholder perspectives in the sections that follow.

Alaska today

Oil and gas activities in Alaska are expanding *in* the Arctic, not *to* the Arctic. The largest oil field in North America is Prudhoe Bay, located on the North Slope of Alaska (Map 7.1). The Prudhoe Bay field was discovered in 1968 and started production in 1977. Over the past 30 years, oil facilities on the North Slope have expanded from one field to an industrial complex, including 23 producing reservoirs, with 115 gravel drill sites and 20 processing facilities, 960 km of permanent roads and trails, 725 km of pipeline corridors, and 353 km of transmission lines (National Research Council, 2003). While oil and gas deposits lie both onshore and offshore, on state, federal and private lands, existing development lies primarily on state-owned land. Active leasing and exploration is ongoing further west in the federal National Petroleum Reserve–Alaska (NPR-A), in state and federal waters to the north, and in the foothills to the south. The eastern frontier, the coastal plain of the Arctic National Wildlife Refuge (ANWR), requires congressional authorization before exploration and development can proceed.

Although indigenous hunters had used the Prudhoe Bay area of northern Alaska for thousands of years, the area contained no permanent human settlements until the construction of the work camp at Prudhoe in 1968. The nearest settlement is the community of Nuiqsut on the Colville River, which defines the boundary between state lands to the east and the 9.4 million hectare NPR-A to the west. The 1998 development of the Alpine field eight miles north of Nuiqsut

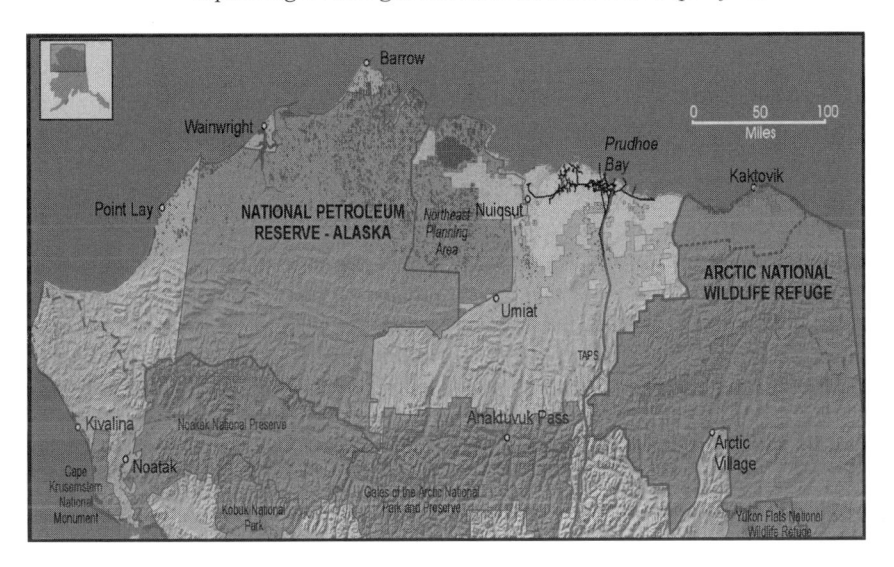

Map 7.1 The North Slope of Alaska.

marked the first time – aside from the gas wells that provide energy to the community of Barrow – that industrial facilities were located close to people's homes. The oil lies under land owned by Kuukpik Corporation, the local village corporation established under ANCSA (see Textbox 7.1).

The discovery of Prudhoe Bay precipitated massive economic, political and social changes for everyone in the state. In the following year (1969) state lease sales in the area elicited bids on 412,548 acres and brought in US$900 million to state coffers – an amount nearly 10 times total state revenues the previous year. But the oil needed a pipeline to an ice-free port, and the pipeline could not be built until long-standing land title disputes were resolved. The Alaska Native Claims Settlement Act of 1971 (ANCSA) awarded rights to 44 million acres and granted about US$1 billion in cash to 13 regional and 216 village corporations organized to benefit Alaska Native shareholders; construction of the trans-Alaska pipeline commenced in 1974. With 800 miles of pipe costing US$8 billion, this was the largest and most expensive human-made structure in the history of the world. During the construction boom, the population of the state grew 25 per cent and per capita income increased more than 50 per cent. Since then, more than US$280 billion dollars worth of oil has been produced from Alaska's North Slope. While the Alaska economy has always centered on resource extraction – particularly furs, gold, salmon and timber – North Slope oil has been worth seven times more valuable than all previous resources combined (Cole and Cravez, 2004).

Alaska's North Slope is the homeland of the Inupiat people. For thousands of years, the Inupiat have extensively used the land and sea for hunting, travel and other subsistence activities. Prior to the development of oil at Prudhoe Bay, the cash economy on the North Slope was quite limited. Before oil, in the Barrow census area in 1969, the annual income was, in 2006 equivalent dollars, US$8204 per person, less than half the income in urban Alaska. In the 1970 census, more than 90 per cent of homes had no flush toilet, nearly 80 per cent had no running water, and 70 per cent had no telephone. Since oil, per capita incomes have more than doubled – though they are still only 80 per cent of the income levels in urban Alaska. All the communities now have modern housing with plumbing, well-equipped schools and Internet access. Despite substantial improvements in the material standard of living and increasing dependence on the cash economy, it remains a mixed subsistence and cash economy. Cash is needed for snowmobiles and rifles, as well as for heating and electricity. Market food costs are high – 220 per cent of urban food costs[2] – and subsistence resources are needed for food as well as for cultural purposes. A total of 57 per cent of households still rely on bowhead whale, caribou, fish, and other subsistence foods for half or more of their diet (NSB, 1999), and villagers plan to continue the mixed subsistence lifestyle in future generations.

Today, nearly two-thirds of the 7253 North Slope residents live in Barrow; the rest live in villages ranging from 200 to 700 people. Access to all North Slope communities is by air. Three-quarters of the North Slope's residents are Inupiat; 44 per cent speak Inupiaq at home and 14 per cent (primarily elders) are not fluent in English (US Bureau of the Census, 2000). Nearly one-fourth of the adult population has not graduated from high school. About 60 per cent are employed full or part time. The Borough government is the largest employer for Inupiat residents and the village corporations are second. Trapping and crafts also provide some earned income. Employment in the oil industry is notably low.

North Slope oil field operations provide employment to over 5000 non-residents, who rotate in and out of oil work sites from Anchorage, other areas of the state, and the lower 48 states. Census figures for the North Slope do not include this transient population.

For additional background to the conflicts and the institutional framework of Alaska and the North Slope, see Textboxes 7.1–7.4, and also Chapter 6.

Stakeholders and coalitions to the oil and gas activities on the North Slope[3]

The history and institutional framework of the North Slope define some of the key stakeholders to the oil and gas activities. The main Native influence is channeled through the Inupiat-majority North Slope Borough government and the regional and village ANCSA corporations, which up till now have been in favor of onshore oil and gas development and against offshore activities. In addition to this, the indigenous people have influence through their tribal governments and the Alaska Eskimo Whaling Commission (AEWC). These three groups of actors were often mentioned as the most powerful locally and they are also deemed to be politically skilled. Regional tribal organizations, such as the Inupiat Commission of the Arctic Slope, do not seem to be as influential.

Textbox 7.1 The history of Alaska land ownership

The United States bought Alaska from Russia in 1867. The land area of Alaska was 375 million acres and it had about 35,000 inhabitants, most of them indigenous people. Alaska was first administered as a territory, and did not gain statehood until 1959.

Close to 1900 the US Congress opened Alaska to homesteading and other programs that allowed individuals and businesses to apply for land. Congress also acknowledged, beginning in 1884, that Alaska's indigenous people had a right to land, but it would take nearly 100 years for land claims to be linked to land title. Big changes in land status began when Alaska gained statehood and got rights to select about 104 million acres of federal land. Alaska could choose from lands that were not reserved for national parks, military bases or other purposes.

In the 1960s, when the state began selecting lands, Native groups saw increasing threats to lands they had traditionally used and they organized statewide to press their land claims. By the end of the 1960s, land claims blanketed Alaska. The Secretary of the Interior responded by halting all land transfers until the claims were settled. In 1971, Congress passed the Alaska Native Claims Settlement Act (ANCSA). It was settled with a grant of 44 million acres and payment of 1 billion dollars. Native regional and village corporations were established to manage these assets.

The years from 1971 to 1980 saw a bitter controversy over how much land would be added to national parks, wildlife refuges, and other conservation units in Alaska. In the end, the 1980 Alaska National Interest Lands Conservation Act (ANILCA) added nearly 104 million acres to the conservation system in Alaska. About 56 million acres were designated as 'wilderness,' the most protected classification. ANILCA generated management issues and conflicts that are yet to be resolved. A major unsettled issue is whether the coastal plain of the Arctic National Wildlife Refuge (ANWR) – known as the 1002 area, referencing section 1002 of the Act – should be designated wilderness or opened to oil and gas activities. This has been a recurring and bitter political fight in the national political arena ever since.

Federal protections for subsistence practices in rural Alaska did not come before 1980 with the passage of ANILCA. Title VIII of ANILCA, entitled 'Subsistence use in Alaska,' focused largely on a priority for subsistence harvests within the fish and wildlife management regimes. However, the statute also included protections for subsistence use practices within the context of federal land use decisions, reflecting the understanding that the opportunity to engage in the subsistence harvest of fish and wildlife resources rests upon secure resource populations and intact productive habitats (McIntosh and Brelsford, 2006).

When the transfer of land is complete, the state will own about 28 per cent, the Native corporations 12 per cent, and other private owners about 1 per cent. The federal government will retain nearly 60 per cent of Alaska, largely in national conservation units.

Textbox 7.2 A new institutional framework

The Alaska Native Claims Settlement Act established 12 Native regional corporations in Alaska and one in Seattle for Alaska Natives living out of state. All Native individuals conceived before the Act was passed and received approximately 100 shares in both their regional and village corporation. These shares cannot be sold, but they can be passed on as inheritance. Several corporations have issued new shares for the so-called 'afterborns.' The regional corporations hold surface and sub-surface rights to their land and the village corporations own surface lands in and around the villages. Under sub-section 7(i) of the Act, 70 per cent of natural resource revenues accruing to the ANCSA corporations must be shared with the other 11 corporations.

The Arctic Slope Regional Corporation (ASRC) is the regional corporation for the North Slope. It owns approximately five million acres of land. Its 2005 net income was 127.5 million dollars. About 400 of its 6000 employees are shareholders. The ASRC has 9000 shareholders, and in 2005 the dividend paid to these was $28.61 per share.

The North Slope Borough (NSB) was organized as a regional government in 1972 and adopted a Home Rule Charter in 1974. It was established by village leaders – the whaling captains – who saw the need to cooperate to manage the impacts and secure benefits from oil development. The oil industry and many of the state elected leaders opposed incorporation of the Borough because it meant another layer of taxation and regulation on the industry. The Borough has powers of taxation, education, planning and zoning, and provides municipal services such as police powers, utilities, infrastructure maintenance, and housing. The income of the NSB is predominantly from taxation of oil and gas infrastructure in the region, but they also receive additional funding from state and federal sources. Constituency to the Borough is not ethnically prescribed. The individual villages all have both a city administration and a tribal village council.

Textbox 7.3 Leasing and permitting

The rights to subsurface mineral resources are transferred from landowners to companies through lease sales. A lease does not authorize subsequent exploration or development: for this the company must obtain a variety of permits from state and federal agencies, including right of way for the pipeline and access corridors, water use, coastal zone management plan consistency review, lease operations, land use, fish habitat authorization, air quality control, storage of drilling wastes, disposal wells, discharge prevention, and contingency planning.

On state lands onshore and in the three-mile band of state waters near shore, the Alaska Department of Natural Resources (ADNR), Division of Oil and Gas, is the agency responsible for the leasing and permitting processes. In the federal National Petroleum Reserve–Alaska (NPR-A), it is the Bureau of Land Management (BLM), an agency within the US Department of Interior that manages leasing and administers the National Environmental Policy Act (NEPA). The NSB must also review and approve the proposed exploration and development activities for consistency with their land use management plan and regulations, with particular concern for fish and wildlife populations, habitats and subsistence activities.

Offshore beyond three miles, the managing agency is the federal Minerals Management Service (MMS). It administers the lease sales and the NEPA review process for the outer continental shelf. The National Marine Fishery Service is also involved to protect marine resources covered by the Endangered Species Act and the Marine Mammal Protection Act. The Borough has far less regulatory power offshore.

The National Environmental Policy Act was signed in 1969, coinciding with the Prudhoe oil field. Under this Act, any activity that involves federal action or approval requires an environmental assessment. If the federal agency in charge, on the basis of this assessment, finds that the activity is likely to have significant environmental impact, an environmental impact statement (EIS) is required. An EIS is a long, public process and it can take several years to complete. The core of the process is the development of different activity alternatives and the environmental, and to some extent social, impacts of these. The process starts with a series of public scoping meetings, a draft EIS is then presented for a new round of meetings and a commenting period before the final EIS is issued and a decision is made. The first EIS was written prior to the 1973 construction-start of the Trans Alaska Pipeline. At this time, there was also an Environmental Studies Program initiated for the offshore by the MMS. Most of the subsequent oil field developments on the North Slope did not require an EIS, as the Army Corps of Engineers made the assessment that there would be no significant environmental impact. The current policy, however, is to prepare a comprehensive EIS for an entire, area-wide lease sale or for a whole set of contemplated developments all at once.

Planning and zoning powers

The Coastal Zone Management Act as implemented in Alaska prior to 2003 gave broad powers to the local coastal policy council – in this case the North Slope Borough – to set environmental standards for development projects. For each project proposal, the local council determines whether the project is consistent or inconsistent with their coastal management plan (CMP) and policies therein. The state Division of Governmental Coordination (DGC) gave great deference to the local council determination, and all state and federal permits for the project had to follow suit. The standards in the local plan are legally enforceable.

In 2003, to streamline the process of CMP consistency reviews, make the standards more uniform state-wide and expedite development, the state administration and legislature overhauled the state's authorizing legislation. The governing Coastal Policy Council, comprised of a majority of representatives from local districts, was disbanded, program authority was transferred from the DGC to the Department of Natural Resources, and regulations of the Alaska Department of Environmental Conservation were removed from the coordinated review process. The new law and regulations narrowed the authority of coastal districts to write enforceable policies restricting development and protecting subsistence and other resources. Specifically, the amendments:

- restricted the area where a consistency review can be conducted to the coastal zone – the effect was to exclude uplands which previously had been included in so far as activities in the watershed might affect coastal resources or uses;
- changed the definition of aquatic habitat from being able to sustain biological organisms to being able to meet water quality standards;
- eliminated subsistence uses as the highest priority; and
- changed the standard for environmental damage that requires mitigation.

The state CMP office also serves a coordination function for all the various state permits that an operator needs to explore and develop a field, such as right of way, water use, lease operations, land use, Title 16 fish habitat authorization, air quality control, storage of drilling wastes, permit to drill, disposal wells and discharge prevention and contingency plan. The state permit review process solicits written comments but does not customarily provide for oral testimony. Federal agency actions such as permitting development in NPRA require public hearings with oral testimony.

The North Slope Borough Planning Commission must also review the development master plan, assess compliance with Borough policies, and reclassify the target acreage for development.

Textbox 7.4 Revenues from oil and gas

Almost all current Alaska oil and gas production occurs on state lands. As the landowner, the state earns revenue from leasing as: (1) upfront bonuses, (2) annual rent charges and (3) a retained royalty interest in oil and gas production. Generally, the state issues leases based on a competitive bonus bid system. It has always retained a royalty interest of at least 12.5 per cent, with a few leases ranging as high as 20 per cent and some as a net profit-share production agreement. Alaska's royalty, rent and bonus bid income in FY 2005 totaled $1.9 billion. The Alaska Constitution provides that at least 25 per cent of the state's royalty, rent and bonus income must be deposited into the principal of the Alaska Permanent Fund, and an additional 25 per cent is funneled into the Fund by statutory law. In 2005 the cumulative value of the Alaska Permanent Fund was 32 billion dollars. Investment earnings from the Fund are used to pay a dividend each year to everyone who has lived in the state for at least one year. The dividend is based on 50 per cent of the five-year average of fund earnings. The amount of the dividend peaked at $1964 in 2000; by 2005 it was back down to 845 dollars, and is now rising again.

In its capacity as a taxing authority, the state levies a production or severance tax based on wellhead value of the oil, as well as an oil and gas property tax and a corporate income tax. These revenues amounted to $863.2, $260.7 and $524 million respectively in FY 2005. Property taxes are shared with local governments: the North Slope Borough share was $192.8 million. The gross value of North Slope oil production in fiscal year 2005 was $13 billion (Alaska Department of Revenue, 2006).

While the state of Alaska manages resources within three miles of the coast, the US Minerals Management Service, Alaska Region, manages offshore energy and mineral resources from 3 to 200 nautical miles. The Outer Continental Shelf (OCS) Lands Act grants the state a 27 per cent share of federal revenues from mineral development in the first three miles of the federal OCS, i.e. 3–6 miles from shore, as designated in section 8(g) of the Act. There is no provision in law for OCS revenue sharing beyond the 8(g) zone. To provide additional incentive for local support of expanded OCS leasing, the Energy Policy Act of 2005 created the coastal impact assistance program to provide grants to mitigate the impact of OCS oil and gas development on coastal communities.

> Alaska's North Slope Borough, for example, has in the past expressed concern that an offshore industry would disrupt North Slope communities without bringing benefits — communities would have to deal with any impacts on subsistence hunting, for example, but the borough would not be able to collect property taxes for outer continental shelf industrial facilities.
> (*Petroleum News* 12 (10), Week of March 11, 2007)

Alaska's allocation for FY 2007 and 2008 will be $848,750; one-third of this – $276,449 – will be available for grants on the North Slope, consistent with a state plan to be approved by the MMS.

On shore, Alaska receives 50 per cent of the federal revenues from oil and gas lease sales located in the National Petroleum Reserve–Alaska. A portion of these revenues is disbursed in grants to mitigate impacts on local communities. Under the Alaska Statehood Act, the state is entitled to 90 per cent of federal revenues from unreserved federal lands in the state, but none of the federal lands on the North Slope falls in this classification. In fiscal year 2006, the federal government received $68 million in mineral revenues from federal lands in Alaska and returned $25.7 million to the state in revenue sharing; $11 million of this was from OCS oil royalties.

The Arctic Slope Regional Corporation (ASRC) and Kuukpik Village Corporation also own land in the Alpine oil field, for which they receive royalties and land rents. Under the section 7(i) provisions of the Alaska Native Claims Settlement Act (ANCSA), 70 per cent of natural resource royalties must be shared with the other 11 regional corporations. In fiscal year 2005 the estimated total value of these royalties was around $100 million, of which ASRC and Kuukpik retained $20.7 and an estimated $9 million, respectively. In addition, Kuukpik received land rents amounting to 600 thousand dollars.

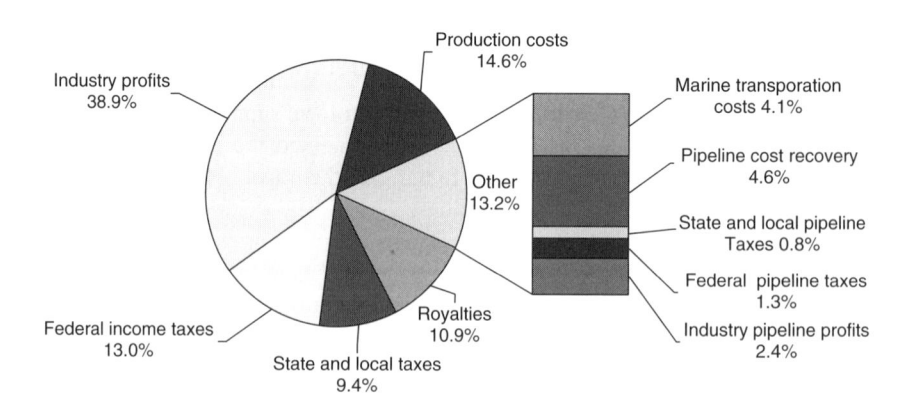

Figure 7.1 Allocation of gross Alaskan North Slope (ANS) production value 2004 (US$13.3 billion total). Source: US Fish and Wildlife Service.

The Borough and the ANCSA corporations depend on revenue from oil and gas. The NSB levies a property tax on oil and gas facilities. Owing to depreciation of the facilities, revenues are decreasing and new activity is needed to get the budget back up. There are no short or known long-term alternatives to oil and gas for this. The power and money of the oil and gas industry is mirrored in the Arctic Slope Regional Corporation (ASRC). The corporation's subsidiaries compete for contracts from the oil companies, and they need the industry to make money. This need is also relevant at the level of the individual residents, because many of them depend on ANCSA corporation dividends being paid out every year.

Yet community development, a subsistence livelihood and the rights of the indigenous people are over-riding concerns voiced at the regional and local level. Subsistence activities are identified as the core identity of the people and the glue of the communities. A representative of a village corporation argued that, since ANCSA corporation stocks are not traded, they have more freedom than larger oil and gas corporations to delay development and wait for the best design to protect community interests.

The state and federal authorities are another important group of stakeholders. Revenues from oil and gas activities seem to be the main interest at the state level, and there is broad support for both onshore and offshore development as long as the state retains authority to regulate. Under the current administration, the federal government is perceived as aggressively pro-development, primarily for reasons of domestic energy supply and national security. The cognizant agencies, however, reflect a more mixed agenda. Every agency in Department of the Interior is tasked with good stewardship of the land and resources under their jurisdiction. The mission of the Minerals Management Service (MMS), to manage outer continental shelf (OCS) mineral resources and revenues 'to enhance public and trust benefits, promote responsible use, and realize fair value,' is most focused on resource development, tempered only by 'responsible use' and compliance with federal environmental laws, such as the National Environmental Policy Act. The Bureau of Land Management, by contrast, manages multiple resources and uses, including: energy and minerals; timber; forage; recreation; fish and wildlife habitat; wilderness areas; and archaeological, paleontological and historical sites. Practices such as re-vegetation, protective fencing and water development are designed to conserve and enhance public land, including soil and watershed resources. The mission of the Fish and Wildlife Service, which administers the Arctic National Wildlife Refuge and also the Endangered Species Act, is the most focused on conservation. Its mission is 'to conserve, protect and enhance fish, wildlife, and plants and their habitats for the continuing benefit of the American people.' Yet oil and gas development may be allowed in a wildlife refuge if it is found to be compatible with the primary purposes of the refuge. Within each agency as well there are sometimes differences of opinion between line staff who work in the field and higher level policy-makers concerning the proper balance between resource development, environmental protection and accommodation of other stakeholder values. Some federal scientists have taken considerable initiative to ensure that their professional opinions are accurately presented in policy circles.

Neither our Native informants nor the representatives of the industry on the North Slope seem to view environmental organizations as particularly important stakeholders. This was sometimes explained as being a result of the different ways the local population and these non-governmental organizations (NGOs) view environmental protection. The environmental NGOs often take a more radical conservationist stand than the local people who depend on the possibilities for hunting and whaling. The Alaska Coalition and allied environmental organizations do participate in the public review of environmental impact statements (EISs) and frequently challenge agency decisions in court, but thus far their success and effect have been limited.

Independent of level or group of stakeholders, the only 'acceptable' opinion on Native Alaskan issues is to support Inupiat subsistence activities related to bowhead whaling and caribou hunting. Everyone agrees on the need to operate industry in a way that does not destroy the environment for subsistence. Indirectly several informants question this position by mentioning all the factors in the past, present and future that make a subsistence livelihood impossible, but they never directly or bluntly say that a subsistence livelihood should cease to exist. However, stakeholders vary on how much they are willing to compromise between development of oil and gas and whale and caribou hunting. The non-profit tribal organizations seem to compromise less than the for-profit organizations, but there are ambiguous attitudes there as well. For example, Kaktovik City Council has declared Shell, who is planning to start offshore activities, a *hostile force*, but the village corporation disagrees with this resolution and the local whaling captains have signed an agreement with Shell through the AEWC. The ASRC is involved in offshore operations, although they have not yet officially stated that they support this kind of activity. A representative of the NSB speaks of how ASRC in reality is an *oil company*, and that the NSB sometimes gets ASRC's position 'thrown in their faces.' To outsiders, the ASRC represents an entity with *Native* backing, so when the ASRC agrees to developments, others can say to the opposing Borough 'but the Native people agree, because the ASRC says so.'

Stakeholder discourses

On the North Slope, several signals of unsustainable global development and threats to sustainable community development are reported. First, it is claimed that the oil activities on the North Slope have displaced wildlife and reduced the Inupiat people's access to their traditional hunting areas. With the expansion to the offshore, it is feared that constructions and noise of the activities will deflect whales and make traditional, subsistence bowhead whaling impossible, and oil spills will harm marine life. Second, modernization and influence from outside by the media, Internet and travels foster a Western style of living and cash economy that erode traditions and cultural values. Dependence on money and modern equipment is also threatening the social structure of the communities. Increased access through roads and airstrips may increase the population impact on the land, exposing it to hard use, which threatens the flora and fauna. Third, global warming is changing the animal and human habitat and is making the traditional knowledge of ice conditions and safe harvesting less valuable. Finally, pollution from energy use in the Western world has already contaminated the Arctic food chain and raises fears about the safety of subsistence foods. Pollution also constitutes a threat to biodiversity.

Environmental sustainability

The Alaska Coalition, being one of the more influential environmental NGOs, ranks the environment as the number one key of the three sustainable development pillars, closely followed by the social aspect. They ask if oil development has a place in a sustainable development: 'it kind of goofs up the whole idea.' An ASRC employee puts it differently: 'I think that if you have the environmental and the economic down, well balanced, the social sustainability follows from that.'

The environmental organizations argue that the industry goes by its bottom line and favors the economics above all. They further argue that the state of Alaska and the federal government are influenced by who is in power. In the federal government, there is some variation in focus with different administrations. The Bush administration is seen as extremely industry friendly, but the former President tried to be more balanced about sustainable development. The state puts the economic benefits above the environmental issues because the politicians want to please the majority of constituents. In total, the current situation seems to be that both state and federal governments are heavily tilted toward favoring industry, not locals or the environment.

Most of the impact assessment that has been done on the North Slope has been environmental. One of the key actors in this has been the NSB. An ASRC spokesman said that the development on the North Slope has been very environmentally aware, maybe more so than anywhere else in the world, but he was more worried about potential offshore developments.

The bureaucrats at the NSB are optimistic about sustainable development of the North Slope seen from a Native perspective. They argue that there has been no lasting damage to the environment. In their view, sustainability means that there has to be a positive relationship between the oil and gas companies and the communities. Some local people are in a way also defending themselves from the outsiders coming to the community and talking about sustainable development. An ASRC representative said: 'sometimes the term sustainable development is held up because of the bad conscience of people from elsewhere – they hold higher standards for this area because they feel guilty about their own dirty highways and stuff.' The informant admitted that the older developments such as Prudhoe Bay and Kuparuk had extensive impacts, but newer developments like the Alpine have small footprints. Several people argue along these lines; new technologies like directional drilling reduce the footprint of development and the environment is almost unaffected.

Are there any cumulative effects?

Environmentalists, governmental authorities and the industry agree that the level of awareness of possible negative (and positive) effects of the oil and gas activities has risen over the years. However, there are different story lines about the situation. One group of people, including regional authorities, their consultants and a committee within the National Research Council, are talking about the need for an integrated management plan. They see problems with the practice of issuing permits project by project, making it difficult to assess possible cumulative effects. Other people, for example, in the state agencies, deny the presence of cumulative effects, and warn about the idea of zero effect: 'We don't live in a world that has zero risk. There has to be a balance between what actually is protecting the environment, and what is protecting a cultural way of life and what just is silly.'

Consultants for the NSB have recently completed a comprehensive land use plan[4] that tries to address all kinds of development, and particularly oil and gas. A core concept in the report is the cumulative impacts and how much of the area can be developed before you reach a tipping point in terms of impacts on air quality, water quality, fish and wildlife, and culture. Several NSB employees also talked about these cumulative impacts. Other informants, for example a representative from the National Academy of Science, would like to see the resource managers in the state and federal governments adopt an integrated management plan for oil and gas that explicitly addresses these cumulative impacts.

On the North Slope, Nuiqsut is often used as an example of the consequences of not having a comprehensive plan for an area. Nuiqsut is on the leading edge of development. The initial development was centralized and not near any communities, but it has continued to expand. Local people claim to recognize changes in animal migration and numbers. This affects the hunting. The road development and increased access makes some supplies less expensive, but it also brings more people in and there is more competition for resources. There has been development of schools, water systems and other infrastructure, and people worry about what will happen to these new services if Borough revenues decline. At the same time, there have also been some very negative impacts and social changes, such as escalating drug and alcohol abuse. The introduction of foods that are not typical in the culture is seen to be the cause of other accelerating health changes, such as diabetes.

Offsite and cumulative impacts are much harder to deal with in regulations or permits than impacts from a specific project. Regulatory officials emphasize how difficult it is to place a costly requirement on the applicant to address cumulative effects and to calculate that applicant's fair share of the cost when there are cumulative effects. The consultants to the NSB described some of the difficulties with a regional perspective for the whole North Slope. When a lease is sold, no one knows what will be found and the industry cannot say what they are going to build. If they may make a find, there is a gap between the broad level of lease sales and very specific projects. There is no intermediate level between an exploration lease stage to a comprehensive overview of the whole area and how a reasonable infrastructure plan can be made to support all of the projects there.

Also, a lot of the information is proprietary, so there are many barriers to developing a coordinated approach. Companies are inclined to think of one project and its own associated infrastructure. Nevertheless, cumulative environmental and social impact assessments of oil and gas activities have been included in the most recent EIS statements for the Northeast and Northwest NPR-A, the Alpine Satellite Development Plan and the various Beaufort Sea lease sales.

Some of the state and federal informants together with some of the industry representatives denied the cumulative effects of the oil and gas activities. One of them said: 'What is the cumulative impact? When you put less than one per cent gravel on the ground? I'm sorry, but cumulative impacts are a figment of the environmentalist imagination. And the idea that there's no comprehensive plan, is also nonsense.'

Are there effects on caribou?

One debate about footprints is whether the caribou have moved away from the infrastructure or not. Widely circulated company propaganda displays pictures of caribou next to the pipeline and (correctly) asserts that the Central Arctic caribou herd is growing. Yet local hunters claim that migrating caribou have been observed to shy away from pipelines. Research biologists find that caribou at times will seek the shade of the pipeline, but that cow caribou at the time of calving avoid infrastructure by 4 km or more (Griffith and Cameron, 1998).

A related issue of debate is hunting access. Hunting and the subsistence way of life will be affected by changes in access to traditional areas, not only decreased access but increased as well. A former top manager in the state administration said that the biggest compromise will be that achieving the economic advantages requires a trade-off of a certain amount of the habitat affected and a certain potential change of growth and population. With the growth of population, subsistence is dramatically affected. With new access roads, people can drive a pickup truck into areas that used to be accessible only by snow machine.

A state representative emphasized the increased access that comes with the roads and infrastructure. He questioned whether changes in the animal populations had happened owing to increased development or if it was caused by additional hunting pressure brought on by the access. According to him, some issues have scientific credibility, but others are emotional or even brought up by environmental groups for fundraising purposes. Research biologists assert that hunting pressure alone will not have a significant effect on herd population, but that the disturbance caused by human activities of any sort can, at critical times, negatively affect caribou feeding and energy balance.

Are there effects on whales?

Whaling is central to Inupiat subsistence, social structure and identity. The prominence of whaling captains in the political and corporate leadership is a testament to this. With arguments of saving marine mammals from extinction, some environmental organizations and the International Whaling Commission (IWC) have objected to the traditional Inupiat bowhead whaling. The indigenous community has been able, particularly through the AEWC, to maintain their whaling quota through the use of modern science and traditional knowledge. Integrating both kinds of knowledge led to a more realistic picture of how numerous the bowhead population is.

The main arguments used against offshore activities are the noise that could deflect the whales and the risk of major oil spills. A state representative said that the probability of a large spill is deemed extremely low, and that the cumulative effects of small spills are also small. Transportation is the riskiest part of offshore oil activities as far as spills are concerned; it is not the exploration, drilling or production. But even transportation risks can be met to a large degree by technology. The Inupiat, however, have a very low risk tolerance for oil spills.

The Inupiat call the offshore their garden, and the debate has been contentious for many years. Representatives of the NSB say that they have opposed offshore activities until now, but that they do not have sufficient power to stop it. They say that they can probably slow it down, but they doubt that anybody has the power to stop it now. Representatives of the Native corporations also explain that the offshore issue is complicated. The corporations depend on growth of the oil and gas activities for their business livelihood, and that might mean doing business offshore too. Local conflicts arise when the Native corporations try to combine political objectives with doing business. The local politicians have to answer the question of 'what do we tell the people?'.

Other informants also recognized that this is a dilemma for local politicians. The first offshore development was called the North Star, and the NSB Mayor negotiated it with the industry. In villages other than Barrow, people thought of the Mayor as being opposed to the project, but in reality he was negotiating with the companies. The people in these positions have to balance the wish to stand up for the regional wishes with the fact that they are facing an 'overwhelming industry.'

A consultant said that companies get permits in spite of protests from the local whaling captains and the research they may present. The law requires making an effort to evaluate the impacts and look at reasonable alternatives, and the agencies may impose some restrictions that try to eliminate or minimize the impact, but they rarely deny a development permit.

The same consultant reports that many people in the NSB think the social and environmental impact assessments are superficial. The industry, the government and the regulatory agencies care more about development than about long-term impacts to the communities. The actors end up making compromises concerning seasonal restrictions, setbacks or geographic offsets. So far, he claims, this seems to be the primary approach between the whalers and the industry on the question of seismic exploration too. There is not a mature body of science on bioacoustics, effects of various types of transportation or seismic activities at this time. State representatives that support offshore drilling argue that the main challenges offshore are technological, but acknowledge that not all the solutions have been found yet: 'In areas with pack ice, the technology has not been proven, and the sub sea completions are just being proven, so you have a question about having sufficient technology to be able to develop successfully.'

Are there technological solutions to environmental conflicts?

Some of the recent conflicts between the Native people, the local and regional authorities, and the industry are about the technological solutions that the companies want to choose. In 2006, Nuiqsut saw a dispute between ConocoPhillips and the local community about a proposed pipeline and vehicle bridge crossing a channel in the Colville River that is very important for subsistence fishing. The local people opposed the design of the bridge because they fear the buttress will cause erosion and silting that will harm fish habitat. They also worry that the pipeline is designed to have only one and not two walls. Kuukpik, the village ANCSA corporation, has managed to halt the process by showing that the environmental conditions demand another alternative, for example, a buried pipeline going under the river.

Another conflict revolved around the offshore prospect McCovey. The main corporate owners tried to get the project through the permitting process, but the NSB strongly objected to the plan to build a giant ice island on which to situate the infrastructure, because it might impact marine life. After a long disagreement that included a court case, the two major partners dropped the project. A small company, which had been a silent partner in the first process, took over as the operator. They decided to use a drilling barge instead of the ice island, and promised to shut it down if whales were spotted in the vicinity and only to do drilling in the winter. The Borough felt much more comfortable with this alternative and the necessary permits were given. This project was brought up by several informants as an example of how a permitting conflict was solved by better communication and technological creativity.

These are both discussions about available technology, price and footprints. For the industry, available technology might mean available at a certain price, whereas for the Native people, it means the technology that makes the least footprint independent of price. One informant claimed that the companies will always try to get the cheapest alternative permitted first, and then improve the environmental protection if they are forced to do so.

The strongest supporters of development had a strong belief in how technology would solve future conflicts. One state representative says: 'Alpine is the future. It is what ANWR development will be like, what development will be like in the NPR-A. Now you see that the wells are very close together, there's very few of them, because you can drill from one common point, and tap the oil in all these different directions.'

Must people choose between traditions and modernity?

Asking about sustainable development for the Arctic also brings about a discussion about the scale of sustainability. Native people are concerned about sustainable communities, and how they can protect their identity and culture. They connect these issues to subsistence activities.

One female Inupiat says:

> Local people very much define themselves and the future they hope for in relation to the traditional subsistence way of life. People talk about jobs and incomes, but the kind of thing that people will talk about in terms of hopes and dreams, their sort of identity and their sense of being the authors of their own future, is very much focused on providing stability and continuity in the subsistence way of life.

These people emphasize the social organizational features of the subsistence livelihood in which generations are integrated, the role of elders and the young, and sharing practices where foods are widely shared between producing households and recipient households.

As an argument for development and against this way of thinking, others are rhetorically asking if this is the kind of life the Native people want to go back to – a community without money and with no modern equipment at all. They speak loudly about all the blessings of the new life: 'Earlier people used blubber for heating fuel, so there have been some benefits from the industry.'

Many local people are looking for some middle ground where there is some adoption of modernity, but they do not want to have the cultural shift forced on them faster.

> The Native people do not want to be a theme park for other people to look at. First and foremost, we want a right to self-determination. We do not want to turn in our snow machines, rifles, whaling bazookas. But if we were to go back – what would we do? Adaptation has always been a part of us. Also, others are allowed to change, why shouldn't we be?

The NSB advocate safe and responsible development and say: 'This is the only economy we have.'

An ASRC employee highlighted the need for a paradigm shift with young people:

> People tend to see a conflict between the Inupiaq way of life and work in the industry – but a blend is desirable. We are victims of that conflict. And we don't want to end up like those disenfranchised people you see elsewhere in the world. There is a collision of cultures. Sometimes it's perceived as oil and vinegar – 'please don't touch our lifestyle'. The demand to stay the same is present both within the community and from the outside.

He continues:

> But what would happen if we stayed the same? We would be something like Disney World where people could come and look at us. There is a desire to protect our way of life, but we need to balance protection with

allowing development. There is conflict around this in the community, just like in any culture.

The mixed economy

The pure subsistence economy does not exist any more. However, as we have described, in terms of culture and individual and group well-being, identities, and social well-being, subsistence is extremely important, even if you have a paid job. With groceries costing twice as much as in urban Alaska, many people need the food from the subsistence activities to maintain a decent living standard. Some of the Natives can live by a combination of corporation dividends, Permanent Fund Dividends and subsistence activities, but this is very little compared to the costs of living. Other people do not receive dividends and rely on social aid. In some villages, as much as 50 per cent of the people need social aid to pay their bills.

The economy that is described by all the informants is a mixed subsistence-based economy where the cash comes primarily from the oil and gas activities through the NSB. A consultant says that what people want is relative stability in this hybrid or mixed economy. Many Natives, both in the NSB and in the corporations, and the local authorities also hold this view. Within this mixed economy, the Inupiat people embrace the new technology, and use it in their hunting and their daily life. An Inupiat said: 'We will use the satellite to find the whale, but we will use our traditional seal skin boats to paddle out to the whale, but then, in order to kill the whale faster so that it doesn't suffer, we will use a bomb.' They want to maintain their subsistence activities in a modernized way, and this has an oil dependency built in to it. Faced with the reality of the mixed economy, many Inupiat people are looking for a compromise with the industry. The industry is also adapting to the need of permits and demands on environmental and social assessments, and mitigating activities for maintaining their license to operate.

What are the social effects of the cash economy?

Introducing corporations and money into the traditional subsistence communities created Native business people, Native shareholders and Native employees. A main story line about this new characteristic of the society was 'you share subsistence food, but you do not share your money.' Some people in the industry talk about the Natives and their ability to adapt to this new way of life, and claim that the influx of cash has led to abuse and westernization – money simply 'isn't good for them.' A representative of a village corporation pointed out another issue, and claimed that money and job competition has introduced new tensions and jealousy. Also, not all Native villagers have shares in the ANCSA corporations. One informant claimed that this has led to new class differentiations.

The success of the ANCSA corporations varies across villages and regions. Some of this variation may be connected to the location of the resources or industry, but it may also be explained by human agency. One academic described the leaders in the ANCSA corporations as exceptional men who live in two worlds. They have urban clothes and travel in jets, but they still have their traditional values, and often combine their corporate jobs with being respected whaling captains.

Owing to the oil and gas activities at the North Slope, the ASRC has been successful and, in terms of cumulative dividends paid, ranks second among the 12 ANCSA regional corporations. At the other end of the spectrum, the Calista Corporation in southwestern Alaska has only paid dividends twice, a cumulative total of US$49 per shareholder. One informant said: 'ANCSA made a community of those who have and of those who have not – it has created a very different social structure.' For the oil companies who come in to these communities, stakeholder engagement is the tool that is used to prevent further negative impacts on the societies.

The village of Nuiqsut illustrates the dilemmas of modernization (see Textbox 7.5).

Textbox 7.5 The two stories of Nuiqsut

Nuiqsut, population 417, is the North Slope village that has been most directly impacted by the oil and gas activities. With the Alpine facility and later the Alpine Satellites, development has moved closer and closer to the village.

The story of Nuiqsut is a contentious one. The Colville Delta has traditionally been a gathering and trading place for the Inupiat and has always offered good hunting and fishing. The old village of Nuiqsut (Itqilippaa) was abandoned in the late 1940s because there was no school. The village was resettled in 1973. A school, housing and other facilities were constructed by federal agencies in the summer of 1973 and 1974; goods were hauled from Barrow by tractor and snow-machine. The City was incorporated in 1975 (see http://www.commerce.state.ak.us/dca/commdb/CF_COMDB.htm). One narrative about the establishment tells of a group of 27 families who traveled from Barrow across the tundra in the middle of the winter to claim their traditional hunting grounds as their own land under ANCSA. They established Nuiqsut and lived there in tents for the first 18 months. Part of the reason why they moved was because they wanted to get back to the old 'Eskimo life.' Other informants hold a different view. They see the establishment of Nuiqsut as a strategic move into an area near Prudhoe Bay where oil had already been discovered. Other aspects of the locale were not taken into consideration, and the result was a village situated at a remote spot where it is hard to provide employment and sustain a local economy.

The past is not the only contentious theme in Nuiqsut. Today, there is a conflict between the local community and the 'oil people' and the impacts they have brought. The impacts, as the community sees them, range from the displacement of caribou, to dramatic increases in administrative work for the City, to increases in substance abuse because ice roads built for the oil facilities make drugs and alcohol more accessible. The oil companies, and to an extent other villages, tell a story of Nuiqsut as a village who has gained a lot from the presence of the oil industry, in terms of financial reward and employment.

The Native village corporation, Kuukpik Corp., has 212 shareholders and a turnover of more than 10 million dollars a year. They own surface rights to land around the village, and have subsidiaries that compete for contracts in catering, transportation, security, and construction work among other things. They pay dividends to their shareholders every year, and also provide scholarships, training programs, and local employment. Through skilful negotiations with the oil companies who needed their land for development, they managed to get deals where they are treated essentially as subsurface owners, meaning that they get royalties from production. This is seen by some as an example of how a small community can stand up to the industry and get advantages from industrial activities.

This success story is again countered by the social problems that Nuiqsut is facing. Unemployment, substance abuse, mental health complaints and the lack of educational opportunities are widespread. Cultural disruption is also an important issue:

> The next thing is a conflict between American culture and Native values. You have satellite TV, you know, you want modern appliances, everything, you know. You want to have a faster snow machine, a bigger boat, a bigger jet boat to get up the river further to hunt. Well, it takes gasoline and it takes money. You've created, what was a subsistence economy, and converted into subsistence and cash economy, both. But without a driving engine to provide cash. The next thing is, you've changed the cultural values of the villages to some extent. It used to be the hunters were, you know, you quit your job and you went off and you went hunting.

Both representatives from the Native population and non-Native employees of the NSB state that the 'communities need to take control of their own lives, they need empowerment.' They need to look at the oil and gas activities as opportunities. These statements are signs of another conflict that Nuiqsut is facing, a conflict between a village that to an extent sees itself as a victim of oil and gas, and others on the North Slope who see them as people not valuing the opportunities they have been given.

Is the cash economy sustainable?

Compared to the issue of balancing subsistence issues against industrial activities, there is little discussion of whether the cash economy is sustainable. The living standard on the North Slope rose dramatically as a result of the oil and gas activities. The level of costs of living rose to a formerly unknown level, and the changes in the economy, livelihood and institutional framework during the last 30–40 years are comprehensive and irreversible.

However, the North Slope communities are not alone in this: the state economy as a whole depends on oil revenues and federal transfers. An environmentalist from the 'Lower 48' States commented on this and claimed that the majority of Alaskans have to get away from being spoiled, reliant on the federal government, the industry and the exemption from income tax.

Can the economy be diversified? Increased access to remote areas in the Arctic could under some circumstances mean increased opportunities for entrepreneurship or other local businesses like tourism, commercial fishing or guided hunting. It can be argued that the North Slope environment is too vulnerable for allowing those kinds of activities. Access can increase the possibilities for local abuse, and create access for tourism to an extent that can be harmful. One informant compared the heavily regulated access for oil companies and access on a general basis, which could mean destroying the carrying capacity and the remoteness of the area.

Fate control

The Arctic Human Development Report frames the issue of fate control as a central concern of Arctic people. This theme was prominent in our interviews in Alaska. Stakeholders on all sides of the oil and gas development seek a compatible mix of subsistence and development, but local stakeholders question how much control they have over the mix. A regional official said that, if development is going to occur, local stakeholders need to do better at taking advantage of the opportunities so that they can continue to engage in subsistence activity. He thought that there is room for both as long as sustainable development has done minimal damage to the patterns of the animals or the land.

How this desire to have development yet maintain some local control over how the impacts play out on the ground is illustrated by three stories: the NPR-A Subsistence Advisory Panel; habitat protection around Teshekpuk Lake in the NPR-A; and conflict avoidance agreements in the Beaufort Sea.

NPRA Subsistence Advisory Panel

During the preparation of an EIS for oil and gas leasing in the Northeast NPR-A planning area in 1998, the Bureau of Land Management (BLM) convened subsistence workshops in the regional centre of Barrow, and in Nuiqsut, the village most directly affected by the proposed leasing program. Local residents expressed frustration with ongoing impacts of oil and gas development on subsistence and on the lack of channels for local interests to be heard. For two decades, public meetings held by the Minerals Management Service concerning offshore oil and gas leasing resulted in widespread public testimony that subsistence resources were being affected, subsistence harvest practices were impacted and local concerns were ignored. Following a process of facilitated public dialog, BLM Northern Field Office staff developed detailed recommendations for a Subsistence Advisory Panel (SAP). The functions of the SAP include reviewing lease plans and providing recommendations on safeguards and mitigation measures, keeping local communities informed of leasing and exploration activities, recommending planning, research, monitoring, and assessment activities to protect subsistence, working with agencies to maintain a repository of subsistence information, and serving as a clearing house for subsistence research in NPR-A. Since its establishment in 1999, many regular meetings have been held, and members express satisfaction with improvements in information exchange, and communication between the communities, agencies and industry. No significant disputes have arisen (McIntosh and Brelsford, 2006). Several informants supported this evaluation but the locals are still not satisfied with the situation for subsistence activities.

An important factor in this story is the actions of federal agencies. In the late 1990s, under the Clinton administration, the Secretary of Interior Bruce Babbitt wanted to preserve ANWR by redirecting attention for oil and gas development to the NPR- A. This resulted in a new leasing program for the central portion of the North Slope in 1998. Informants assess consultation with the communities on subsistence hunting issues to be a distinct feature of the environmental review under Secretary Babbitt. It was recognized as innovative and very much beyond the norm in federal land management at the time. It is seen as the high point of consultation and a very serious effort to avoid impacts on subsistence users. However, in 2003 and 2005 under the new administration, Secretary Norton decided to go back and revisit that leasing program and to change some of the measures that were intended to protect environmentally sensitive areas and important subsistence areas.

Teshekpuk Lake

Teshekpuk Lake is an important wildlife habitat area, including calving grounds for the Central Arctic caribou herd and wetlands for migratory waterfowl. The geography of the area is shown in Map 7.2. The Borough land use management plan designated Teshekpuk Lake as a protected area owing to its high value as wildlife habitat and its importance for subsistence for people from Nuiqsut, Atqasuk, and Barrow. The BLM plan for NPR-A initially followed suit and classified Teshekpuk Lake as a special protection area where no leasing was allowed. A decade later however, the new federal administration, over the unanimous protests of people in the region, cut the protected area in half and opened the rest to development under performance-based environmental standards. The lease sale was temporarily blocked, however, by a federal judge who, in response to a lawsuit brought by a coalition of environmental organizations, ruled that BLM had not considered the cumulative effects of development.

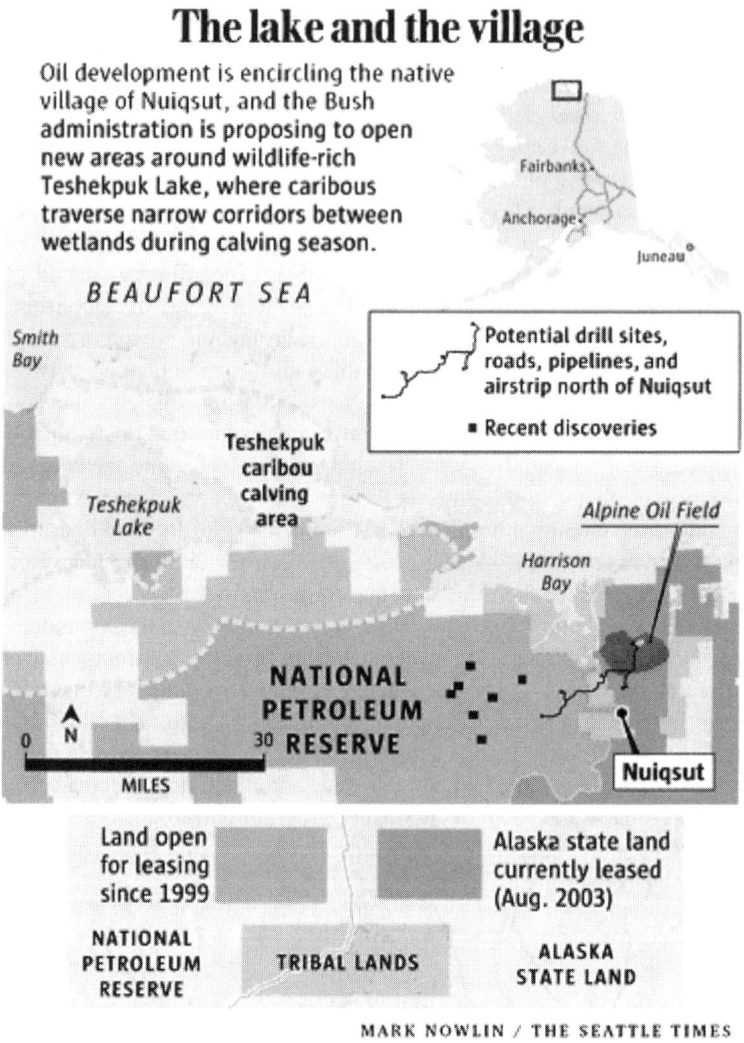
The lake and the village
Oil development is encircling the native village of Nuiqsut, and the Bush administration is proposing to open new areas around wildlife-rich Teshekpuk Lake, where caribous traverse narrow corridors between wetlands during calving season.
Fairbanks
Anchorage
Juneau
BEAUFORT SEA
Smith Bay
Potential drill sites, roads, pipelines, and airstrip north of Nuiqsut
Recent discoveries
Teshekpuk caribou calving area
Teshekpuk Lake
Alpine Oil Field
Harrison Bay
NATIONAL PETROLEUM RESERVE
N
0
30
MILES
Nuiqsut
Land open for leasing since 1999
Alaska state land currently leased (Aug. 2003)
NATIONAL PETROLEUM RESERVE
TRIBAL LANDS
ALASKA STATE LAND
MARK NOWLIN / THE SEATTLE TIMES

Map 7.2 Teshekput Lake.

The communities felt that the federal government was going back on its word: the areas that had been excluded were suddenly up for renegotiation. This was a matter of great frustration and concern in the North Slope communities. The NSB, for instance, had not been fully satisfied with the 1998 plan, and were strongly opposed to the inclusion of the areas they saw as environmentally sensitive and central to subsistence. The informants' conclusion on this is that the legislative protections are fairly weak. Section 8–10 in the Alaska National Interest Lands Conservation Act does ensure public notice and analysis of impacts, but it does not go further to guarantee protection. The protection of these subsistence uses really comes from the political mobilization by the local communities, if and when they develop their own scientific and technical arguments.

The NSB is recognized as a capable actor in doing this. According to an NSB employee, the battle for what alternatives should be considered is an important part of the planning and permitting process. The fate of the McCovey-project and the conflict over the Nuiqsut bridge have already been described. Another example is the collaboration with the AEWC in opposing the whaling ban.

Conflict avoidance in the Beaufort Sea

The village of Kaktovik, as all the other villages on the North Slope, is strongly opposed to exploration in the Beaufort Sea. During the summer of 2006, eight companies shot seismic in the Chukchi Sea and the Beaufort Sea. This is in federal waters; Native people and the NSB do not have authority to stop the activity. However, the environmental and subsistence protection system does give the Native people a position to negotiate. The executive director of the AEWC says: 'we have to take advantage of the situation and – because you cannot stop the offshore activity we have to use the situation and be specific on the demands to the industry.' The tools used are Conflict Avoidance Agreements (CAA) and Good Neighbor Policies (GNP). Representatives from the NSB also argue that they will use their position to negotiate:

> Offshore the development is going to happen, there is no moratorium on the ocean here, and we don't want to fight it in court where we would probably lose anyway, so we try and influence it before it happens – make development that is considerate of our concerns.

According to an NSB attorney, a CAA is not mandatory (although the AEWC claims that the federal government *requires* it), but is encouraged by the governmental agencies in charge of the permitting process.

Companies like Shell, ConocoPhillips, and GX Technology Corporation are active and are trying to get the permit to shoot seismic. To do this they are negotiating a CAA with the villages. An academic comments on this situation saying: 'The CAAs are an example of the stakeholders themselves negotiating with industry, and the agreements are good.'

A CAA is an agreement where whaling and oil activities are scheduled in such a way that they don't interfere with each other. When there are several operators in the water at the same time, these agreements become very complicated. In 2006, six different companies signed a programmatic agreement. According to the AEWC, factors included in the agreement are charity contributions to the AEWC for the establishment of communication centers with 24 hours on duty dispatchers in the villages, funding for a search and rescue services, and VHF radios for the communities, whaling boats, and seismic ships. A contentious issue is how extensive the marine mammal monitoring done by the operating companies should be. Every ship should have an observer on the deck so that activities can cease if mammals are seen. Shell has agreed to do monitoring from the air as well. The AEWC wants hydrophones to be included as part of the monitoring.

By April 2006 the AEWC had reached a CAA with the whalers in Kaktovik, Nuiqsut, Barrow, Wainwright and Point Hope. By May, however, Kaktovik was feuding with Shell over its plans for offshore seismic exploration. Kaktovik's City Council passed a resolution calling Shell a 'hostile and dangerous force', and authorizing the mayor to take legal or other actions necessary to defend the community. In a news release issued with the resolution, Mayor Lon Sonsalla said Shell had failed to address village concerns about how it would keep seismic testing scheduled for this summer from disturbing migratory bowhead whales and how the company would operate safely in unpredictable sea ice. 'Instead of technical staff or people with authority, they send public affairs. Instead of true consultation with community leaders on substantive issues, they try to schedule superficial social gatherings where they can talk at us' (*Anchorage Daily News*, May 24, 2006). Kaktovik's May 9 resolution said Shell 'has ignored our advice' and called on all North Slope communities to oppose Shell's work on its offshore leases until it establishes a respectful relationship with the indigenous people. One informant without specific ties to Kaktovik commented on how Shell had contacted the village for consultation very late in the process, and that this might be an important background for the skepticism.

Balance of power

A representative from the Alaska Native Science Commission (ANSC) says:

> ANCSA came into being because the US wanted access to the oil on the North Slope. The good thing is that it brought us, the Native community, two things – land and money. And if you have land and if you have money, you have power.

Locals question whether the parties are on an equal footing in negotiations. An NSB bureaucrat says defensively: 'People are going through a transition, and oil holds them hostage all the time.' A more offensive attitude toward the industry is held by a village corporation manager: 'The threats that the companies will leave the state when the Native companies negotiate on finding environmental and society friendly solutions – when these are not the cheapest ones, get people frustrated.' Philanthropy can also be used as leverage, such as in Nuiqsut: ConocoPhillips has not given money to their annual festival this year, and people in Nuiqsut are convinced it is because they are in conflict over the bridge issue. A Native informant said that the people on the North Slope tend to be non-confrontational. 'It's a cultural trait. So, when people push, [we] have a habit of giving. But then we push too.'

There are strong and emotional responses to the questions on environmental effects of the oil and gas industry. NSB bureaucrats say that the environmental community uses the legal system to force the industry to follow the environmental laws. If the industry fails to meet the requirements in environmental laws, however, the laws are changed! They hold the opinion that it partly comes down to economics – industry claims it's too expensive to follow environmental requirements. The Nuiqsut bridge was used by informants as an example of this.

Some informants do not see that the Inupiat can have a more central role in the oil and gas activities: 'Some of the Natives will not appreciate the accommodations they will have to make. Since they do not participate in the cash economy by paid work, they will not see the benefits.' A consultant to the NSB did not see this as a 'system failure' but as a private individual choice: 'This non-participation is a personal lifestyle choice.'

All together, it might be hard to convince local stakeholders that they have much of a choice. A mayor says: 'The oil and gas activity is closing us in – Alpine is becoming a spaghetti and it is barricading us. It is limiting our access to hunting ground.'

Arctic National Wildlife Refuge

To open ANWR or not?

The ANWR debate has a unique constellation of vocal stakeholders. The question of opening the coastal plain of ANWR to oil and gas activities has generated national and state-wide debate for decades. The Alaska Coalition is a major player on the conservation side of the debate. Arctic Power, an NGO funded with a mix of private and state money, voices the position to open ANWR. They get their main support from the Republicans, and they report that their 'case is dead if the Congress is Democratic.' A consultant for Arctic Power differentiates between offshore and onshore activities for the possibility for having support of the Native people. 'If you're talking about ANWR ... you get a lot of support.'

The debate on opening ANWR does not only contain arguments on impact on subsistence living. One of the main arguments is that oil from ANWR will increase the energy supply, supply security and price stability for the United States. On the other side, the argument that the oil will go to the Far East, to Japan and China, has been used against developing it. A consultant for Arctic Power says:

> There have been a lot of false arguments on both sides in this debate. In my view this effect on gas prices is also false. What it would do, it might change the prices of gasoline in pennies, no more. But, what it would do, I think, would be to get rid of the real spikiness.

Another informant also talked about distorted facts:

> ANWR is used by the environmental groups. They push the ANWR button whenever they want to raise money. I personally think that they have

distorted the facts; they've even distorted the pictures. I mean you see pictures ... of this huge Brooks Range mountain range, ... taken with a telephoto lens [so] the coastal plain is foreshortened. Remember that once you get to the mountains you're not on the coastal plain anymore. That area is protected and it will remain protected. The area we're talking about is the coastal area.

There are also different visions of development. One informant says: 'If you went out on the street here in Anchorage or anywhere in United States of America, and asked people about this, and they'll say, "oh well the whole area is going to be developed." It's nonsense. If you develop it to the same degree as the North Slope, it is much less than 1% [of the surface area].'

There is uncertainty on how much oil there is in ANWR. The US Geological Survey (USGS) estimates that there is an excellent chance (95 per cent) that at least 11.6 billion barrels of oil are present on federal lands in the 1002 area, and a small chance (5 per cent) that 31.5 billion barrels or more are present. But the proportion that would be technically and *economically* recoverable depends upon the state of technology and the price of oil. The USGS estimated that, with 1996 technology and prices (US$24/barrel in 1996 dollars), there was a 95 per cent chance that 2.0 billion barrels or more could be economically recovered, and a 5 per cent chance of 9.4 billion barrels or more. Roughly one-third more oil may be under adjacent state waters and Native lands (US Department of the Interior, 1999).

Another argument for the opening of ANWR has been that production from Prudhoe Bay is declining: from its peak of 2 million barrel a day, it is now down to 700,000 barrels a day. The pipeline from Prudhoe to Valdez, the Trans Alaska Pipeline System still has plenty of capacity left to take ANWR oil. Some of the supporters of development argue that this 'would give people the confidence to come here and continue to explore because they know that their delivery system is going to be viable, and [it] will continue into the future.'

Consultants and oil and gas representatives argue that the industry is much less interested in ANWR and its potential than they are in National Petroleum Reserve and the offshore areas. Perhaps the most significant evidence of this is in the lobbying effort in front of the federal Congress. The state of Alaska has provided millions of dollars per year for a lobbying effort, but in the last two or three years, the oil industry has not contributed to this lobbying effort. Actors such as Arctic Power are not happy about this, and say that it is probably because the companies know that it would be seen as very partial. Several state-connected informants also argue the Alaskans are in desperate need of additional development, ANWR being one of them.

Kaktovik, situated on the coast of the Beaufort Sea in the eastern ANWR, is the village closest to the prospective oil development. Most of the people in Kaktovik support the opening of ANWR, although a substantial minority has recently spoken out to oppose it.[5] The village corporation owns surface rights to 92,000 acres in the refuge, and the ASRC owns subsurface rights as well.

Representatives from the village, and from Barrow, lobby for the opening of the refuge. Since the Inupiat own some of the land, the former Mayor of the NSB, George Ahmaogak, also strongly supports this cause. His argument is that, unlike in NPR-A where the federal jurisdiction supersedes the Borough, the local people would have more control over how development happens on their own land.[6]

The Gwich'in are Athabaskan Indians, and live in both Canada and Alaska. In Alaska, they mostly live in Arctic Village and Venetie, south of ANWR. They call themselves the people of the caribou. Their major subsistence resource is the Porcupine caribou herd, which migrates thousands of kilometers each year from its winter habitat south of the Brooks Range to its summer calving area on the coastal plain of ANWR. The Gwich'in oppose the opening of ANWR because they fear it will harm the caribou herd. According to one oil and gas consultant, however, the Gwich'in have lost some legitimacy as bearers of 'the Native voice' in this case since the other Native tribes support the opening. However, one informant pointed to the fact that, in 1984, the same people who now represent the Gwich'in in the battle against the opening of ANWR, leased their own land in the area to a company for exploration drilling. The operators found no potential for development on these lands. The informant accused the Gwich'in of resisting the opening of ANWR just because they themselves will not benefit from it.

Many stakeholders complain about the flux in environmental and developmental policies. ANWR is often used as the example. The coastal plain, the §1002 area, requires Congressional action to be opened for development. It actually happened in 1995, but President Clinton vetoed it. At this writing in 2008, the out-going Republican president and key Republican leaders in Congress are still pushing to open the ANWR to oil and gas development, but the Democratic-controlled Congress is against it and the two leading Democratic candidates for president have also gone on record opposing it.

Relations with environmental organizations

In the Native fight for protection, it might have been expected that environmental organizations could have been allies. However, the most important argument for their position is the value of lands as habitat for wildlife, for people to know that there are wild places left like they were before Western humanity arrived. In spite of the fact that many environmentalists do not object to subsistence hunting, they are perceived by many Native people as so extreme on conservation that they do not see them as partners.

A representative for a local corporation said:

> We try to stay away from the environmental groups. We think that, if you look at the ANWR model, what has happened with the Gwich'in, is that they fundamentally bought in and were utilized by the environmental lobby. We don't want to be utilized by anyone's lobby other than our own. But to that extent, it creates more work.

A Native woman said:

> I think the environmental groups have a difficult relationship with Native communities because we see things in a very different light. For environmental groups it's mostly about protecting at all costs, and for the Native community it really is having respect for the environment. That means using what the creator has given us to use in a respectful way. We see ourselves as the first stewards of the land and of the environment. It's been our job to do that for centuries. Whaling is for instance a matter of cultural survival. And we've had many, many disagreements with environmental groups who would rather see a people die than a whale die.

The Alaska Coalition holds the position that everyone should have a say as to what happens to the land and is opposed to the expansion of oil and gas activities in the Arctic. However, their representative adds that it is not a simple question. Some places are more suitable for development than others and some places are more important to protect than others. The answer has to be given case by case. They argue that the United States has an unsustainable energy policy, it revolves around oil and gas only, and this needs to change. The coalition looks at the local people as legitimate decision makers, but that this means that all local people in a democratic sense should have a say, not just the leaders. The next tier for them is that the lands are a part of the greater United States, they are connected environmentally, and they are a part of our common, national heritage. The conflicts that are built into this position are seen in the debate about opening ANWR.

Conclusion

As we have seen over and over again, the primary controversy over expanding oil and gas development in Alaska concerns sustainability of the subsistence economy and culture. While there is broad agreement that a mix of development and subsistence is desirable, there is little agreement on the nature of the trade-offs involved, what the best mix might be, or how to achieve it. Can science and technology resolve these disputes? Must some areas be off limits, or do environmental performance standards and conflict avoidance agreements provide adequate protection for subsistence values? If zero risk is impossible, how much risk is acceptable? How much cost can the industry afford to pay for a design that minimizes risks for local stakeholders? Can a process of consultation avoid animosity and mistrust, clarify the core issues and build a respectful cooperative approach to problem-solving? We have no answer to these questions.

Owing to the strong position the indigenous people hold on subsistence activities, we might expect them to be allied with the environmentalists. There is agreement between indigenous people and environmentalists on issues like oil spill, air pollution from outside, but on others they still disagree. The Natives' primary goal is to utilize the natural resources to survive and make a living, while the environmentalists hold a pro-conservation position. Conservation in this sense sometimes means untouched, which for non-Alaska environmental groups may mean prohibiting subsistence hunting. This position is more visible as oil and gas, climate change and other human activities displace animals from traditional hunting areas and force local people to use larger motorized vehicles to travel further to find game. The environmentalists' idea of the pristine wilderness is opposed to the Natives' view of Nature as the resources you use to survive. This may also explain why most Natives support the onshore oil and gas activities that are not defined as a threat to their living. The supporters of both onshore and offshore development accept that they must be as careful as possible to not interfere with the subsistence activities of the locals, but they use arguments of a secure national oil and gas supply as the main argument for having to use all the resources in the area. No stakeholder group objects to the right of the Natives to sustain their culture and traditional lifestyle, but the long-term effect of the oil and gas activities on subsistence and the realism in this position is not openly debated. The dependency on oil and gas revenues for modern technology, welfare and living standards blur the clean-cut positions of those who are in favor of and those who are against further oil and gas development in the North Slope of Alaska.

The most persistent lines of conflict arise from structural differences in stakeholder interests. Locals will always prefer technologies that entail lower risks to the environment, while companies will always prefer technologies that are cost lower. Whalers will always be less tolerant of risk to the marine environment (approaching zero tolerance) than companies who find the current EIS risk projection (10–11 per cent risk of a spill of 1000 gallons or more over the life of the field) acceptable. Federal and state decision-makers will always give more weight to the interests of national and state-wide constituents than will local and regional officials who must advocate local interests. Inupiats who retain their cultural values will always adopt a longer-term time horizon, past and future, than company officials that face short-term shareholder demands and high discount rates. Arctic residents will always care more about local control than their counterparts. The art of negotiation and conflict resolution entails finding enough interests in common to forge a politically acceptable compromise among these competing interests.

Other sources of conflict are avoidable. For example, our key informant interviews turned up some disparaging attitudes that would, if openly expressed, only serve to balkanize stakeholder groups and preclude constructive dialog. For example, the stakeholder who dismissed cumulative effects as 'nonsense' and a 'political ploy' because he or she couldn't imagine how covering 'one per cent' of the surface area with infrastructure could possibly affect wildlife was displaying closed-minded ignorance and animosity toward other stakeholders. Wildlife biologists have no problem explaining how linear structures might fragment habitat, or how industrial noise or low-flying air craft might disturb wildlife and affect populations. Economists have no problem explaining how industrial infrastructure for one project will lower the development costs for adjacent projects, so the spatial impacts will tend to expand.

Notable in the discourse is what is not discussed. While there is a great deal of attention to the sustainability of subsistence activities as the foundation for Inupiat identity, cultural values, social well-being, and the economic base for communities, there is less attention paid to the sustainability of the cash side of the mixed economy. NSB personnel are acutely aware of the reality that Prudhoe Bay production is declining, NSB tax revenues are declining, and the North Slope economy is already shrinking in terms of population and employment. Perhaps this bedrock reality is so compelling that the need for additional development is a given, such that mitigating adverse impacts on subsistence becomes the policy variable and subject for debate. Or perhaps, as expressed over and over, local people see the development as unstoppable, and that their only hope is to slow it down, minimize the adverse impacts and maximize the local benefits. Still, there was remarkably little discourse regarding avenues to maximize local benefits and economic sustainability through increasing revenues, employment, business income or diversifying the economy for the near and long term. A few stakeholders that we did not manage to interview do take the position that tourism would be a preferable economic activity for Kaktovik than oil and gas development in ANWR.

Other notably absent issues include climate change and pollutants. While climate change is of major concern in Alaska, it is not linked in local discourse to the question of expanding oil development. Inupiats are active stewards of the environment and all forms of pollution are of concern to them, but not necessarily on their list of urgent priorities to warrant much discussion in our interviews. Contaminants in the food chain are less of a problem in Alaska than in other regions of the Arctic. The primary source of these contaminants is long-distance atmospheric transport from Europe and northern Russia, or old, abandoned military sites, not the local oil and gas industry. The prospect of an oil spill offshore is a significant concern for potential contamination of marine life – the lingering effects of the Exxon Valdez oil spill in the south of Alaska are an omnipresent reminder of these risks. But development to date has been mostly onshore, and the risk of contamination from onshore spills was, at the time of our interviews, perceived as small. Spills of drilling muds, waste water, hazardous materials and even oil pan drips from vehicles are closely monitored and, while a perennial concern, are not a major source of contention. There is some air pollution from industrial activities and flaring of gas but, except for Alpine, these are remote from human settlements and not a major concern at this time.

However, attitudes are changing in the wake of a major pipeline leak at Prudhoe Bay in March 2006. With 4790 barrels of oil covering two acres of tundra, this was the largest spill ever on the North Slope: five times larger than then next largest spill in 1989. The pipeline had leaked for at least five days before the snow-covered spill was discovered, and clean-up was hampered by bitterly cold temperatures with wind-chill approaching 50° below zero [Celsius]. Investigation revealed that inadequate monitoring and maintenance of the pipeline failed to detect and correct unexpected levels of internal corrosion, and that many miles of pipe in the older sections of the field were similarly vulnerable. Indeed, a second, smaller leak followed in August. Partial shut-down of the field cost the state treasury millions of dollars per day in lost production taxes and royalties. The incidents resulted in criminal and civil investigations by federal and state environmental regulators, federal prosecutors, pipeline regulators, and members of Congress.

While the oil spill surprised and shocked some people, it did not dampen the enthusiasm and support by others for expanding the oil and gas activities. The incident, however, is a reminder of risk and unpredictability that is inherent in oil and gas activities, and incidents like this may also change the attitudes of important stakeholders in the oil and gas industry. In the years to come, Alaskan Northern villagers may face the problem of increased burdens from oil and gas where they want it least – offshore. To reconcile the traditional ways of life will not be less challenging. The question is what alternative development paths there are for these settlements. This is the paradox of the situation of sustainable development in this area.

Notes

1 In this chapter, we use the terminology preferred in Alaska, where Native with a capital 'N' refers to Alaska Native, which includes seven major indigenous culture groups: Inupiat, Yupik, Aleut, Athabaskan, Tlingit, Haida and Eyak. In the context of Alaska's North Slope, Native and Inupiat are used interchangeably.
2 Bret Luick, March 2007 Food Cost Survey, University of Alaska Cooperative Extension Service.
3 Our informants for this chapter will not be identified since, in conjunction with their consent to be interviewed, we pledged anonymity. See the Alaska methodology in Chapter 3.
4 The North Slope Borough Comprehensive Plan, 2005. (Available online at www.co.north-slope.ak.us)
5 http://www.anwr.org/features/kaktovik.htm
6 Uiniq Magazine, Fall 2005, published by the North Slope Borough.

References

Alaska Department of Revenue (2006) *Spring 2006 Revenue Sources Book*. Online. Available HTTP: http://www.tax.state.ak.us/sourcesbook/2006/spr2006/oil.pdf (Accessed 19 November 2007).
Cole, T. and Cravez, P. (2004) *Blinded by Riches: The Prudhoe Bay effect*. UA Research Summary, no. 3.
Griffith, B. and Cameron, R. D. (1998) *Shifts in the Distribution of Calving Caribou: Developing a Model for Assessing the Impacts of Development*, Poster presented at (1) Annual Conference, Alaska Chapter of the Wildlife Society, Girdwood, Alaska, January 16, 1998; (2) 8th North American Caribou Workshop, YT, April 21–25, 1998.
McIntosh, S. and Brelsford, T. (2006) 'Raising the Profile Of Subsistence Uses': Section 810 Analyses And Subsistence Protections, in *BLM Alaska's North Slope Energy Initiatives*, presented at Invited Session: Subsistence Research, Resource Management, and Public Policy in Alaska, 66th Annual Meeting of the Society for Applied Anthropology, Vancouver B.C., March 31, 2006.

Minerals Management Service, US Department of the Interior (2003) *Beaufort Sea Planning Area Oil and Gas Lease Sales 186, 195, and 202*, Final Environmental Impact Statement MMS 2003-001. Online. Available HTTP: http://www.mms.gov/alaska/ref/ EIS%20EA/BeaufortFEIS/beaufortfeis.pdf (Accessed 19 November 2007).

National Research Council (2003) *Cumulative Environmental Effects of Oil and Gas Activities on Alaska's North Slope*, Washington, DC: National Academies Press. Online. Available HTTP: http://www.nap.edu/books/0309087376/html/

North Slope Borough (NSB; 1999) *1998 Economic Profile and Census Report*. Barrow, Alaska.

US Bureau of the Census (2000) *Population and Housing Characteristics*. Washington, DC: GPO. Online. Available HTTP: http://www.census.gov/main/www/cen2000.html.

US Department of the Interior, Geological Survey (1999) *The Oil and Gas Resource Potential of the Arctic National Wildlife Refuge 1002 Area, Alaska (1999)*, 2 CD set. USGS Open File Report 98-34.

8 Oil and gas activities at the Mackenzie Delta, in Canada's Northwest Territories

Aldene Meis Mason, Robert Anderson and Leo-Paul Dana

Introduction

Events in Canada have paralleled those internationally. Indigenous people in Canada have been leaders in the struggle for recognition of traditional lands and resources and access to their socio-economic benefits. The Canadian Government has been one of the more progressive in recognizing these rights. In addition, the governments in Canada have built sustainable development requirements into all planning, policies and programmes since the Bruntland principles were introduced.

This chapter focuses on the building of the Mackenzie Valley Pipeline running from the natural gas fields in the Mackenzie Delta and Prudhoe Bay in Alaska to markets in Southern Canada and the United States. This has been one of the grand industrial projects of the twentieth century. In 1974, a consortium of multinational oil companies (called Arctic Gas) made application to the Canadian government to build the pipeline. At the time, most believed the application would be approved. However, a moratorium was declared. After 30 years of events and interactions, the parties are again engaged in a regulatory review process to determine whether the current Mackenzie Gas Pipeline Project (MGP) should go ahead. Since the regulatory review process began in 2002, the MGP projected costs have doubled from CAD$6 billion to CAD$17 billion.

The chapter begins with an overview of Canada today with regard to the current oil and gas industry in the Arctic. It then discusses the status of indigenous populations located in the North and their indigenous rights. The Mackenzie Pipeline has been planned to cross the traditional lands of four distinct groups of people and the impact will not be the same on all groups. The third section presents the discourse. It describes the history of the Mackenzie Valley since the Berger inquiry in the 1970s to the present. Three sub-stories are identified: (1) land claims and other rights recognition; (2) socio-economic development and (3) environmental protection. It then examines these main story lines including international, corporate, state and civil sectors. Finally, the chapter concludes by using the framework to summarize the discourse among the parties. It then uses the key principles and concepts from corporate social responsibility (CSR) and sustainable development to draw conclusions about the evolving modes of social regulations (MSRs), resulting modes of development and the degree to which these are sustainable. In the current arena, we see that sustainable development concepts have become part of the emerging MSRs because the parties (communities, state and civil) demand it. In addition, strategic CSR has taken on more depth as a necessity for corporate success in this new flexible regime of accumulation. This deeper strategic CSR is shaped by and includes sustainable development principles and practices.

Canada today

This section will first present a brief overview of the oil and gas industry in Canada, especially in the Arctic. Attention then shifts to indigenous matters, especially the demographics of the Arctic, its population and socio-economic circumstances. It then provides a brief look at indigenous rights.

Oil and gas activity in the Northwest territories

The Northwest Territories (NWT) covers approximately 1,171,918 km^2. Oral traditions confirm Aboriginal peoples of the Arctic used oil and gas seepages for fire and light and to caulk their canoes before the arrival of Europeans. The explorer Alexander Mackenzie noted these in the eighteenth century. Although non-Aboriginal people claim they were the first to find oil in Norman Wells in 1919, the Sahtu Dene say that Francis Nineeye was the first to find oil. Taking a sample in a lard pail to Tulita, he gave it to Gene Gaudet, the Hudson's Bay Manager. The sample was sent out on the boat. The Sahtu said they never heard of that oil again and there is no record. The Sahtu Dene also guided the geologists exploring the region around Fort Norman to Legohli (where the oil is; Auld and Kershaw, 2005: 60).

Oil and gas activity in the NWT to date has primarily centred on the Beaufort Sea in the Western Arctic. The Beaufort–Mackenzie Basin contains large volumes of discovered oil and natural gas resources, and has high potential for future discoveries. Petroleum exploration for oil and gas began in the mid-1960s, with the bulk of exploration drilling activity occurring in the 1970s and 1980s. 'A total of 183 exploration wells (and 66 development wells) have been drilled in the region, resulting in the discovery of 53 oil and/or gas fields' (Government of the Northwest Territories, 2001: 1), quoting a National Energy Board report from 1998 in the Marine EcoSystems Overview of the Beaufort Sea Large Ocean Management Area by North/South Consultants and Inuvialuit Cultural Resource Centre (2005: 9). According to the World Wild Life Federation – Canada (WWF), 1 million hectares of oil and gas leases are located in the Beaufort Sea (WWF, 2005b).

However, NWT's energy resources are largely untapped:

> Discovered reserves are estimated at 2.8 billion barrels with an undiscovered recoverable resource of up to 10 billion barrels; the estimate for discovered natural gas reserves is almost 11 trillion cubic feet (Tcf), with an ultimate recoverable resource of at least 60 Tcf.
>
> (NWT, 2006: 59)

The Canadian Potential Gas Committee's (2001) *Report on the Gas Potential in Canada* suggests that these are too low – marketable reserves are more likely in excess of 60 Tcf in the NWT: 5 Tcf in the Mackenzie Valley, 30 Tcf in the Beaufort Basin and 26 Tcf in the High Arctic.

The NWT is positioned to play a significant role in the North American market. Canada's total conventional oil and natural gas proven reserves are estimated at 4.3 billion barrels and 58 Tcf, respectively, at the end of 2004. To date, the most significant development has been the Norman Wells oil field. This field is Imperial Oil's single largest producing source of conventional crude oil for the North American market. Natural gas production is currently on a scale of about 70 million cubic feet per day (NWT, 2006: 59).

Although vast reserves are here, oil and gas development in the NWT is still in the early stages. The challenge is to get the product to market. If the pipeline is approved, a boom will occur. The preferred solution has been a CAD$17 billion 12,220-km pipeline up the Mackenzie River Valley, a route that crosses the traditional land of four indigenous groups: the Inuvialuit, the Sahtu Dene, the Gwich'in and the Deh Cho. The anticipated value of the oil reserves from the Mackenzie Valley Gas Project hovers around CAD$3 trillion.

Overview of NWT socio-economic indicators

The NWT total population in 2005 was 42,982 (males 22,093 and females 20,889). The Aboriginal and non-Aboriginal population were almost balanced – 21,413 versus 21,569. (In Canada, Aboriginal people include the Inuit, Métis and First Nations. 'Indigenous' is not a term which is commonly used.) The average annual population growth rate was 0.3 per cent. Households with more than six people were 7 per cent in the NWT, and 12.7 per cent in smaller NWT communities, compared to 3.1 per cent of Canada. About 79 per cent of families in 2001 were couple families and 21 per cent were lone parent families. This rises to 28.3 per cent in smaller NWT communities and compares with 15.7 per cent for the rest of Canada (NWT Bureau of Statistics, 2006).

About 44 per cent of Aboriginals 15 years and older speak an Aboriginal language; in smaller communities this rises to 61.9 per cent. Although 67.5 per cent of people have a high school diploma or further education, in smaller communities only 36.8 per cent do. Fifty per cent have graduated from high schools as compared to 75.9 per cent for Canada. Labour force participation is about 75.6 per cent and the unemployment rate is 10.4 per cent; 69.7 per cent of males and 65.7 per cent of females are employed. Of those employed, 85.9 per cent worked full time and 11.6 per cent part time, and 75.6 per cent worked more than 26 weeks. Average family income is CAD$91,362 while 16.2 per cent of families earned less than CAD$25,000 and 61.1 per cent earned more than CAD$60,000. The cost of living index compared to Edmonton, Alberta, ranges from 147.5 per cent in Inuvik to 162.5 per cent in Tuktoyaktuk. The 2004 Food Price Index compared to the Yellowknife NWT is 140.5 per cent and 206.4 per cent in Tuktoyaktuk.

Violent crime has been increasing steadily from 49 (per 1000 persons) in 2000 to 68.7 in 2004 (compared with 9 for Canada). Property crime has also increased from 59.1 per cent (per 1000) in 2000 to 74.1 per cent in 2004. Heavy drinking (the percentage of persons age 12 and over drinking five or more drinks per occasion more than once per month) was 40.4 per cent for the NWT compared to 21.4 per cent for Canada. In 2003, 36.7 per cent were engaged in hunting and fishing; 5.9 per cent in trapping. The percentage of households eating harvested meat and fish was 17.5 (32.7 per cent in the Beaufort Delta; 17.7 per cent in Inuvik and 49.5 per cent in Tutoyaktuk). In smaller communities the consumption of country foods is 43.7 per cent. The proportion of households in unsuitable, unaffordable or inadequate housing was 6.3 per cent for the NWT, 35.3 per cent for smaller NWT communities, versus 14 per cent for Canada (NWT Bureau of Statistics, 2006).

Indigenous or Aboriginal rights

In the 1973 Calder decision, the Supreme Court of Canada recognized that Aboriginal people have an ownership interest in the lands that they and their ancestors have traditionally occupied. In this landmark decision, the Court held that this right had not been extinguished unless it was specifically and knowingly surrendered. Following this decision, the federal government was forced to rethink its position on Aboriginal title. In doing so, the federal government accepted the legal concept of Aboriginal title as outlined by the Supreme Court. Ottawa also created a negotiating structure to settle land claims of lands under Aboriginal title. These two concepts – Aboriginal title and a negotiating structure – are both complex and interrelated.

The existence of Aboriginal legal rights to lands other than those provided for by treaty or statute is known as Aboriginal title. Until a settlement is reached, these public lands remain under the ownership of the federal and provincial governments. They are legally known as Crown lands. Aboriginal title is rooted in Aboriginal peoples' historic 'occupation, possession and use' of traditional territories. Aboriginal title is obtained after proof of continued occupancy of the lands in question at the time at which the Crown asserted sovereignty. Aboriginal title is held collectively by all members of an Aboriginal nation, and decisions regarding the use of the land and resources are made collectively. At first, Aboriginal title was restricted to the right to hunt, trap and fish within the traditional subsistence economy. Later, the rights expanded to include certain commercial rights.

Discourse section

This section will discuss the chronology of the Mackenzie Gas Project, describe the key Aboriginal groups, identify the key sub-story lines, then take each of these to capture the discourse among the relevant parties – international, state, civil and corporate sectors.

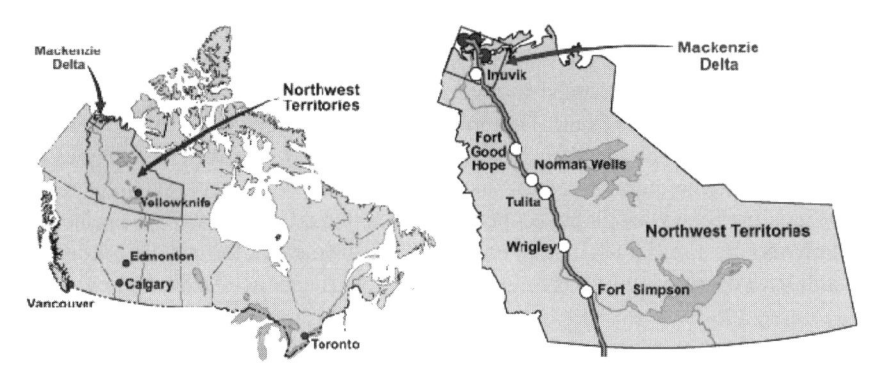

Map 8.1 The Mackenzie Valley Gas Pipeline Project and its inquiry.

In March 1974, Justice Thomas Berger was appointed to head an inquiry that would consider the social, environmental and economic impacts if pipeline construction went ahead on all the people of the Mackenzie valley – Dene, Sahtu, Métis, Inuvialuit and non-Aboriginal. Justice Berger took the hearings directly to 35 communities and the people most affected if the pipeline proceeded. The presentations fell into two camps – those opposed to the project and those favouring the project. Many opposed were local residents who felt that they would bear the social costs of the project and they felt threatened. Environmental organizations from outside the region saw the pipeline as one more example of industry's attack on the environment. Arctic Gas and other proponents of the pipeline argued that industrialization and modernization in Northern Canada were 'inevitable, desirable, and beneficial – the more the better' (Usher, 1993: 105). They did not deny that the process would have negative impacts on traditional Aboriginal society. In their view, development 'required the breakdown and eventual replacements of whatever social forms had existed before' (Usher, 1993: 104). They agreed that the process would be painful for Aboriginal people, but from it would emerge a higher standard of living and a better quality of life. Project proponents held the view that 'all Canadians have an equal interest in the North and its resources' (Usher, 1993: 114). This view was based on the 'colonial' belief that title to all land and resources had passed from Aboriginal people to the Crown.

During the Berger Inquiry, Aboriginal leaders challenged Arctic Gas spokespersons. The Aboriginal argument was that the pipeline project would introduce 'massive development with incalculable and irreversible effects like the settlement of the Prairies' (Usher, 1993: 106). Aboriginal spokespersons saw the project as destroying their culture, and leaving their people with few economic benefits and many social costs.

During this time, the Supreme Court of Canada had rendered the landmark Calder decision. Aboriginal people have an ownership interest in the lands that they and their ancestors have traditionally occupied. This right could not be extinguished unless it was specifically and knowingly surrendered. Following this decision, the federal government was forced to accept the legal concept of Aboriginal title as outlined by the Supreme Court. Thus, during the Berger inquiry, the Dene and other Aboriginal leaders played their trump card: Aboriginal title to the lands across which the pipeline must proceed.

Within the context of global economic development, Usher (one of the principal researchers for the Berger Inquiry) argued persuasively that the pipeline proposal would have disastrous impacts on Aboriginal peoples and their traditional culture.

> This massive assault on the land base of Native northerners threatened their basic economic resources and the way of life that these resources sustained ... when all the riches were taken out from under them by foreign companies, Native land and culture would have been destroyed and people left with nothing.
>
> (Usher, 1993: 106–107)

In the context of Aboriginal society and economy in the 1970s, Judge Berger recognized that the Aboriginal peoples of the Mackenzie Valley were not ready to participate and therefore benefit from the project. In fact, great harm might come to their culture. Berger recommended a ten-year delay. 'Postponement will allow sufficient time for native claims to be settled, and for new programs and new institutions to be established' (Berger, 1977: xxvii).

From 1977 to 1994, a moratorium existed on the issuance of exploration rights for oil and natural gas in the Mackenzie Valley. Exploration licences issued before the moratorium were honoured, resulting in several significant oil and gas discoveries in the NWT in the late 1970s and the 1980s (Brackman, 2001: 7).

Berger's decision ushered in a new era in the relationship between Aboriginal people, the federal government and corporations that wished to develop resources on traditional Aboriginal lands. A key characteristic of this new era has been the emergence of Aboriginal business development based on financial capacity provided by land claim settlements and by the decision of Aboriginal leaders to participate in the market economy. This shift in attitude towards industrial projects later resulted in the formation of the Aboriginal Pipeline Group.

The Inuvialuit

The Inuvialuit people are Inuit, and therefore in Canada they are not referred to as First Nations. They have more in common with their relatives in Alaska than with other peoples of the Canadian Arctic. The Inuvialuit had never signed treaties with the Government of Canada. In May 1977, the Committee of Original Peoples' Entitlement (COPE) submitted a formal comprehensive land claim on behalf of approximately 4500 Inuvialuit living in six communities in and around the mouth of the Mackenzie River on the Arctic Ocean.

Negotiations between the Inuvialuit and the federal government continued through the late 1970s and early 1980s, culminating in the Inuvialuit Final Agreements in May 1984. The goal of the Inuvialuit negotiators was to maintain their traditional way of life and, at the same time, venture into the market economy (Bone, 2003: 193). Under the terms of the *Inuvialuit Final Agreements*, the Inuvialuit retained title to '91,000 square kilometres of land, 13,000 square kilometres with full surface and subsurface title; 78,000 square kilometres excluding oil and gas and specified mineral rights'. The agreement included the communities of Aklavik (*Aklarvik*), Sachs Harbour (*Ikaahuk*), Holman (*Uluksaqtuuq*), Paulatuk (*Paulatuuq*), Tuktoyaktuk (*Tuktuuyaqtuuk*) and for the purpose of the claim, Inuvik (*Inuuvik*). The claim settlement also included an offshore area and the North Slope of the Yukon Territory over to Victoria Island. In the mid-1960s, the Inuvialuit in the Inuvialuit Settlement Region was about 1580, in 1991 it was 2890 (Usher, 2002)

The Sahtu Dene

Coates and Morrison (1986) suggested that the Federal government was not interested historically in pursing treaties with the people of the North unless Indian concerns and priorities required them. With the discovery of oil in Norman Wells in the early 1900s, the Department of Indian Affairs was advised that the oil and gas licences in the area were outside the law. Treaty negotiations that allowed for adhesions to Treaty 8 (now called Treaty 11) were quickly concluded during the summer of 1921. Dene and Métis people ceded their title to 599,000 km^2. The Native people, concerned that their harvesting practices would continue, were promised complete freedom to hunt, trap and fish.

In 1966, the Indian Brotherhood of the NWT launched an oral history project to determine the Dene understanding of the Treaty 11 process. Oral testimony showed the Dene people did not understand the Treaty had ended title to their traditional lands. In addition, many Native people had been away from their communities during the summer when the negotiations were taking place so were not included in the treaty. These people had to be added in subsequent years.

In March 1973, 16 Dene chiefs legally claimed an interest in land covering more than one million square kilometres. Justice William Morrow concluded that the Dene people did indeed have Aboriginal rights in the area. This meant no development such as the Mackenzie Valley Pipeline could proceed until land title was established. The Dene and Métis came forward with a single Denedeh land claim. By 1988, Agreement in Principle was reached; however, the deal fell apart.

The Gwich'in

The Gwich'in Tribal Council (GTC) represents the interests of approximately 2900 Gwich'in people of the Mackenzie Delta of the NWT. The Gwich'in were the original inhabitants of the area and the Gwich'in Comprehensive Land Claim

Agreement (GCLA) was signed in 1992. The Gwich'in Settlement Area (GSA) is located immediately south of Inuvik and occupies an area of approximately 57,000 km². The Arctic Circle crosses at the lower third of the GSA. The majority of the approximately 3000 participants in the GCLA live in four communities found within the GCLA: Aklavik, Fort McPherson, Inuvik and Tsiigehtchic. The agreement gives the Gwich'in fee simple, or private ownership of the surface of 8658 square miles of land (22,422 km²) in the NWT, which includes 2378 square miles of land (6158 km²) where the Gwich'in own the subsurface as well, and the surface of 600 square miles (1554 km²) of land in the Yukon (Indian and Northern Affairs Canada, 1993).

The Sahtu Dene and Métis

In September 1993, the Sahtu Dene and Métis Comprehensive Land Claim Agreement (SCLCA) was signed. Settlement legislation known as the *Sahtu Dene and Métis Land Claim Agreement Act* received assent on June 23, 1994 and provided constitutional protection as a modern-day treaty under section 35 of the *Constitution Act 1982*. The Sahtu Dene and Métis Settlement Area (SSA) covers 280,238 km² (approximately 108,200 square miles) including the area of Great Bear Lake. Less than 3000 people live in the Sahtu settlement area. Five communities exist in the SSA: Colville Lake, Deline (formerly Fort Franklin), Fort Good Hope, Norman Wells and Tulita (formerly Fort Norman). The Mackenzie River runs through the Eastern portion. One-third of the proposed Mackenzie Gas Project will run through the SSA. The Sahtu Dene and Métis have title to 41,437 km² of settlement lands in the NWT (about 16,000 square miles – an area slightly larger than Vancouver Island). Of this, 1838 km or 22.5 per cent includes the ownership of subsurface resources (petroleum and minerals; Sahtu Secretariat Inc., 2005).

The Deh Cho

In contrast, the Dene First Nations group of Deh Cho have a different story. The Deh Cho represent about 4500 people from ten communities. The First Nations in this tribal council include Acho Dene Koe First Nation (Fort Liard), Deh Gah Gotie Dene First Nation (Fort Providence), Tthe'K'ehdeli First Nation (Jean Marie River), Katl'Odeeche First Nation (Hay River Reserve), K'a'agee Tu First Nation (Kakisa), Liidlii Kue First Nation (Fort Simpson), N'ah adehe First Nation (Nahanni Butee), Pehdzeh Ki First Nation (Wrigley), Sambaa K'e First Nation (Trout Lake), Ts'uehda First Nation (West Point) as well as Fort Liard Métis Nation, Fort Providence Métis Nation and Fort Simpson Métis Nation. Fort Simpson is the oldest continuously occupied trading post on the Mackenzie River (which they call the Deh Cho River). Fertile soil, coupled with moderate climate, allows for the cultivation of vegetables and the raising of livestock.

The Deh Cho have been trying to settle their land claim for 40 years. They are the only tribe along the pipeline that has not reached a final land claim settlement. The Deh Cho regional and Aboriginal leadership have been opposed to the pipeline until their land claim is settled with the Canadian government. They are using it for some leverage. They feel the land claim will give them certain rights.

The Mackenzie Valley Pipeline: Act 2

The end of the twentieth century saw a rebirth of interest in the energy resources of Northern Canada and Alaska, and a pipeline to bring these resources south to the Alberta Tar Sands, and then to the Canadian and American markets. Reasons for this were: (1) constantly increasing demand and record-breaking prices accompanied by declining production from existing wells in Canada and the United States; (2) significant cost reductions in bringing oil and gas to market as a result of technological advances and (3) the resolution of native land claims coupled with the increased recognition that people should have employment and income streams that provide for self-reliance and dignity, rather than government welfare (Bergman, 2000). Since 2001, the need has increased to establish sovereignty over an assured North American oil and gas supply because of the wars in the Middle East and the acts of terrorism.

In July 1999, the Government of the Northwest Territories (GNWT) and TransCanada Pipelines signed a Memorandum of Understanding to encourage the timely development of the natural gas reserves of the NWT and the construction of an economic, competitively priced, natural gas transmission infrastructure. In February 2000, four of Canada's largest energy companies – Calgary-based Imperial Oil Resources Ltd, Shell Canada Ltd, Gulf Canada Resources Ltd and Mobil Oil Canada – launched a joint study into the feasibility of developing and transporting Mackenzie Delta gas through a pipeline to Southern markets. This prompted proponents of an alternate route – Westcoast Energy Inc. and TransCanada Pipelines to announce their re-evaluation of the Foothills Pipeline Project first proposed in the 1970s that would take Alaskan natural gas southwards along the Alaska Highway route through the Yukon, British Columbia and Alberta into the United States.

Following the announcement by Imperial, Shell, Gulf and Mobil Oil Canada, 30 Aboriginal leaders (representing the Inuvialuit, the Sahtu, the Gwich'in and the Deh Cho) met in Fort Laird and Fort Simpson. As a result of these meetings, the Aboriginal Pipeline Group was formed in June 2000.

In October 2004, Imperial Oil Resources Ventures Limited filed an application with the National Energy Board of Canada for approval of the Mackenzie Valley Pipeline, which is part of the Mackenzie Gas Project. The owners were Imperial Oil Resources Limited, the Mackenzie Valley Aboriginal Pipeline Limited Partnership, ConocoPhillips Canada (North) Limited, Shell Canada Limited and Exxon Canada Properties.

The Aboriginal Pipeline Group (APG) also owns 30 per cent of the proposed Mackenzie Valley Pipeline to secure future benefits. According to Bob Reid, President of APG in 2007, this ownership allows APG to participate in the board of directors for the Mackenzie Gas Project and directly shape the pipeline's future (APG, 2007). At that time the full cost was expected to be CAD$1 billion.

APG reported a new model for Aboriginal participation in the developing economy would maximize ownership and benefits from the proposed Mackenzie Valley Pipeline and support greater independence and self-reliance among Aboriginal people (APG, 2004: 1).

Negotiations between the APG and the corporations culminated in an agreement announced on June 19, 2004. According to Claudia Cattaneo writing in the *National Post*, APG would receive an annual dividend of CAD$1.8 million for the next 20 years if no new reserves are found and the pipeline carries 800 million cubic feet of natural gas a day, increasing to CAD$8.1 million after 20 years, when the debt is paid off. If significant reserves are found and the pipeline is able to move 1.5 billion cubic feet a day, the APG would receive an annual dividend of CAD$21.2 million, increasing to CAD$125.8 million after 20 years. APG goals would also have a say in the pipeline's development and receive the highest possible Aboriginal participation in its construction and operation (Cattaneo, 2004).

While negotiating for almost three years, the corporations never had any objection to the APG becoming a full partner. The companies actively courted them, considering their participation key to a successful project – so very different from the corporate attitude at the time of the Berger Inquiry. All the parties sought a business-to-business relationship of equals. APG's inability to finance their share of the CAD$250 million cost of the project's first phase caused the delay. The group needed to raise CAD$80 million but Ottawa refused to help. APG turned to the private sector and TransCanada PipeLines Ltd (TCP), which agreed to loan APG CAD$80 million, which was to be repaid later from pipeline revenues. The way in which the CAD$80 million was finally secured also serves to further illustrate a fundamental change from the 1970s.

TransCanada PipeLines Ltd is a proponent of the Alaska route and is also a supporter of the Mackenzie Valley route. Gas from the Mackenzie Delta will feed into the company's existing pipeline network, increasing utilization and reducing costs to shippers (Cattaneo and Haggett, 2003). The company also has a long-standing and sophisticated interest in, and history of, working with Aboriginal groups as captured by Hope Henderson:

> With pipeline and power facilities now within 50 km of more than 150 Aboriginal communities, TransCanada realizes a significant business advantage by nurturing long-term relationships with its 'First neighbours'. In 2001, a Corporate Aboriginal Relations Policy was adopted, which outlines commitments to employment, business opportunities and educational support through scholarships and work experience.
>
> (Henderson, 2003: 10)

Consistent with this approach and in its own interest, TCP received an option to buy 5 per cent of the other companies' equity in the pipeline (Cattaneo and Haggett, 2003). The agreement negotiated between the APG, the pipeline corporations and TransCanada is another reflection of the changing relationship between

Aboriginal communities, corporation and governments in the new economy. The federal government's Indian Affairs minister also was excited by the private sector corporations agreeing to work with the Aboriginal community. In addition, they approached Washington to indicate that the proposed Alaska line should not receive government subsidy (Cattaneo and Haggett, 2003).

At the same time as APG and the corporations were negotiating their agreement, the Deh Cho was seeking a land claim agreement with the federal government. Almost 40 per cent of the proposed pipeline route is on lands claimed by the Deh Cho. Because of the Aboriginal title, the developers were required to deal with the Deh Cho. The pipeline corporations asked Ottawa to reach a land claims agreement, thus resolving the pipeline corridor issue. On April 17, 2003, the Federal government and the Deh Cho reached an interim Five-year Resource Development Agreement until reaching a final land claims agreement (Deh Cho First Nations and Government of Canada, 2003.) Under its terms, each year the federal government will set aside on behalf of the Deh Cho a certain percentage of the royalties collected from the Mackenzie Valley. The amount will be paid out to the Deh Cho when a final agreement is concluded. In the interim, up to 50 per cent of the total each year (maximum CAD$1 million) will be accessible for economic development. Seventy thousand square miles of Deh Cho claimed lands will be set aside as part of a system of protected areas, while '50 per cent of the 210,000 square kilometres with the land with Aboriginal title will remain open to oil, gas and mining development, subject to terms and conditions set out by the aboriginal group' (Canadian Press, 2003).

Environmental groups praised the deal. The World Wildlife Federation called it a 'tremendous achievement'. The group awarded the Deh Cho and the federal government the Gift to the Earth, an international conservation honour for environmental efforts of global significance.

With the interim agreement in place, the pipeline project was ready to move forward to the next stage – environmental review. But by November 2003, the Deh Cho were threatening to seek a court injunction to halt the review approval process unless the government included Deh Cho representation. Keyna Norwegian, chief of the Liidlii Kue band in Fort Simpson, suggested they should also have input into the decisions (VanderKlippe, 2003). Deh Cho have continued to feel strongly that protecting traditional areas is more important than transporting the gas across their land. Chief Norwegian has repeatedly expressed that the Deh Cho can live without the pipeline.

The Deh Cho have formed allies among the environmental groups having concerns about the risks to the Bathurst caribou, to the stability of the pipeline from melting permafrost, risk to the 500 rivers that the pipeline must cross, and a general resistance to ongoing reliance on petrochemicals. The dispute remained resolved but collapsed on June 6, 2004. The Deh Cho thought they had reached an agreement in May, which gave them a seat on the review board. However, the federal negotiator's understanding differed and the agreement reached suggested ways in which the Deh Cho could participate. Chief Norwegian of the Deh Cho accused the regulators of reneging on an agreement and the impasse continues.

The Mackenzie Valley Pipeline: Act 3

On January 25, 2006, the National Energy Board (NEB) commenced its public hearings into the application of Imperial Oil Resource Ventures Ltd for permission to construct and operate the Mackenzie Valley Pipeline. The NEB is concerned about the project's economic safety and technical issues. Two weeks later, on February 14th, a seven-member Joint Review Panel commenced parallel hearings into environmental, socio-economic and cultural issues. The Joint Review Panel concluded the public hearings on November 29, 2007. This public hearing phase had been extensive: 115 days of hearings, more than 5000 submissions, over 11,000 pages of transcripts. The JRP had travelled to 26 communities to hold its hearings (MGP Exchange, 2008:1). The Inuvialuit, Sahtu and the Gwich'in have been involved in these hearings. As part owners of the proposed gas pipeline, their past efforts to establish a partnership with the developers succeeded. As owners of the AGP, their relationship with government has been similar to that of their non-Aboriginal corporate partners, namely a relationship of applicant to regulator.

As of 2008, negotiations are still ongoing with the Deh Cho. Both the proponents of the pipeline and APG have been putting considerable pressure on the Deh Cho. APG gave the Deh Cho a deadline of December 31, 2006 to formally join the pipeline group, saying this was essential in order to secure timely funding. Chief Norwegian responded by saying this would jeopardize the Deh Cho if it supported APG and the pipeline's construction before settling the land claim. Since then, the APG has put the 34 per cent participation for the Deh Cho in trust.

Land claims and other rights recognition

The Aboriginal Groups recognized that they wanted to ensure their full participation into future decisions affecting their lands and people. As previously indicated, the Inuvialuit land claim was settled in 1984, followed by the Gwich'in in 1992 and the Sahtu Dene and Métis in 1994. Although the Deh Cho have been negotiating with the Federal Government for more than 40 years, an interim agreement has been in place only since 2003.

Under these land claim agreements, the legal title over their territory was transferred from the Canadian Government to the respective Aboriginal group. The transferred lands are privately owned in fee simple and are not reserves under the Indian Act. In addition, significant payments were made. For example, over a 20-year period, the land claim payments to the Inuvialuit Regional Corporation totalled CAD$169.5 million. This included a CAD$7.5 million Social Development Fund (SDF) and a CAD$10 million Economic Enhancement Fund settlement (Inuvialuit Corporate Group, 1997: 4). On the other hand, the Sahtu Denis and Métis would receive financial payments of CAD$75 million (in 1990 dollars) over a 15-year period and share the royalties from resource development paid to Canada's governments each year in the Mackenzie Valley. The Gwich'in also received a tax-free capital transfer of CAD$75 million (1990 dollars) paid over a 15-year period.

Each of the agreements guarantees participation in institutions of public government for renewable resource management, land use planning and land and water use. They also allow for participation in environmental impact assessment and reviews in the Mackenzie Valley. The agreements provide for negotiation of self-government agreements to be brought into effect through federal and territorial legislation. They also have extensive and detailed wildlife harvesting rights; and rights of first refusal to a variety of commercial wildlife activities. They will receive a share of annual resource royalties in the Mackenzie Valley.

Each Aboriginal group formed a special company to receive the lands and funds on behalf of its beneficiaries. For example, the objectives and goals of the Inuvialuit Regional Corporation address preserving their specific culture, identity and values within a changing Northern society; ensuring equal and meaningful participation in the Northern and national economy and society; protecting and preserving the Arctic wildlife, environment and biological productivity; stewardship over their lands; preserving and growing the financial compensation from the initial funds and distributing accumulated wealth to the beneficiaries. The corporations also represent and advance their interests in areas of external relations such as federal, territorial and municipal governances; international, circumpolar and other Aboriginal organizations and private sector and special interest groups (Inuvialuit Regional Corporation, 2007a, b).

Between 2000 and 2002, the Canadian Association of Petroleum Producers (CAPP), the Government Regulatory Agencies and the Indian and Northern Affairs Canada (INAC) partnered with the various Aboriginal settlement corporations to develop regulatory roadmaps for oil and gas exploration and development in the NWT. These will facilitate sustainable economic development in the North by helping companies fulfil the regulatory requirements that safeguard human health and safety, protect the environment and encourage economic benefits particularly as both the authorities and the processes for regulating these activities are relatively new (Regulatory Roadmaps Project, 2004).

According to the Mackenzie Gas Pipeline Proposal, 40 per cent of the pipeline would go through territory of the Deh Cho First Nations. The pipeline would pass about 17 km (10 miles) from Fort Simpson a town of about 2500. Also, three gas compressor facilities would be located within the Deh Cho territory. As far back as 2001, the Deh Cho had refused to sign on to participate in the APG. The Deh Cho regional and Aboriginal leadership opposed the pipeline until their self-government and land management negotiations were concluded with the Canadian government. According to Grand Chief Norwegian, they felt the land claim settlement would then establish their rights and grounds for interaction with various levels of government and with the oil and gas companies, thus allowing for maximum benefit from the project (Norwegian, 2006).

By November 2003, the Deh Cho were threatening to seek a court injunction to halt the review Joint Review Panel approval process unless the government renegotiated the terms of the process to include a Deh Cho representative on the Review Board along with those of the Sahtu Dene, Gwich'in and Inuvialuit.

In September 2004, the Dene First Nations group of Deh Cho filed two federal lawsuits to stop the Joint Review Panel of the Mackenzie Valley Pipeline Project until the Deh Cho were included in the Environmental Impact Review process. They also wanted to nominate their own Deh Cho representative to sit on the Joint Review Panel in a similar manner to the Sahtu, Gwich'in and Inuvialuit. They were also concerned that several appointed members of the Joint Review Panel had publicly expressed their support for the pipeline and thus undermined the panel's impartiality.

The Deh Cho Process negotiations on land resources and governance were halted and out-of-court negotiations began. In July 2005, the Deh Cho and the Canadian government reached a settlement. The agreement included: establishment of a Deh Cho Resource Management authority, as a body of public government, in a Final Agreement; consultation between the two parties before Canada issues any authorizations or makes any decisions relating to the MGP that might adversely affect their section 35 rights, and, if appropriate, will accommodate Deh Cho First Nations concerns; consultation with the Deh Cho First Nations on terms and conditions for new oil and gas exploration in its territory; completion of interim land withdrawal negotiations in the Wrigley area to protect environmentally sensitive areas and identify a pipeline study corridor; implementation of an approved and favourably considered land use plan as soon as possible after the Plan's completion; CAD$817,352 guaranteed funding for 2005–2006 to be provided to complete the Deh Cho Land Use Plan; Canada will provide CAD$15 million in economic development funding over three years to assist in exploring and developing business opportunities; Canada will provide CAD$6 million for participation in the environmental and regulatory process; establishment of a process to monitor and recommend mitigation measures regarding environmental assessment and socio-economic impacts of an approved MGP; and Canada will provide CAD$10.5 guaranteed funding over three years to increase capacity to complete the Deh Cho Process negotiations in an expedited manner (Deh Cho First Nations and Government of Canada, 2005).

The APG had held a 34 per cent stake for the Deh Cho in the one-third equity position that had been acquired in the Mackenzie Pipeline. This offer was held in place until June 30, 2006.

In June 2006, Indian and Northern Affairs Minister Jim Prentice tabled an offer from the Canadian government. The package was the first tabled since negotiations opened in 2001. It included cash payouts of CAD$104 million plus interest paid out to the 4500 Deh Cho residents over the next 15 years; a resource-sharing component, consistent with other agreements along the Mackenzie Valley, that would include 12.5 per cent of the first CAD$2 million and 2.45 per cent of any additional royalties; a land and self-government agreement covering 10,000 square miles; Deh Cho control over surface and subsurface rights, allowing them to set royalty rates and collect 100 per cent of royalties and the right of the Deh Cho to make their own laws (Park, 2006).

In the Fall of 2006, the Conservative federal government repeatedly stated that it would move ahead with the pipeline with or without the Deh Cho's approval.

Overall, the self-government and land claims agreements have resulted in the Aboriginal governments having significant power and influence. The communities have established co-management boards. These boards still have the power to 'say no to' or hold up issuing water, land use and other permits even if a decision is made to go ahead with the pipeline. The oil and gas companies approach the chiefs and reserves, mayors and towns to determine what needs to be done to get the permits. By making corporate donations to community projects, the companies are establishing goodwill and hoping to get more favourable permit decisions. Communities are participating in the National Energy Board and Joint Review Panel hearings to ensure certain terms and conditions are met and negative impacts are mitigated. The Aboriginal groups have also formed alliances with each other to increase their power. However, the oil and gas companies have undertaken to negotiate separate agreements with each Aboriginal group.

The Inuvialuit and Gwich' also work through international organizations. The Inuvialuit works through the Inuit Tapairitt Kanatamai (ITK), the Inuit Circumpolar Council (Canada) and the Arctic Council. The Gwich'in participate in a steering committee where they meet with representatives from Russia, Finland and the United States. The information sharing helps to anticipate issues - 'if it's occurring there what's to stop it from occurring in Canada'.

Overall, various stakeholders felt that, compared to the 1970s with the Berger Inquiry, the Aboriginal people have matured and are able to speak for themselves. Their leaders are well educated and can participate as equal partners in negotiations and decision-making.

Socio-economic development

Overall, most Northerners saw the pipeline as positive because the economy was so dependent on government jobs, military jobs or transfer payments. People needed opportunities to become self-reliant and have the dignity that it provided. Resource-sharing and land use agreements negotiated with the companies would provide additional income. The biggest impacts would come in the construction phase of 3–5 years with some work in the clean-up period of years 4 and 5. For the first three years, they will have an opportunity to participate in businesses constructing and servicing the pipeline. After that, about 50 people would be in operations and another 200 in maintaining the pipeline. However, Northerners and Aboriginal people were concerned about their ability actually to get these jobs. In the past, the Southern companies and employees had dominated.

The pipeline is being designed to carry 10 cubic feet per day. Initially, this would carry gas from the three fields. In the longer term, substantial development of oil and gas from other fields (proven and unproven) could occur. Aboriginal community and women's organizations have expressed concerns about the cumulative impacts of these new developments. Positive effects seen to come from the pipeline included: transfer of knowledge, employment skill sets and technology; increased linkages with Southern companies and educational institutions; more service jobs; potential for more local energy supply thus substantial reduced costs of living; improved human welfare as people have incomes, learn to manage their money and lifestyles better and reduce alcohol and drug use to keep their jobs.

The Inuvialuit were seen to be most mature with regard to business development, as they had benefited from the oil and gas development of the Beaufort Sea in the 1970s and had settled their land claim earliest. The Gwich'in were next most mature, as they had benefited from their land claim settlement and used this to form partnerships with and invest in Southern companies. The Sahtu and the Métis had more than 100 years of experience with the fur trade, and so have experience as small business people. Some communities like the Deh Cho in Fort Liiard have developed all sorts of businesses. However, business development varied greatly between communities and between land claim settlements. The interplay among the actors was very different from that during the Berger Inquiry. For example, since its land claim settlement, the Inuvialuit have created or acquired more than 30 companies operating in eight industry sectors within the Arctic, Southern Canada and internationally. The Inuvialuit Regional Corporation's performance over a ten-year period has doubled the beneficiaries' equity while distributing CAD$11.6 million in dividends to the beneficiaries. Between 1996 and 2004 the Inuvialuit Corporate Group contributed CAD$83 million to communities and individuals.

The Aboriginal groups have spent considerable time and energy researching and defining the specific terms and conditions under which the current pipeline project would be supported. Both the federal and territorial governments have provided funds to assist in this process.

The Inuvialuit and Gwich'in have developed several partnerships and joint ventures with key companies from Southern Canada with international reputations that are involved in pipeline construction and servicing. This will maximize the opportunities to the beneficiaries of their settlement areas if the project were to go ahead. These partnering companies are involved in airline and helicopter transportation, camp operations, heavy equipment and construction for roads and highways, construction-based services, such as pipeline, infrastructure and facilities construction, transportation and logistics services; land-based drilling and well servicing; stationary camp accommodation, housekeeping and catering services.

The Pipelines Operations Training Committee (POTC) was formed in 2002 to develop and implement training programmes that would prepare Aboriginal and other Northern workers for the long-term employment opportunities associated with the Mackenzie Gas Project. The committee is made up of Aboriginal, industry, territorial and federal government representatives. A Training and Employment Framework was developed, which included job descriptions and competency profiles. Several targeted training programmes have been developed and implemented in cooperation with Aurora College, Bow Valley College and post-secondary institutes of technology in Alberta to deliver the required technical certifications. They also provide job readiness and in-class and on-the-job skills training. Adults and youth in all Aboriginal settlement areas are being encouraged to complete high school with appropriate maths, science, literacy and other essential skills. The Federal government has assisted by providing increased funding through the Aboriginal Human Resources Development Agreement. The oil and gas companies have also partnered to provide career fairs, job placements and internships.

A variety of stakeholders from industry, government, education and training providers and diverse serving agencies worked together federally on the Petroleum Human Resources Sector Council of Canada. The Council released in 2003 *A Strategic Human Resources Study of the Upstream Petroleum Industry: The Decade Ahead*. They indicated the North was in a start-up phase of the industry cycle. Key focus for industry and local communities would be to minimize the 'boom and bust' economic impact for the region and ensure employment for Aboriginal peoples. Several of the sector council's recommendations include: attracting and providing access to employment for non-traditional workers, such as women, Aboriginal peoples and new immigrants, as they are underutilized; addressing present and upcoming skill shortages; better management of the difficult progression from education and training to long-term employment needs; greater appreciation of the differences in Aboriginal and non-Aboriginal cultures is needed (targeted prerequisite and bridging programmes must meet regional and Aboriginal needs); ensure better portrayal of role models and provide more information about careers and entry-level requirements. They noted that Aboriginal peoples as well as Northern residents will likely be key resources, as they are already acclimated and geographically proximate. Industry should continue to strive for lasting partnerships with Aboriginal communities to support the creation of a related business development such as a service industry.

Considerable attention has been paid to small business and local economic development in all the settlement areas. For example, the Deh Cho have provided training to communities regarding joint venturing, impact/benefit agreements, resource royalties, financial management, preparing economic development plans, establishing businesses, marketing, etc. (Deh Cho Business Development Centre, 2004: 5). A custom-designed two-day workshop 'Aboriginal Oil and Gas Management' covered an overview of the petroleum geology of the Deh Cho, GNWT career development programmes and services; history of mineral rights in the North, Northern industry economics, regulating the Northern seismic project; finding and producing oil and gas; producing operations in the North; Northern public safety, cultural and environmental aspects; Northern lands administration; Northern royalties, joint venture agreements and Northern employment and business opportunities in oil and gas (Government of the Northwest Territories, 2002: 2).

For businesses to develop and participate meaningfully in the regional economic development activities associated with the pipeline, they must have access to the opportunities. Thus priority access to procurement opportunities for businesses in the settlement areas have been built into negotiated agreements, permits and licences by the territorial government as well as the governments for each of the Aboriginal settlements. A registry of Aboriginal owned businesses has also been developed. For example, the Inuvialuit Regional Corporation (IRC) noted that, since the return of the oil and gas industry in the winter of 1999/2000, approximately 70 per cent of all contracts have gone to Inuvialuit businesses (IRC, 2006: 14). On the other hand, the Sahtu Dene and Deh Cho commented they have much fewer and smaller businesses in their regions, and many of these are in the start-up phase.

Business support centres have been developed within the key communities to assist with training, identifying opportunities, developing business plans and applying for government funds. The government of the NWT has increased its funding to support these business support centres as well as for business start-up loans and grants. For the first time, a training manual is now available on how to develop a small business. This manual has been translated into each of the Aboriginal languages.

In 2004, the GNWT in partnership with the NWT Association of Communities held a three-day conference 'Community Government Leaders Conference: Preparing for the Pipeline'. The conference was held in Inuvik and attended by more than 50 mayors, chiefs and councillors, senior administration officers, band managers and other key staff from 20 communities (GNWT, 2004). In 2005, the NWT Government also offered a series of two-day community-specific workshops to identify the socio-economic concerns associated with the pipeline and also suggestions for mitigating the negative impacts and accentuating the positive (GNWT, 2005a,b,c). These impacts were divided into four areas: health and social, housing, justice, employment and income.

Key socio-economic issues include: (1) the inability of already stretched community infrastructures to handle the increased usage and demand resulting from the activities associated with the exploration and development; (2) cumulative impacts on the quality of life in NWT communities, including overcrowding, increased rents, homelessness, crime, substance abuse, domestic violence, unwanted and teen pregnancies, family break-ups, sexually transmitted diseases, gambling, gangs, etc. and (3) loss of government and other skilled employees to take jobs with resource extraction companies offering higher salaries and benefits. Community impacts of the proposed pipeline included water supply, sewage, solid waste sites, land development, roads, granular materials, fire response, emergency response, hazardous materials response, community service personnel, community staffing, secondary industry demand, population growth/social impacts, local inflation, governance, municipal facilities and municipal contracting and mobile equipment.

The Status of Women's Council (SWC) held focus groups, conducted individual interviews and held meetings in 12 communities along the proposed MGP. The SWC then made a submission and appeared before the Joint Review Panel with regard to the Environmental Impact Statement (EIS) Terms of Reference. They also provided comments on the Proponents Gender Analysis Report. SWC developed a *'How To Guide for Participating in Socio-economic Impact Assessment (SEIA) in the NWT* (SWC of the NWT, 2006a)' to assist women in gaining a voice and have more meaningful participation.

In the EIS, the companies had concluded the residual project effects for the MVP on the community well-being would not be significant, as these would be short-term for the three-year period. The SWC in its presentation to the Joint Review Panel disagreed, saying the induced effects would be long-term, such as teen births, human immunodeficiency virus (HIV) infection, increased drug use or increased family dysfunction, are long-term. If individuals are victimized through a project-induced increase in family violence or sexual abuse, they will suffer long-term impacts (SWC of the NWT, 2006b: 2). The SWC recommended the Panel apply the precautionary approach in evaluating information in the EIS regarding residual effects on community wellness conditions. These viewpoints about the negative social impacts were reiterated many times by respondents in the interviews.

In 2005, the SWC of the NWT received CAD$1.7 million from Human Resources and Social Development Canada under the Pan-Canadian Innovations Initiative for the Northern Women in Mining and Oil and Gas Project to increase the attraction, training and retention of women for trades and industry opportunities in mining, oil and gas (SWC, 2005).

Overall, Aboriginal, civic and government sectors had legitimate expectations of the oil and gas companies for corporate social responsibility and sustainable development. Oil companies should not think that Canada is the rightful land owner. They needed to comply with the protocol. They also need to visit the community and start dialogues before doing anything. They should conduct themselves in ways that minimize impacts on the land and the environment. They also need to ensure that local people benefit from the jobs and contracts. Contractors and sub-contractors must be required to hire locally. Jobs and procurement requests should be openly and fairly posted, without deals on the side or in the South. Contracts should be broken up if necessary to assist smaller businesses to qualify. Job requirements should be identified well in advance so people can receive appropriate training. Companies should partner to provide training programmes, practical experience and apprenticeships. Training scholarships were needed. Employees in the work camps should have courses in money management; access to families through emails and phone calls so relationships are maintained and access to the Internet for bill payment on-line. No alcohol and drugs should be allowed in the camps. The companies need to work along with the local communities and Aboriginal groups to provide alcohol and drug-abuse prevention and treatment programmes, and to contribute to women's shelters. Safe and healthy workplaces must be provided and support Aboriginal people. Burial grounds and heritage sites should be preserved. Royalties and revenues should flow to the people. The companies need to ensure adequate infrastructure is in place and be prepared to pay for what is associated with the industry's development.

Environmental protection

Environmental protection is a key concern for all parties involved in the Mackenzie Pipeline. Northerners were very concerned about the potential negative social and environmental impacts, particularly because the environment cannot be separated from the culture. Some respondents clearly felt that the pipeline should not go ahead because of the potential to harm the fragile Northern environment and landscape.

The pipeline companies have developed environmental impact assessments, which go beyond the requirements outlined in the many pieces of regulatory legislation and guidelines. By having extensive consultation and communication with the impacted communities, the companies have listened to the concerns and suggestions. They have developed detailed plans for a variety of scenarios to minimize and mitigate the harmful effects. The Aboriginal leaders also felt that the negotiated land use and resources agreements with the companies would place terms and conditions on the proposed pipeline development, and subsequent exploration and drilling activities that would minimize and mitigate the negative aspects.

The Government of Canada and the NWT have well-developed sustainable development policies, associated regulations and guidelines, strategies and action plans. Inuvialuit and First Nations people have practised sustainable development for thousands of years. They also have received responsibility for sustainable development as part of the land claims settlements. They are very concerned not only about the building of the pipeline and the development of existing exploration and drilling licenses, but also with the other projects this would allow for in the future and the cumulative impacts. All parties clearly recognized that oil and gas were non-renewable resources – once these resources are removed from the land, they are gone and will not return. Thus, they are concerned with the impacts on the futures of their children and grandchildren.

Most Northerners felt that one aspect could not be considered more important than another. The three dimensions of sustainable development: the environment, and the social and economic aspects, were likened to a three-legged stool – if you remove one leg, the rest are unbalanced. Others commented that Aboriginal and Northern culture, and the social aspects are not separable in the North. The Western economic culture is very young. Aboriginal and Northern cultures are much more rooted in tradition. Several observed that the pipeline proponents were more concerned about environmental than social impacts. Others commented 'the Southern stakeholders weight the economic factors much higher' – it is an important dimension but one of the key differences.

Another key difference was that Northerners were concerned not only about the development of the current project but also its cumulative effects. The Joint Review Panel directed Imperial Oil Resources to submit cumulative impact mapping for the MGP. The Canadian Arctic Resources Committee commissioned a series of scenarios. Cizek Environmental Services (Cizek, 2005) suggested that the three anchor fields could only supply the pipeline for between 9 and 14 years. The MGP would require the connection of other existing and numerous undiscovered gas fields to keep it filled for the following 50 years.

The biggest environmental threats identified by respondents included climate change, pollution, landscape degradation, species reduction (particularly for species already at risk and those the people depend on for food or culture) and culture/traditions. There is potential for spills and mismanagement of waste. Habitat preservation and remediation is also very important.

The ecosystem was seen as very robust but at the same time shallow. The NWT has one of the severest climates in the world: it can range from minus 40°C in winter to plus 30°C in summer. As components in the environment still survive, this makes it very robust. However, it is fragile because the recovery takes so long if there is an impact on an Arctic system. Because very few levels exist between the highest and lowest levels in the food chain, and the food chain is so highly interconnected, removing one component owing to the environmental impacts can make it vulnerable. In the past, the government has had to supply Northerners with food because caribou or musk ox herds declined. Today, traditional foods still make up to 60 per cent of the diets of Northerners.

The Inuvialuit, Gwich'in, Sahtu Dene and Deh Cho were very concerned about the potential of impact of a spill or pipeline rupture on the fish, whales, marine plankton, plants, water and animals. Although a marine spill is much less easy to contain, the effects of a spill on land would be severe and long lasting in this fragile environment. Oil and gas companies operating in the North commented that safety is absolutely critical in all their operations, and the ability to meet and adhere to safety standards would be a criterion for successful bidding on contracts and also for employment. The governments have also introduced stricter regulations and zero pollution for the pipeline construction, operation and exploration, and drilling. Most stakeholders viewed this as necessary, however, many commented that zero pollution is unattainable.

The effects of climate change are already in the NWT. People who spend time on the land and the elders see changes, as they have something to compare it to. Warmer temperatures have occurred over the past few decades. Comments included lakes and rivers not freezing as fast; lakes drying up; patterns of animal movements changing; fewer fish, caribou, ducks and birds; more bears and wolves; other animals using the water for travel; whales in a lake; species from the South expanding their ranges; more and faster growth of all the trees, willows, plants and different types of flowers.

Warmer weather is also melting the permafrost. Permafrost is the permanent ice in the ground, in the ocean and on the land. In the NWT, the ground below 1 m is permanently frozen. Melting permafrost is affecting building foundations, all-weather roads, winter roads and infrastructures. Operational seasons for winter ice roads are becoming uncertain. As many NWT communities are only accessible by air or winter roads, this is increasing transportation costs and thus costs for goods as well as decreasing supplies. Shifting ground can also cause roads to crack or slip, landslides and can disrupt community water supplies.

Several respondents were quick to point out the oil and gas companies' activities in the North were neither a direct cause of the climate change nor the cause of the reduction in the size of caribou herds. However, the MVP is supplying non-renewable energy, which creates greenhouse gases.

Aboriginal groups and Northerners have aligned themselves with environmental groups like the World Wild Life Federation, Canadian Wildlife Federation, and world conservation groups who look at the big picture. They have been strong advocates and allies. They are seen not as anti-development but as sustainable development. Although the Sierra Club has been intervening in the pipeline project, it is seen more in opposition. The Canadian Arctic Resources Committee is viewed as fairly influential, as it is based in the seat of federal government, Ottawa, Ontario. Ecology North and Alternatives North are locally based in the NWT. As interveners during the Joint Review and National Energy Board hearings, they raised good questions about sustainability and also concerns about greenhouse gas emissions and Arctic warming (Ecology North, 2006a, b, c).

Conclusion

Much has changed over the past 40 years. First, many land claims have been settled. Second, as a result of these settlements, Aboriginal organizations that emerged have elected to engage in the market economy. Third, companies are perfectly prepared – maybe even eager – to have the Aboriginal groups participate as equal equity partners in the Mackenzie Gas Project. Nellie Cournoyea, Chair of the Inuvialuit Regional Corporation, expressed this new business climate as:

> the biggest change since the 1970s is that the oil and gas industry realizes aboriginal people are an integral part of development, and that they must receive a fair share of resource revenue and have the opportunity to invest directly in pipelines and offshoot businesses.
>
> (Bergman, 2000)

Fred Carmichael, President of the Gwich'in Tribal Council and Chairman of the APG summed it all up by saying 'We're ready,' at the opening of the National Energy Board Hearings (Jaremko, 2006). Fourth, the environmental and social concerns about pipeline development trampling fragile Northern environments and Aboriginal settlements that, in the 1970s, lacked the strength to protect traditional livelihoods and lifestyles no longer exist. As Thomas Berger stated in his remarks to *Edmonton Journal* reporter Gordon Jaremko (2006): 'The recommendations I made have been carried out. How events unfold in an area as dynamic as the Mackenzie Valley will depend on the people of the Mackenzie Valley. I'm confident they'll decide what's in their best interests.'

The shift in relationship between the Aboriginal groups, especially the Inuvialuit, Sahtu and the Gwich'in, and environmental groups deserves attention. Aboriginal groups and environmentalist were strong allies during the Berger days. Now the Inuvialuit, Sahtu and the Gwich'in are proponents of the project. Environmental representatives are not. The position of the Deh Cho is somewhat ambivalent. On the one hand, this First Nations has accepted the interim agreement, which permits the Mackenzie Gas Project to proceed to the National Energy Board Hearings. On the other hand, the Deh Cho are still very focused on environment issues as part of their land claim and still seek the environmental groups as allies in their land claim negotiations. As the National Energy Board hearings proceed, the position of the Deh Cho will be critical. While they are not opposed in principle to economic development, they wish to control such development within the framework of their comprehensive land claim agreement. If the Deh Cho becomes dissatisfied with the Resource Development Agreement, the gas pipeline project could be in trouble. Finally, with the costs climbing to CAD$16 billion since 2002, the pipeline companies are seriously reconsidering whether to continue and, if the return on the investment will be realized, other alternatives may become more attractive, particularly if the federal government remains firm in its unwillingness to take an equity position.

Sustainable development requires a balance between the needs of people, nature's other species and future generations. Sustainability is also about recognizing and working within limits. It is also about complex interconnections and interdependencies. Concepts such as cumulative and adjacent impacts, incorporating externalities, precautionary principles, mitigation and risk minimization, intergenerational equity, interconnected and nested layers of ecosystems, environmental, socio-economic and cultural impacts, as well as human diversity are being addressed in this current decision whether to proceed with the Mackenzie Gas Pipeline in the NWT. Aboriginal groups, companies, civil, state and communities are clearly interacting and participating in the decision-making process.

The business sector is clearly being required to demonstrate corporate social responsibility. Some examples discussed include the following: extensive community consultation, communication, dialogue and negotiation. Detailed plans are being developed in conjunction with stakeholders to maximize benefits, to minimize risk and damage, and to remediate proactively. Companies are going beyond meeting the minimal legislation requirement to provide healthy and safe work places that are substance free. They are encouraging diversity and planning to allow for active participation of Aboriginal peoples and women in education, training, employment and business opportunities. They are also providing ongoing benefit through revenue sharing and equity partnerships. Strategic corporate social responsibility of more depth has become a necessity for corporate success in the new flexible regime of accumulation. By using sustainable development principles and practices to shape corporate social responsibility, the result will be a sustainable mode of development. The Mackenzie Gas Pipeline decision has clearly demonstrated how modes of social regulation are also evolving over time. The corporate sector will continually have to monitor and proactively meet changing stakeholder expectations.

References

Aboriginal Pipeline Group (2004) *Aboriginal Pipeline Group Brochure*, Inuvik: Aboriginal Pipeline Group.

Aboriginal Pipeline Group (2007) *An Exceptional Deal*, by Bob Reid, President of APG. Online. Available HTTP: http://www.mvapg.com/page/page/2501121.htm (Accessed 14 January 2007).

Auld, J. and Kershaw, R. (eds; 2005) *The Sahtu Atlas: Maps and Stories from the Sahtu Settlement Area in Canada's Northwest Territories*, Norman Wells, NWT: Sahtu GIS Project, p. 60 (quoting Blondin, J. from *Dehcho: 'Mom, We've Been Discovered!'*, Dene Cultural Institute, 1989).

Berger, T. (1977) *The Probable Economic Impact of the Mackenzie Valley Pipeline in The Arduous Journey: Canadian Indians and Decolonization*, Ponting, J. R. (ed.). Toronto: McClelland and Stewart.

Bergman, B. (2000) From new pipe dreams, *MacLeans*, July 17, p. 34.

Bone, R. M. (2003) *The Geography of the Canadian North: Challenges and Issues*. Toronto: Oxford University Press.

Brackman, C. (2001) *The Northwest Territories Petroleum Industry*. Online. Available HTTP: http://www.bmmda.nt.ca/outgoing/NWTpetroleumindustry.pdf (Accessed 17 November 2007).

Canadian Gas Potential Committee (2001) *Report on Natural Gas Potential in Canada – 2001*, Calgary: Canadian Gas Potential Committee. Available HTTP http://www.canadiangaspotential.com/.

Canadian Press (2003) 'Natives sign land deal', *Regina Leader Post*, April 19, p. D7.

Cattaneo, C. (2004) *National Post*, Don Mills, July 26, Ont., p. FP.1.

Cattaneo, C. and Haggett, S. (2003) *National Post*, Don Mills, June 19, Ont., p. FP.3.

Cizek, P. (2005) *A Choice of Futures: Cumulative Impact Scenarios of the Mackenzie Gas Project*, prepared for Canadian Arctic Resources Committee, 24 October 2005. Online. Available HTTP: http://www.carc.org/2005/A%20CHOICE%20OF%20FUTURES%20final.pdf.

Coates, K. S. and Morrison, W. R. (1986) *Treaty Research report Treaty No. 11 (1921)*, Treaties and Historical Research Centre, Indian and Northern Affairs Canada. Online. Available HTTP: http://www.ainc-inac.gc.ca/pr/trts/hti/t11/index_e.html (Accessed 17 November 2007).

Deh Cho Business Development Centre. *Annual Report 2004–2005*. Online. Available HTTP: http://www.dehchobdc.ca/Main/About/DCBDC%20annual%20reports/Annual%20Report%202004%20-%202005.pdf.

Deh Cho First Nations and Government of Canada (2003) *Interim Resource Development Agreement*. Online. Available HTTP: http://www.dehchofirstnations.com/documents/agreements/dehchoirda_e.pdf (Accessed 17 November 2007).

Deh Cho First Nations and Government of Canada (2005) *News Release and Backgrounder: Government of Canada and Deh Cho First Nations Reach Agreement, 11 July 2005*. Online. Available HTTP: http://nwt-tno.inac-ainc.gc.ca/dehcho/pdf/NR_DCFNAgreement-Jul05_e.pdf (Accessed 18 November 2007), and http://nwt-tno.inac-ainc.gc.ca/dehcho/pdf/Bkgrndr_DCFNAgreementJul05_e.pdf (Accessed 18 November 2007).

Ecology North (2006a) *Documents – 17 October 2006 -Yellowknife-Technical Hearing – Topic 4 – Physical Environment – Land, Water and Air Greenhouse Gas Emissions and Air Quality*.

Ecology North (2006b) Powerpoint presentation for the technical session. Online. Available HTTP: http://www.ngps.nt.ca/Upload/Interveners/Ecology per cent20North/EcologyNorth_Hearing_Presentation_for_Oct_17_2006_technical_session.pdf (Accessed 18 November 2007).

Ecology North (2006c) Intervener presentation of the Arctic Council Impact Assessment (2004) *Impacts of a Warming Environment*, Cambridge University Press. Online. Available HTTP: http://www.ngps.nt.ca/Upload/Interveners/Ecology per cent20North/EcologyNorth_Arctic_Climate_Impacts_Assess_ per cent20Impacts_of_a_Warming_Arctic_Highlights.pdf (Accessed 16 November 2007).

Oil and gas activities at the Mackenzie Delta 197

Government of the Northwest Territories (2001) Department of Resources, Wildlife, and Economic Development, *The Northwest Territories Petroleum Industry* (by Brackman, C.), quoting National Entergy Board Publication (1998) *Probabilistic Estimate of Hydrocarbon Volumes in the Mackenzie Delta and Beaufort Sea Discoveries.* Online. Available HTTP: http://www.bmmda.nt.ca/outgoing/NWTpetroleumindustry.pdf.

Government of the Northwest Territories (2002) Departments of Education, Culture and Employment Wildlife, Resource and Economic Development, *Oil and Gas Update* 1 (8): September. Online. Available HTTP: http://www.iti.gov.nt.ca/mog/pdf/oilgas_update2002.pdf.

Government of the Northwest Territories (2004) Departments of Municipal and Community Affairs and Resources, Wildlife and Economic Development in partnership with the NWT Association of Communities. *Community Government Leaders Conference: Preparing for the Pipeline. Conference Report.* December 6-8, 2004. Online. Available HTTP: http:www.maca.gov.nt.ca/pipelineresource/documents/MGPConferenceReportWebVersion.pdf (Accessed 15 December 2006).

Government of the Northwest Territories (2005a) *Social Impacts of the Mackenzie Valley Project. GNWT Beaufort Delta Regional Workshop. Inuvik, NT, June 7 and 8, 2005.* Online. Available HTTP: http://www.hlthss.gov.nt.ca/pdf/reports/social_health/2005/english/mackenzie_valley_gas_project_beaufort_delta.pdf (Accessed 15 December 2006).

Government of the Northwest Territories (2005b) *Dehcho Regional Workshop on the Social Impacts of the Mackenzie Valley Gas Project, Fort Simpson, NT, May 30 and June 1, 2005.* Online. Available HTTP: http://www.hlthss.gov.nt.ca/pdf/reports/social_health/2005/english/mackenzie_valley_gas_project_deh_cho.pdf (Accessed 15 December 2006).

Government of the Northwest Territories (2005c) *Sahtu Regional Workshop on the Social Impacts of the Mackenzie Valley Project., Norman Wells, NT, September 30, 2005.* Online. Available HTTP: http://www.hlthss.gov.nt.ca/pdf/reports/social_health/2005/english/mackenzie_valley_gas_project_sahtu.pdf (Accessed 15 December 2006).

Henderson, H. (2003) *Canadian HR Reporter*, December 15, Toronto, 16 (22): 10.

Indian and Northern Affairs Canada (1993) *The Gwich'in (Dene/Métis) Comprehensive Land Claim Agreement, January 1993. Comprehensive Claims Process.* Online. Available HTTP: http://www.ainc-inac.gc.ca/pr/info/info22_e.html (Accessed 18 November 2007).

Inuvialuit Corporate Group (1997). *Inuvialuit Corporate Group: 1996 Annual Report*, Inuvik: Inuvialuit Corporate Group.

Inuvialuit Regional Corporation (2006) *Board Summary and also Newsletter*, March/April: 14. Online. Available HTTP: http://www.irc.inuvialuit.com/publications/pdf/2006-01 percent20March.pdf (Accessed 13 December 2006).

Inuvialuit Regional Corporation (2007a) *Inuvialuit Final Agreement. About.* Online. Available HTTP: http://www.irc.inuvialuit.com/about/finalagreement.html (Accessed on 30 October 2006).

Inuvialuit Regional Corporation (2007b) *Mandate, Goals, Philosophy.* Online. Available HTTP: http:www.irc.inuvialuit.com/about/sbout/mandate.html (Accessed 30 October 2006).

Jaremko, G. (2006) 'Bring on the pipeline, job-starved Arctic says', *The Edmonton Journal*, January 26. Online. Available HTTP: http://www.canada.com/edmontonjournal/news/story.html?id=b15a83f5-ec58-4070-baae-0a7c9d9871a9&k=78474 (Accessed 18 November 2007).

Lee, T. (2006) *Interim Report: Northerners Presenting to the JRP about Looking after the Land.* Prepared for the World Wildlife Fund Canada. Online. Available HTTP: http://www.wwf.ca/AboutWWF/WhatWeDo/Initiatives/RESOURCES/PDF/WWF_InterimHearingReport.pdf (Accessed 18 November 2007).

Mackenzie Gas Project (2008) 'Joint Review Panel hearings phase concludes', *The MGP Exchange*, issue 14, February. Available HTTP: http://www.mackenziegasproject.com/moreInformation/upload/MGPNewsletter_2008_Final.pdf (Accessed 2 April 2008).

North/South Consultants and Inuvialuit Cultural Resource Centre (2005*) Marine Ecosystem Overview of the Beaufort Sea Large Ocean Management Area (LOMA)*, Prepared for Fisheries and Oceans Canada. March 2005. Available HTTP: http://www.dfo-mpo.gc.ca/Library/320674.pdf.

Norwegian, H. (2006) Phone interview with Aldene Meis Mason on November 2.

Northwest Territories Bureau of Statistics (2006) *NWT Social Indicators – 2006 Census Data.* Online. Available HTTP: http://www.stats.gov.nt.ca/Social/home.html (Accessed 12 June 2007).

Northwest Territories (2006) *Energy for the Future.* Online. Available HTTP: http://www.iti.gov.nt.ca/energy/pdf/Whitepaper.pdf (Accessed 18 November 2007).

Park, G. (2006) 'Dealing with the Deh Cho: Canada tables offer to end Mac conflict; Prentice positive, Deh Cho hesitant', *Petroleum News,* 11 (24): 11 June. Online. Available HTTP: http://www.petroleumnews.com/pnarchpop/060611-02.html (Accessed 18 November 2007).

Petroleum Human Resources Sector Council of Canada. (2003) *The Strategic Human Resources Study of the Upstream Petroleum Industry: The Decade Ahead.* Online. Available HTTP: http://www.petrohrsc.ca/pdf/finalreport_april2004.pdf.

Regulatory Roadmaps Project (2004) *Oil and Gas Approvals in the Northwest Territories: Inuvialuit Settlement Region, Sahtu Settlement Area, Gwich'in Settlement Area, Southern Mackenzie Valley, and Beaufort Sea.* Online. Available HTTP: http://www.oilandgasguides.com/aguides.htm.

Sahtu Secretariat Inc. (2005) *The Sahtu Métis Comprehensive Land Claim Agreement. Comprehensive Land Claim Objectives.* Online. Available HTTP: http://www.sahtu.ca (Accessed 18 November 2007).

Status of Women Council of the Northwest Territories (2004a) *Review of the Draft Agreement for the Environmental Impact Review of the Mackenzie Gas Project*, July, 2004. Online. Available HTTP: http://www.statusofwomen.nt.ca/download/review_eir_mgp.pdf.

Status of Women Council of the Northwest Territories (2004b) *Review of Draft Environmental Impact Statement (EIS) Terms of Reference for the Mackenzie Gas Project*, July 2004. Online. Available HTTP: http://www.statusofwomen.nt.ca/download/review_tor.pdf.

Women's Participation in Mining Oil and Gas. Accomplishments in 2004–2005 and Project Report 2005–2006.

Status of Women Council of the Northwest Territories (2005) *The Northern Women in Mining, Oil and Gas Program.* Online. Available HTTP: http://www.statusofwomen.nt.ca/women_mog.htm.

Status of Women Council of the Northwest Territories (2006a) *A 'How to Guide' for Participating in the Socio-Economic Impact Assessment (SEIA) in the NWT.* Online. Available HTTP: http://www.statusofwomen.nt.ca/download/SEIA_How_To_Guide.pdf.

Status of Women Council of the Northwest Territories (2006b) *Presentation to the Gas Project Review Panel General Hearing 16 February 2006 by Gerri Sharpe-Staples, President*. Online. Available HTTP: http://www.ngps.nt.ca/Upload/Other%20Hearing%20Participants/060206_SWC_Presentation.pdf.

Usher, P. J. (1993) 'Northern development, impact assessment and social change', in Waldram, J. B. and Dyck, N. (eds) *Anthropology, Public Policy and Native Peoples in Canada*, Montreal: McGill–Queens University Press.

Usher, P. J. (2002) 'Inuvialuit use of the Beaufort Sea and its resources, 1960–2000', *Arctic*, 55 (Suppl. 1): 18–28.

VanderKlippe, N. (2003) 'Bickering could kill pipeline, warns aboriginal group leader', *Edmonton Journal*, Edmonton, Alta., November 20, p. H.1.

World Wild Life Federation – Canada (2005a) *2005 Annual Report*. Online. Available HTTP: http://www.wwf.ca/Documents/WWF/annualreport2005.pdf (Accessed 18 November 2007).

World Wild Life Federation – Canada (2005b) *Map: Conservation and Industry Activities*. Online. Available HTTP: http://www.wwf.ca/AboutWWF/WhatWeDo/Initiatives/RESOURCES/PDF/PA_HCV_pipelines_prospecting_040205_L_web.pdf (Accessed 18 November 2007).

9 Going North

The new petroleum province of Norway

Ove Heitmann Hansen and
Mette Ravn Midtgard

Introduction

The petroleum industry in Norway is a mature industry, but there have been few discoveries and little activity in the Norwegian Arctic. Some of the reasons for a late start in the Arctic include the lack of suitable technology, minor field discoveries and high costs, as well as strong political tension around the environmental issues of expanding oil and gas activity. Ever since oil and gas in the Arctic first appeared on the political agenda during the 1970s, it has been a divisive issue in Norwegian politics. This conflict cuts across the traditional left–right groups of Norwegian politics.

New technological developments, particularly sub-sea techniques and horizontal drilling, high oil prices and a rapidly increasing, worldwide demand for energy have brought the Norwegian Arctic into a new focus. There is still an ongoing 'political battle of the North', but the present debate has shifted focus from whether or not to explore and exploit oil and gas, to 'where', 'when' and 'how'. Local politicians demand petroleum land facilities to create new jobs and new regional businesses. At the same time, environmental organizations, fishermen and politicians are worried about potential accidents and toxic waste from the oil industry in a vulnerable area. The discussions about expansion of the oil and gas activities to the Arctic Norway can be separated into four main discourses: the environmental; the regional economic; the rights and situation of the Sámi minority and the international.

Introduction to the Norwegian arena

Norwegian society is a modern, market economy with a centralized system of governance. Because policies for the utilization of natural resources are adopted at the national level, Norway's development of oil and gas activity is, in many ways, a success story about how Norway ensures its petroleum resources are managed for the benefit of the Norwegian society as a whole (see Textbox 9.1). The Norwegian petroleum sector has contributed significantly to the economic growth and to the Norwegian welfare state.

Textbox 9.1 A brief overview of Norwegian petroleum history

The first licence round was announced on April 13, 1965, and 22 permissions were assigned for extraction in 78 blocks. The permission for extraction gave exclusive rights to investigate, drill and extract. The first wildcat was drilled on the Norwegian Continental Shelf (NCS) in the summer of 1966. It proved to be dry. The oil adventure began in the summer of 1969 when Phillips Petroleum Company Norway completed its last exploratory well bore at the Ekofisk field. The company was close to giving up after several futile attempts, and was getting ready to pack up and leave when they made a massive oil discovery. The oil production started in the North Sea from the Ekofisk Field in 1971. The first three extraction permissions north of sixty-second parallel were given in 1980. A number of oil and gas accumulations have since been discovered. Several pipelines have been built to marked areas in Europe. Since 1980, 39 extraction permissions have been given in the Barents Sea.

The Norwegian Government wanted moderate development and, therefore, only a limited number of blocks were opened at each announcement. In the beginning, there was a dominance of foreign companies that carried out the activity, but key goals for Norwegian oil and gas policies since the early 1970s have always been national management and control. The White Paper No. 76 (1970–71) stated the most important premises for a national petroleum policy, and the Norwegian state moved from being a tax-collecting organ to an oil producer. The Government wanted to build up the Norwegian oil industry and state participation. The Norwegian oil company Statoil was established in 1972 as a state-owned company, and the principle of 50 per cent state participation in each production licence was established. Later, the Storting decided that the level of state participation could be higher or lower than 50 per cent, depending on circumstances. Statoil was partially privatized in 2001, and now operates on the same terms as every other player on the NCS. In 2007, the two Norwegian oil companies, Statoil and Hydro, merged into one company, named StatoilHydro.

The political context of oil and gas in the Arctic

The environmental movement in Norway developed during the 1960s and early 1970s, and the discourse has centred on the exploitation of hydroelectric power and nature conservation. During that time, the fusion of populist and environmental ideas led to the emergence of a new political split in Norway: economic growth versus environmental concerns.

Later, the growth versus environmental protection division re-emerged in the context of the petroleum sector. Whereas the question of state control over oil activities fell along the left–right dimension, questions about the pace of construction and extraction, as well as the controversy over proposed drilling off the coast of the Norwegian Arctic, grouped the parties along the same axis as the growth versus protection issues of the 1970s (Knutsen, 1997). The issue of extraction rates, however, more or less disappeared from the political agenda in the 1990s: During the 1980s, 1990s and thereafter, the extraction rates proposed in the 1970s have been dramatically exceeded (Willoch, 1996).

The concept of sustainable development broadened the environmental agenda in the late 1980s. Many of the issues and problems addressed by the World Commission, however, touched upon existing debates and conflicts within Norway, and sustainable development was to some degree adapted to already existing political ideologies (Aardal, 1993). While the follow-up to the World Commission reduced the tension concerning economic growth to some extent, the energy split continued to play out on the issues of oil and gas in the Barents Sea, energy consumption and CO_2 emissions. The importance of this split is illustrated by the political dispute over the domestic use of natural gas and by the building of gas-fired power plants in Norway. In 2000, the Bondevik I Government resigned from Office on the issue of two proposed gas-fired power plants that would increase Norway's CO_2 emissions by approximately 6 per cent per year. This is probably the only government in the world that has left office on the issue of climate change.

The political split over energy cuts across the traditional political divisions in Norway, especially along the left–right axis. The Progress Party have most consistently argued for domestic use of natural gas and the opening of the Barents Sea for oil and gas production. The Conservative Party and the Labour Party are also in favour of development. The Centre Party occupies the middle ground, followed by the Christian People's Party, Liberal Party and Socialist Left Party, who are at the opposite end of the spectrum.[1] The Centre party and the Christian People's party support oil and gas activities, so long as the activity is within the Integrated Management Plan (IMP) and its principles. So does the Liberal party, but they also insist on petroleum-free zones, both in the Barents Sea and outside Nordland. The Socialist Left are against any petroleum activity in principle, but so long as they are part of the coalition government, they are engaging in compromises.

The current resource management model

The Petroleum Act of November 29, 1996, No. 72, provides the general legal basis for the licensing system that regulates Norwegian petroleum activities (see also Textbox 9.2). This act determined that the Norwegian government owns the sub-sea petroleum deposits on the Norwegian Continental Shelf (NCS). The government announces a fixed number of blocks for which the oil companies can apply. The companies must carry out an impact assessment, covering aspects, such as the environmental, economic and social effects of such activities on other industries and adjacent regions. Exploration permission gives rights to perform geological, geophysical, geochemical and geotechnical examinations, including small drillings.

Although competition is desirable, cooperation between the players in the petroleum industry is also beneficial. Therefore, the authorities award production licences to a group of companies rather than to just one company. In each group, one company is responsible for the management of the operation. This management company is responsible for ensuring the procedure complies with the regulations. The authorities judging who is eligible to receive the award must assess the geology of the area proposed for development, along with the technical expertise, financial strength and the experience of the oil company. Based on the applications, the Ministry of Petroleum and Energy establishes a licence group. Within this group, the oil companies exchange ideas and experience, and share the costs and revenues associated with the production licence. The companies compete but must also cooperate to maximize the value of the production licence they have been awarded.

The production licence is awarded for an initial exploration period. A specified work obligation must be met during this period, including seismic surveys and/or exploration drilling. An area fee is charged per square kilometre, according to detailed rules. Provided that all the licensees agree, a licence can be relinquished once the work obligation has been fulfilled. The main rule is that, once petroleum activity ceases, everything must be cleared and removed. To date, the Ministry of Petroleum and Energy has approved more than ten decommissioning plans. In most cases, it decided that abandoned facilities were to be removed and taken ashore.

Textbox 9.2 National organization of the petroleum sector

Norway has a hierarchical structure. At the top we find the Storting (Parliament) and the government. The government has a number of different ministries with different responsibilities in relation to oil and gas activities, and the ministries have directorates for management and control.

The Storting

The Storting, the Norwegian parliament, creates the framework for Norwegian petroleum activities. The methods used include passing legislation and adopting propositions, as well as discussing and responding to white papers concerning the petroleum activities. Major development projects or matters of great public importance must be discussed and approved by the Storting. The Storting also supervises the government and the public administration.

The government

The government holds the executive power over petroleum policy and is responsible vis-à-vis with the Storting for this policy. In applying the policy, the government is supported by the ministries and subordinate directorates and agencies. The responsibility for executing the various roles within the petroleum policy is shared as follows: the Ministry of Petroleum and Energy is responsible for resource management and the sector as a whole; the Ministry of Labour and Social Inclusion is responsible for health, the working environment and safety; the Ministry of Finance is responsible for state revenues; the Ministry of Fisheries and Coastal Affairs is responsible for oil-spill contingency measures and The Ministry of the Environment is responsible for the external environment.

The Ministry of Petroleum and Energy

The Ministry of Petroleum and Energy holds overall responsibility for management of petroleum resources on the Norwegian continental shelf. This includes ensuring that the petroleum activities are carried out in accordance with the guidelines drawn up by the Storting and the government. The Norwegian Petroleum Directorate is administratively subordinate to the Ministry of Petroleum and Energy. In addition, the ministry holds a particular responsibility for monitoring the state-owned corporations, Petoro AS, Gassco AS and Gassnova, and the partly state-owned Statoil ASA.

The Norwegian Petroleum Directorate

The Norwegian Petroleum Directorate (NPD) plays a major role in the management of the petroleum resources, and is an important advisory body for the Ministry of Petroleum and Energy. The Storting established the NPD in 1972. The main task for the Petroleum directorate is to be an organ for management and control. The Norwegian Petroleum Directorate exercises management authority in connection with exploration for and exploitation of petroleum deposits on the Norwegian Continental Shelf (NCS). This also includes the authority to issue regulations and make decisions according to the rules and regulations for petroleum activities.

The Norwegian government's participation at the NCS

The participation in petroleum activities is divided between the State's Direct Financial Interest (SDFI) and its shareholdings in Statoil ASA, Norsk Hydro ASA, Petoro AS and Gassco AS. SDFI is an arrangement whereby the state owns interests in a number of oil and gas fields, pipelines and onshore facilities. The state's interest is decided when production licences are awarded and the size varies from field to field. As one of several owners, the state pays its share of investments and costs, and receives a corresponding share of the income from the production licence. Petoro AS was established in May 2001 as a state-owned limited company to manage the SDFI on behalf of the government, while Gassco AS operates the integrated gas transport system on the NCS. As the majority shareholder, the state owns 70.9 per cent of Statoil. The remaining 29.1 per cent of the shares are divided between private investors in Norway and abroad. About 44 per cent of the shares in Norsk Hydro are owned by the state.

Source: Norwegian Petroleum Directorate.

The government receives a large share of the value created through taxes, charges/fees, direct ownership in the fields (through the State's Direct Financial Interest; SDFI), and dividends from ownership in Statoil. The CO_2 tax, introduced in 1991, is an instrument for reducing CO_2 emissions from the petroleum sector. According to the Kyoto Protocol, Norway is obliged to ensure that average emissions for the years 2008–2012 do not increase by more than 1 per cent compared to the level of emissions in 1990.

The Petroleum Tax Act from June 13, 1975 regulates the taxation system. Petroleum taxation is based upon the rules for ordinary corporation taxes. However, because the oil business is more profitable, there is an additional tax on this business sector. The ordinary tax rate is 28 per cent, and the special tax rate is 50 per cent, for a total tax rate of 78 per cent. All state petroleum revenues are put into a fund called The Government Pension Fund.[2] The Government Pension Fund was established in 2006 and consists of two parts: 'the Government Pension Fund – Global', a continuation of the Petroleum Fund; and 'the Government Pension Fund – Norway', which was previously known as the National Insurance Scheme Fund. The Ministry Finance is responsible for the management of the Government Pension Fund. The purpose of the Government Pension Fund is to facilitate government savings, which are necessary to meet the rapid rise in public pension expenditures in the coming years, and to support the long-term management of petroleum revenues.

Geographical context

Norway has no formal definition of its Arctic areas, but for the purposes of this chapter, it is defined as the Norwegian Sea and land areas north of the Arctic Circle (see Map 9.1).[3]

The size of this sea area is approximately 1,400,000 km², which is four times larger than the Norwegian land area. The area is bounded by the Norwegian Sea in the Southwest, the Arctic Ocean in the North and the Russian part of the Barents Sea in the East.

The Norwegian Petroleum Directorate has calculated that the Barents Sea and the area outside Lofoten may hold 35 per cent of the country's undiscovered oil and gas (see Textbox 9.3). This is estimated to be 1215 million Sm³ oil equivalents

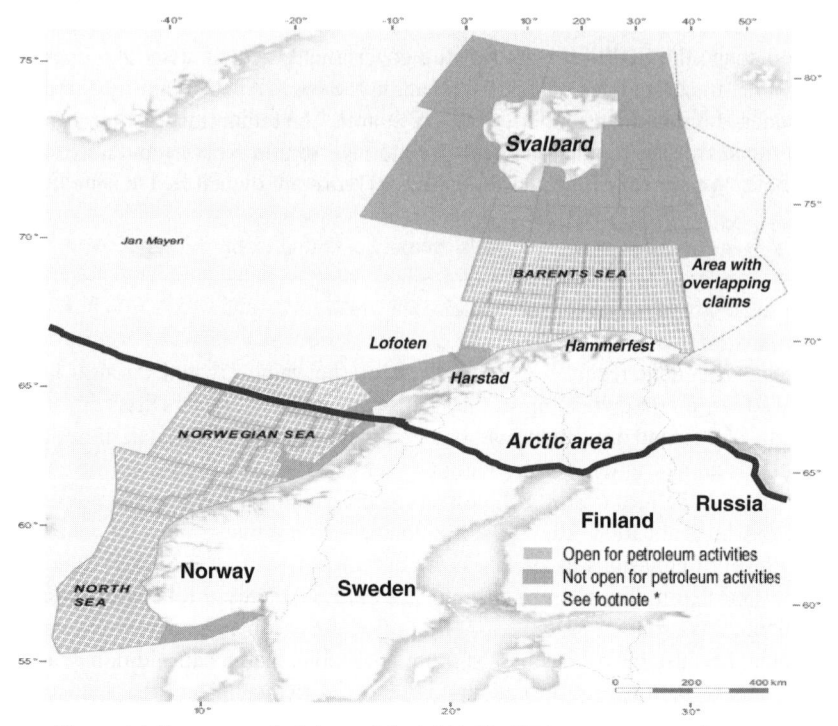

Map 9.1 Open and closed areas on the Norwegian Continental Shelf within the Arctic area. (Source: Norwegian Petroleum Directorate.) *Area excluded from year-round petroleum activities in light of the ULB (government report on the impact on the Lofoten–Barents Sea of petroleum activities).

in total, with 485 million Sm^3 as liquid and 730 million Sm^3 as gas. This estimate does not include the area of overlapping claims between Russia and Norway, a disputed area. The uncertainty linked to this number is greater than that for the rest of the NCS, where there has been more activity. There are higher expectations of finding gas rather than oil in this area.

In Norway today, approximately 80,000 people are employed in the petroleum sector. In 2005, the state's net cash flow from the petroleum sector amounted to approximately 33 per cent of total revenues. In 2005, crude oil, natural gas and pipeline services accounted for 52 per cent of Norway's exports in value terms. Measured in Norwegian krone (NOK), the value of petroleum exports was 445 billion, 35 times higher than that of the export value of fish. The value of the remaining petroleum reserves on the Norwegian Continental Shelf was estimated at 4210 billion NOK in the National Budget for 2006.

Textbox 9.3 The Barents Sea and Lofoten area

The Norwegian Continental Shelf (NCS) in the Barents Sea and outside Lofoten constitutes two different geological territories (provinces). The Barents Sea is shallow water with an average depth of 230 m. The Barents Sea covers 7 per cent of the Arctic Ocean, yet the largest harvestable marine resources of the Arctic are in this area. This is due to the fact that most of the Northeast-Atlantic fish resources have part of or all their life cycle in the Barents Sea. There are about 5.4 million nesting seabirds in the Norwegian part of the Barents Sea, and most of these birds migrate south in the winter.

The Shelf outside Lofoten

The Continental Shelf by Lofoten is geologically very narrow and, therefore, a limited area. Parts of the Shelf outside Lofoten (Nordland VI) were opened for exploration in 1994, and one wildcat was drilled in 1999. The assessed potential of this area is very high, and the source rock shows that gas is most likely to be found. Also, oil is likely to be found in shallower water. This area is temporary closed for oil and gas activity.

The Barents Sea Shelf

The Barents Sea Shelf is divided into many test-drilling areas, which vary considerably and have very different geological histories. These areas have had little or no investigation but some (the Hammerfest basin, the Tromsø basin and parts of Lopphøgda) have been better explored. Parts of the Tromsø basin were open for petroleum activity in 1979 and the first permission for production was given in 1980. The Tromsø basin was expanded in 1985, and the South Barents Sea was opened for exploration in 1989. A total of 39 licence permissions have been granted in the Barents Sea (2006), 64 wells have been drilled, and several small- and middle-sized gas deposits have been discovered.

Source: Integrated Management plan for the Barents Sea.

Activity in the Barents Sea Region

All Norwegian petroleum activity takes place offshore. Onshore reserves or reservoirs have not been identified and the probability of any reserves is regarded as low. Exploration in the Barents Sea has yielded a number of discoveries. One project is the Snøhvit gas field, northwest of Hammerfest, a city in the county of Finnmark. The project is under construction and involves sub-sea-completed wells, a process plant for gas liquefaction (liquefied natural gas; LNG) and a closed system for separating and re-injecting the carbon dioxide into the reservoir. The field came on-stream in 2007. This is the first time Norway has signed agreements to sell gas to markets outside of Europe. Another commercial oil field is Goliat, where the production licence process will begin in 2007. Goliat will face serious challenges due to the conflicting indigenous, social, environmental, biological and fishing-related issues. Norwegian Hydro made the latest discovery in the Barents Sea in an oil field called Nucula.

The first Integrated Management Plan in Norway

In 2002, the Norwegian parliament decided that the government should prepare an Integrated Management Plan (IMP) for the Barents Sea. This decision, the follow-up to the 'Protecting the Riches of the Seas' report,[4] resulted in the first comprehensive management plan in Norway. The management plan should assess the considerations related to the environment, the fishing industry, the petroleum industry and sea transportation arising from increased activity in the Norwegian Arctic. Furthermore, the plan states: 'These plans must have sustainable development as a central objective, and management of the ecosystems must be based on the precautionary principle [World Commission on the Ethics of Scientific Knowledge and Technology, 2005] and be implemented with respect for the limits that nature can tolerate'.[5]

The IMP for the Barents Sea and the Lofoten area is the starting point for management plans for other Norwegian Sea areas. The Storting approved the plan on March 31, 2006. As a result, the Lofoten area and the Northern Barents Sea remains closed to petroleum activity, which includes the area around Bear Island, waters further North, the Tromsø Patch and the edge of the pack ice (see Map 9.2). An additional 50-km wide zone from Troms II and east along the coast of Finnmark would also be protected. The exceptions are where activity has already started and already announced blocks in the area, which range from 35 to 50 km off the coast.

Oil production in Norway is expected to fall gradually after 2008. Gas production is expected to increase until 2010. From representing around 30 per cent of Norwegian petroleum production in 2005, gas production is likely to continue to increase and may come to represent a share of more than 50 per cent by 2014.

The environmental aspect and arguments in the Arctic petroleum debate

The Arctic region is characterized by its pristine nature and often regarded as the frontier. A common assumption is that the Barents Sea is biologically vulnerable, more vulnerable, but also cleaner than any other oceans. Although biological to some degree, this impression of its nature is related to the traditional main employer in the coastal areas, the fisheries. So far it is either the fear of harming nature, or the lack of that fear, that involves people the most and generates the most heated debates. Facts are uncertain; the problem has nothing to do with the discovery of a particular fact, but with interpretation of a complex reality. And values are

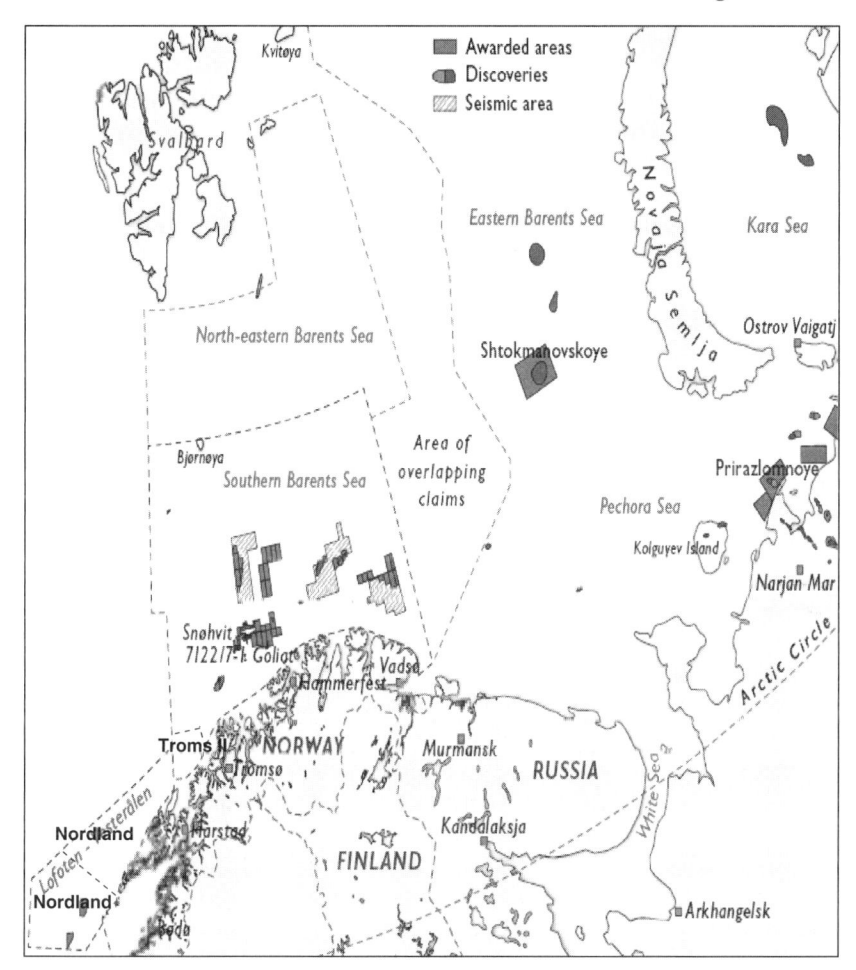

Map 9.2 Norwegian and Russian parts of the Barents Sea.

in dispute; what is at stake is of huge importance to several stakeholders: the costs, the benefits and so on. The IMP for the Barents Sea and Lofoten area identifies some areas as especially vulnerable. The physical, chemical and biological characteristics vary from area to area, and one area is not equally vulnerable the whole year round or to different types of impact. Therefore, many arguments in the environmental debate have different perspectives, and often the stakeholders use many different stories to underline their point of view.

Is the Arctic environment vulnerable or robust?

The concern for nature has been the most highlighted argument for not opening the Norwegian Barents Sea to oil and gas exploitation. Scientists, to different degrees, claim that the environment is too fragile in the Barents Sea for this type of economic activity. On the other side, this debate has many stakeholders, and we find politicians, people in the petroleum industry and scientists who are sceptical of the vulnerability argument: They argue that there is no evidence for excluding petroleum activity from the Barents Sea.

Environmental organizations are most emphatic about the vulnerability of the environment in the Barents Sea and Lofoten area. They are all against oil and gas activities in the Barents Sea and the Lofoten area, and claim that this region is threatened by oil and gas and related activities. Both Bellona and the World Wildlife Fund (WWF) Norway have chosen to emphasize the vulnerability of the Norwegian Sea and the Barents Sea in the fight against oil and gas activities in the Arctic. In fact, Bellona's position is that there should be no oil and gas activity at all in the Barents Sea and Lofoten area. They argue that oil activity represents a great threat to natural and renewable resources in this region and that an oil spill incident will result in very big losses for the ecosystems, as well as economic losses for the fisheries and aquaculture industry. Coldness, ice and sometimes extreme weather make both oil activity and clean-up after an incident very difficult and risky.

The arguments about the Barents Sea as one of the world's most important ecological regions are taken further by the WWF, which has highlighted it as an area with extraordinary biodiversity value, being the world's highest density of migratory seabirds, some of the richest fisheries in the world, diverse and rare communities of sea mammals and the largest deep water coral reef in the world. Oil and gas development may result in discharges of drilling chemicals, radioactivity and produced water, and will certainly result in habitat destruction and a risk of medium to large oil spills through blowouts, and pipeline leaks when loading on to tankers or other accidents.

> Oil spills in the sea ice, in polynyas or along the ice edge will have particularly dramatic consequences. The existing response system and procedures for dealing with oil spills are in this region of little effect, particularly in rough weather.[6]

The Norwegian Pollution Control Authority (SFT), the Institute of Marine Research (IMR) and the Directorate of Fisheries are all warning against oil exploration and extraction in the Arctic. This is because of the risk of accidents and the lack of knowledge about the consequences resulting from spills. It is especially the lack of knowledge that these institutions underline. SFT is concerned about the large gap in knowledge about the effects of increased activity. The section of the oil and gas industry in SFT has said: 'The utmost care and precautions must be taken in these matters. It is important for us to ensure that central parts of the ecosystem are not damaged or that vulnerable resources are affected in a negative way'.[7] The head of the IMR argues that the area is vulnerable because it is an important spawning area for several fish species.

Other scientists have entered the debate and argued against the conception of the Arctic as vulnerable and even argued against this premise. Four professors from the University of Oslo wrote in the Norwegian newspaper, *Aftenposten*, in 2005: 'The crude oil is a natural product, which is not especially dangerous to the environment, and the Barents Sea is not particularly vulnerable compared to other naval areas in the world'.[8] Their expertise is in biology and geology. The IMR responded:

> The professors give the impression that the ecology in the Barents Sea is robust enough to stand oil extraction ... . Our knowledge indicates that the ecosystems in the northern area are very easily affected by human activity such as fisheries, military operations and also industrial activity on the continent further south through long haul contamination.[9]

The professors from Oslo argue that history has shown us that oil spills in sensitive areas have not caused long-term environmental damage. They refer to oil pollution from tankers or oil platforms in the last 30–40 years, which has been very well studied. In particular, they studied the *Exxon Valdez* accident in Alaska in 1989, when 37,000 tons of oil leaked from an oil tanker. They claim that after one year or so, 90 per cent of the ordinary fauna and flora was back, and that integrated studies in the last 15 years have further shown no negative impacts on the salmon, sea lions, seals or sea birds.

An employee of the Norwegian Petroleum Directorate who shares the view of a robust Barents Sea said:

> The Arctic is constantly changing. Both nature and our planet are quite robust, and the planet has faced several changes over time. The Barents Sea is very robust and can handle discharge from crude oil. The ecological consequences of oil and gas activities would be marginal.[10]

This attitude is also common among oil companies, who refer to either 'the four professors' or other official channels. Additionally, they often say that technology will provide solutions that will take care of the environment. A presentation from the Norwegian oil company Statoil stated: 'technology will solve environmental challenges'.[11] The oil company, ENI Norge, who discovered the Goliat oil field in the Barents Sea, has stated: 'Most biologists consider the Barents Sea not to be a sensitive area; it is more politically rather than biologically sensitive'.[12]

Another oil company, BP, said:

> It is possible to enter the Arctic today, there are no technological restrictions against that ... . The challenges in the Arctic can be solved ... . There is a lot of insecurity among people generally and politicians, but how much do people outside the oil branch know about today's possibilities for a safe production? The big oil companies do not want to go into areas where the environment can be threatened; this can have serious consequences for their reputation.[13]

The United Nations Environmental Programme (UNEP) warns against oil and gas activity in the Norwegian Arctic area and is worried about the increasing pressure in Norway for development of the Northern resources. Their Yearbook 2005 stated:

> We know that the Arctic areas are very vulnerable, and the environment is shifting constantly. We can see the climate changes around the Arctic and the Antarctic. Oil and gas activity in the Arctic will be gambling, no matter how strict the safety regulations are.[14]

In the political landscape, we find opinions both supporting and not supporting oil and gas activity; in between are those who are supportive, given severe precautions and heavy restrictions. The right side of the political spectrum is generally more positive to petroleum activity than the left. But all stakeholders, with their different stories of the vulnerability of the area, emphasize that there are differences between the Barents Sea and the area outside Lofoten. There are three drilling areas outside of Lofoten that the discourse highlights: Nordland VI, VII and Troms II (see Map 9.2). The Directorate of Fisheries recommends that neither exploration nor extraction should be allowed in the area east of 400 m sea depth in Nordland VI, VII and Troms II. Compared to other areas outside Norway, Greenpeace says that environmental values at stake in the Barents Sea and Lofoten are quite different from those in the Norwegian Sea, especially from a global perspective. The climate is exposed and short, and simple food chains make this sea environment more vulnerable to oil and chemical pollution than the more southern Norwegian Sea.

In a statement about the Integrated Management Plan, SFT said:

> This area (Nordland VI, VII and Troms II – areas outside Lofoten) is of such great value seen from an environmental- or business perspective that currently any risk of damage is unacceptable. Therefore we have to consider Petroleum-free zones.[15]

Petroleum-free zones: the solution for co-existence

More stringent environmental requirements have been set in the later licence rounds for areas considered more vulnerable. In addition, impact assessment studies were initiated for petroleum recovery in the Barents Sea and in the area off the Lofoten Islands before the seventeenth round. The discussion about petroleum-free zones started in 2002 when Einar Steensnæs of the Christian Party, the Norwegian Petroleum and Energy Minister at the time, said:

> The impact study will give us a better basis for assessing continued operations in an area that is environmentally sensitive. Zero discharges form a basic requirement here. Moreover, we shall consider establishing petroleum-free zones if it appears impossible to achieve co-existence between petroleum and fishery activities.[16]

The Chief Executive Officer of Hydro, Eivind Reiten, commented that there was no doubt that petroleum-free zones would make it considerably more difficult to achieve the government's goal of continued growth on the Norwegian shelf up to 2050, as laid down in White Paper No. 38.

The debate about which areas should be protected and where the petroleum-free zones should be returned to the forefront when the Soria Moria Declaration of 2005 was presented. The environmental organizations, together with the Social Left Party politicians, have been advocates for no petroleum activity in areas they consider vulnerable. The WWF declared that it was the Social Left Party's responsibility to get petroleum-free zones incorporated into the Integrated Management Plan. Greenpeace's point of view was that there is no hurry to extract oil and gas, but it is urgent to protect vulnerable areas. Furthermore, they said that these areas should be kept free from oil and gas activity, and that the IMP should minimally define some of the areas as permanent, petroleum-free zones. Bellona said that the only way to keep areas closed to petroleum activity was to establish petroleum-free zones, and that 'by giving the vulnerable areas outside Lofoten and the Barents Sea the status of petroleum-free zones, some of the world's most unique areas could be secured for the future'.[17]

Politicians in the Socialist Left Party and the Central Party also fear that, if oil companies discover oil in the Barents Sea, it will be next to impossible to deny exploitation. They and the environmental non-governmental organizations (NGOs) consider the only solution to this potential situation is to establish petroleum-free zones.

The environmental groupings were disappointed when the IMP did not define any petroleum-free zones in the Barents Sea. The arguments and accusations were at times very heated and culminated when a draft of the IMP was accidentally presented on a stakeholder's web page while the plan was still under hearing. The Norwegian Oil Industry Association (OLF), an influent stakeholder for increased activity, commented: 'No petroleum-free zones. Perhaps most surprising in the draft of the Integrated Management Plan is that the government is avoiding the term "petroleum-free zones" about the areas that shall have special assessment'.[18]

The Integrated Management Plan was presented to the parliament on March 31, 2006 and there was still an absence of terms such as 'petroleum-free zones' in the official version. The IMP provided a temporary protection of the area outside Lofoten and Vesterålen (Nordland VI, Nordland VII and Troms II). In addition there would be a 50 km wide zone from Troms II and east along the coast of Finnmark that would be protected. The exceptions were where activity had already started and where there were blocks already announced in the area, which ranged from 35 to 50 km off the coast. Therefore, the Goliat project was allowed to continue. At the presentation of the IMP, the Environment Minister said it was a balanced compromise. She also said it would be updated in 2010 based on the knowledge gained in the meantime.

It was important to the environmental organizations that the other parties in parliament supported this temporary protection. This assured them that protection could last longer than just the course of this parliamentary period but, in any case, the debate will surface again in 2009. Regardless of the result, Friends of the Earth will fight to reduce the activity in the North; they will say that a temporary solution is better than nothing. On their homepage we can read:

> It is the fourth time we have managed to keep the platform away from Lofoten and Vesterålen. Although no area was granted permanent protection this time, we know that professional environmental consultations have concluded time after time that these fields cannot withstand oil and gas drilling. Sooner or later the Storting and the Government must listen to their own professional departments.[19]

The WWF's opinion is that petroleum-free zones are the only solution for safeguarding the 45,000 working places, which depend upon a clean and productive ocean. Furthermore, they say that petroleum-free zones are not a direct cost, but an insurance that Norway must afford to secure working places for its Northern population and as an area of world-class nature.

The climatic influence: from local to global

Offshore oil and gas activities entail considerable outputs of gases into the atmosphere from power generation, flaring, well testing, leakage of volatile petroleum components, supply activities and shuttle transportation. Air emissions have effects on the climate. OLF says that the oil and gas industry is willing to buy the quotas needed for Norway to fulfil its climate commitments. At the same time, the industry wants to allocate funds for environmental measures. The environmental organizations and the United Nations Environment Programme (UNEP) are present in the debate over the role of oil and gas due to climate change. The UN has concluded that: 'Increased commercial pressures on the Polar regions bring with them increased threats to the ecosystems and, in the Arctic, increased challenges for the sustainability of local economies and ways of life'.[20] UNEP especially warns against oil drilling in the Barents Sea, as it considers the Arctic areas to be extremely vulnerable to rapid climate change.

The debate about the effects of oil and gas activity on climate change is primarily based on global concerns. More oil and gas activity will not reduce CO_2 emissions; therefore, Norway will have problems with reaching its emission reduction targets. Research is showing that the sea ice in the Arctic, with a doubling of the CO_2 discharges, may not only completely disappear in the summer by 2050, but may also be reduced by 20 per cent in the winter. Under this scenario, the whole Barents Sea would be ice-free in the winter. More stormy weather along the coast would present major negative consequences for sea transportation, oil and gas activities, coastal fishing and fishery and the inhabitants of the coast. The ice in the Barents Sea will be pushed North and East because of increasing Southwestern winds and warmer weather. This will expand the fishing area and possibly make it easier for the oil and gas industry to operate in the winter in a larger part of the Arctic, raising new geopolitical questions.

In 2004, the WWF said they were very disappointed with the Norwegian Foreign Minister, Jan Petersen, from the Conservative Party, who had not taken the opportunity to promise new cuts in CO_2 emissions after the first Kyoto period. Bellona is also using the climate argument to stop oil and gas activity in the Barents Sea and Lofoten area, saying: 'It should not be possible to open up the Barents Sea for exploration because of the climate- and environmental changes'.[21] Bellona claims that the Barents Sea should not be open for exploration owing to climate–environmental changes, as this would lead to it becoming an even more exposed area than earlier assumed. Evidence of accelerating climate change in the Barents Sea and Lofoten area, including notable shrinking of the Arctic sea ice, continues to accumulate, while current and planned expansion of commercial exploitation in both regions has raised concerns for sustainability. Bellona says that oil and gas activity in the Barents Sea and Lofoten area could raise the Norwegian CO_2 emissions by 8 per cent. The government says that the establishment of a CO_2 value chain is high on their agenda, and they are using this to potentially further increase oil recovery, which would contribute towards Norway meeting its international commitments concerning greenhouse gas emissions. But will this extra oil recovery cause more CO_2 from the used oil at the same time?

Another issue in the Arctic is that these climate changes will bring more and heavier storms, which again may result in erosion of coastal areas and invasions of new species to a biological environment that has been protected previously by stable frost. New invading animals may bring new diseases to the area. Researchers also fear that CO_2 and methane contained in the permafrost and seas would be released to the atmosphere as a result of melting ice and a warmer sea. Changes in the climate in the Arctic area could have important effects on the climate in other parts of the world too. But a warmer Arctic will also have positive effects regionally, as greater possibilities for sea transportation, enlarged fisheries and increased petroleum activities in and around Spitsbergen.

It's our turn now

The inhabitants in Northern Norway have been optimistic and positive at several stages during the debate about petroleum development in the province. The last time the petroleum companies operated in the Barents Sea was in 1993, but they soon left when they found no oil, and they considered the rather small gas discoveries not profitable enough for exploration. Since then, new technology solutions and market demands (delivery security and price) have made the Barents Sea area noteworthy again. Statoil were given permission to start constructing the Snøhvit LNG plant in Hammerfest in 2002, and production started in 2007. This gas field has meant the start of a prosperous petroleum age in the Arctic Norway. ENI Norge discovered far more oil in the Goliat field in 2005 than anticipated, which was another major step forwards for further oil activity in the Barents Sea.

The gas project Snøhvit, the Goliat plans, the postponed giant gas field Shtokman on the Russian side and the Soria Moria Declaration have once again caused optimism in the North. This time, at least, construction work is under way and a defined amount of oil has been found in the field. This is an upswing, mentally as well as economically. The mental reorientation is expressed by the report, *2025 Rings in Water*, initiated by the NHO,[22] and by the Soria Moria Declaration:

> The government regards the Northern Areas as Norway's most important strategic target area in the years to come. The Northern Areas have gone from being a deployment area for the security policy to being a power centre for the energy policy, as well as an area that faces great environmental policy challenges.[23]

This mental reorientation has led to optimism and the belief that Finnmark, and to some extent Troms, will be the next growth regions of Norway. Because the Integrated Management Plan has closed Lofoten/Vesterålen for oil activity until the 2009 national election, Nordland has experienced less activity, optimism and attention.

The construction by Statoil of the Snøhvit LNG plant in Hammerfest has meant economic revitalization of businesses in the municipalities of Hammerfest, Kvalsund and Alta. This project has proved that gas activity provides not only direct profits, but also many unexpected, indirect, spin-off effects. From being a municipality in decline, Hammerfest is now one of the most prosperous municipalities in the North. The remaining parts of Finnmark, except South Varanger and Alta, have not participated in this development, and therefore its politicians have raised proposals for how larger areas could benefit from the ongoing and forthcoming petroleum development. The strong wish for participation in what is often called 'the adventure in the North' can be understood in the light of the condition of the basic industry along the coast, the fisheries.

The coastal societies in the North have relied heavily upon the fisheries and fish processing, whose structure and employment were reduced to 6 per cent of the total workforce in 2006. This is the main reason why people in the North claim, 'it's our turn now': the need for economic revitalization and new employment opportunities along the coast are essential for their survival and for strengthening the viability of the industry.

The new significance attributed to Finnmark has also led to a new belief in the value of the county, thus making the politicians and the whole Northern population think they have a good basis for negotiation. This is clearly seen in the local demands; however, the constructors and the operators show less understanding for Finnmark's viewpoint. Additionally, the former policy instruments have disappeared, which again may cause difficulties for the expected and long-desired industrial transformation of the northernmost region of Norway.

Unemployment in the North and shortage in the South

One of the main arguments presented by the Storting for giving Statoil the production licence for the Snøhvit field was that the project will have influence on industry and commerce, employment, scope of expertise and so on, regionally and locally (the Hammerfest area).[24] The same argument is expressed in the Soria Moria Declaration: 'The oil and gas activity will provide more working places and economic growth in Northern Norway'. Before the 2005 election, the pro-petroleum parties were using petroleum development as a tool for employment and modernization of the 'old-fashioned' industrial structure, while those opposed were using environmentally based arguments in the political discourse.

Oil and gas activities have positively affected Norway's employment and economy: In total, Norway has 3 per cent unemployment on average, while eight municipalities in Finnmark, all situated along the coast (Statistics Norway, 2006), have an average of 8 per cent unemployment. The domination of the Northern coastal communities on the list can have a two-fold explanation: Firstly, the collapse of the fish-processing industry, and secondly, it is the larger towns that experience an economic boom – but the coastal municipalities are not part of these growth centres. These types of events pull the youngsters away. People are hopeful that petroleum development will lead to more employment, as Finnmark's own petroleum strategy states: 'The petroleum activity in the Barents Sea is probably the industry that will provide the largest positive effects for both Finnmark and the region as a whole'.[25]

According to all past experiences, oil and gas activity in a region generates more jobs and synergy effects for other types of business. In 2005, during the Snøhvit construction period, the unemployment rate in Hammerfest was low, 3.2 per cent. However, even more remarkable has been the increased participation of local women in the workforce. These are women who had fallen out of the employment statistics, mainly because of long-term unemployment. The construction period opened a window of opportunity for these women, just as in the Kårstø and Tjeldbergodden cases (Olsen, 1988; Kotte, 1997; Gvozdic, 2001). The Snøhvit LNG plant in operation will directly employ 180–200 people, and indirectly another 300 for support services and in public services (Agenda, 2003; Finnmark County 2005). The municipality will receive property taxes from the plant in addition to increased personal taxes, which will lead to further public spendings. This is good news for the Hammerfest area (defined as Alta, Kvalsund and Hammerfest), but almost of no importance to the rest of the county.

Disagreement about the number of jobs

There is some scepticism regarding the estimates of the number of workplaces the oil and gas industry will generate, and of the basis of Northern Norway's optimism. These discussions arose in the autumn of 2005 and involved all the stakeholders identified: scientists, politicians, the petroleum industry and environmental organizations.

On behalf of the WWF, the Nordland Research Institute (NRI) worked out a prognosis regarding employment development and opportunities over the next 35 years in Northern Norway (Bay Larsen, 2005). This report stated an enormous growth potential in the tourism sector and opportunities for real advances in the aquaculture industry. The employment numbers in the report for the three northernmost counties raised a debate both on radio and in newspapers. The report was criticized by the International Research Institute of Stavanger (IRIS) for reasons such as, 'The note seems to be a systematic unbalanced description ... . The extent of mistakes and asymmetrical presentation is so wide, that it cannot be explained by lack of data or uncertain material'.[26] Some considered that this process was initiated by the oil industry and that the aim was to show its power: 'Think about a number; kindergarten places, population numbers or jobs that the adventure will bring. Everything can be said, every positive argument is bought. It all increases the euphoria'.[27] Alternative reports[28] claimed different numbers.

The importance of these numbers relates to politics and to the present discussions about opening up the Lofoten, Vesterålen and Barents Sea for petroleum exploration. Petroleum activity will mean a risk of incidents and an oil spill will result in fishing fields closing down for at least some periods. If the people in the North, who are sceptical of petroleum development,[29] are to change their attitude; then something must be given in return. The answer has been new job opportunities and economic growth, but most employees in the oil and gas industry in the North will commute from other places in Norway. A leader of the Labour Youth Party said:

> There must be an end to the argument that oil and gas activity shall save North Norway. People in Stavanger and foreign countries will get most of the jobs if the Barents Sea and Lofoten area are opened for full activity.[30]

Well-known sceptics, who argue about the number employed by the Snøhvit gas field (an investment of 56 billion NOK), say that we need to invest 1000 billion NOK and build at least 20 Snøhvit plants to create 8000–10,000 new jobs in the region. The only fields used for commercial operation are the Snøhvit gas project and, in 2011, the Goliat oil project in the Barents Sea. A report originating from the consulting firm, Barlindhaug, in the summer of 2006 calculated that 4000 people would be employed in the petroleum sector by 2020. This report also states that the future business development in Northern Norway should be based upon tourism and fisheries.[31] The choice of an exploitation and production solution is an important element in these discussions. Many Northern Norwegians claim that landing the resources will produce the largest benefit. Let us examine the different positions.

Landing of oil and gas, or ...

ENI Norge will deliver their Plan for Development and Operation (PDO) for the Goliat oil in 2008. This oil company has met the demand from the local political authorities to land the oil in Finnmark. The county of Finnmark agreed upon this statement in 2005: 'In the case of oil discoveries, the oil shall be transported ashore in all those cases where it is technically possible'.[32] The Prime Minister, Jens Stoltenberg, sounded a similar note when he said: 'I feel confident that oil- and gas activity in North Norway will contribute to employment and value creation'. He continued, 'we have succeeded in having large land-based facilities. We wish to explore the possibilities of other land-based terminals'.[33] Influential Northern

Norwegian actors, such as Johan P. Bardlindhaug[34] and Arvid Jensen,[35] have also expressed clearly that landing is the most favourable solution for the region and its business life. ENI Norge is considering these questions and has announced three alternative solutions for development. In debate after debate, ENI Norge claims that landing the oil does not necessarily create more jobs locally and that other solutions might benefit Finnmark better. 'Without regard to a chosen technical solution, ENI will make sure that the oil activity in the north results in as many positive spin-offs as possible for the region and for the local community'.[36]

The use of political tools

In the early stages of construction of the Norwegian petroleum manufacturing industry, the oil companies that wanted to position themselves for licences were obliged to offer some benefits to Norwegian firms and regions. The time for goodwill contracts and technology deals are formally over, but the oil companies still recognize that they have corporate responsibilities, as indicated above by the ENI Chairman's statement. Some conditions do exist, such as when the parliament accepted Statoil's application for the Snøhvit production licence. The words of interest in the resolution are: 'The operator must in the planning and construction of the processing- and LNG plant arrange for he disposal of gas and cooling water'.[37]

A new industrial structure

The expectations are huge and, for the businesses in Northern Norway, the petroleum development is more important than ever before. The possibilities that the petroleum sector offers regional businesses are based on the condition that the oil and gas companies are bringing the resources to land. The chairman of the Finnmark County has said that this must be a premise for the industry and that a single-point buoy mooring in the Barents Sea (where tankers just have to fetch the oil) will not be accepted. A member of parliament from the Norwegian Labour Party, Karl Eirik Schøtt-Pedersen, said: 'The oil and gas in the Barents Sea must be pipelined to petroleum installations on land. This is the way we can create jobs and new regional businesses'.[38]

Finnmark – the most prosperous colony in Norway

There are large expectations in Northern Norway that oil and gas activity will generate economic growth, extended effects on other industries, new working places and population growth. Local politicians have several times stressed that the oil and gas activity must generate working places along the coast. Even local politicians from the Social Left Party favour increased activity if the development generates more working places. A local member of the Social Left Party said:

> One thing must be clear. If the public and private working places generated from the oil and gas industry do not appear along the coast of Finnmark,

I will be against oil and gas exploration and production in the North. We must get the working places that are generated by the oil and gas activity.[39]

The Troms County Council for Communication (Synnøve Søndergaard) also supports these statements: 'we will not be a raw material deliverer'.[40] The Northern Norway media, which has been very pro-development, reacted in several chronicles when the merged StatoilHydro revealed in their organization plan that the headquarters for the Northern activities were to be located in Stjørdal, which is not in Northern Norway. The main newspaper, *Nordlys*, as well as the right-wing paper, *Harstad Tidende*, used the words 'colony' and 'colonialism' about this plan and Northern Norway's position. *Harstad Tidende* wrote, 'it is a strengthening of Northern Norway's role as a colony, that's the hard reality' (HT, 2007). People in Northern Norway demand petroleum industry onshore in this part of the country; because they fear that the industry will monitor, control and manage the petroleum exploration from other places in Norway. Technically, the development can take place by means of computers in Stavanger, Oslo or Stjørdal. Accordingly, most politicians in Northern Norway demand that the government must set out clear guidelines for the oil and gas companies. They want research, development, production and operation of the oil industry, oil protection, sea safety and supply services to be present in Northern Norway.

Those environmentalists from Oslo

The environmentalists are the one opponent to the positive atmosphere in the North. When the Snøhvit gas project was about to get its production licence, a few environmentalists in Finnmark and Hammerfest attempted a counter-mobilization. In the debate concerning the Integrated Management Plan for the Barents Sea, however, there were hardly any local voices heard against petroleum development. In the debate preceding the approval of the IMP, the leader of the Nordland Labour Party expressed his opinion: 'It is not Nordland's job to be a nature park for the rest of Norway',[41] and this statement pretty much sums up other expressed opinions. The environmentalists' argument about climate change has so far not been adopted; in fact the most frequently expressed view is 'we like the climate change' (e.g. the summers have become longer and warmer). Many stakeholders from Northern Norway trust technology,[42] or as a bureaucrat expressed it; 'they (e.g. the environmental organization) play perhaps too much on feelings'.[43] The Northern Norwegian environmentalists have, to a very small extent, been recognized in this debate, although the counter-forces have been strong.

In a television debate the evening before the IMP was approved, the mayor of Vesterålen and the chairman of Petro Arctic/leader of the consultancy firm Bedriftskompetanse, used two main arguments against the four representatives from the different environmentalist groups. The first argument was that the environmentalists' views were based on feelings: '...a monopoly on being environmentally-friendly whilst sitting around coffee tables in Oslo'.[44] The second argument was that they were trying to influence development in an area geographically distant from them. They were attacked for being from the Southern part of Norway and without real knowledge about Northern Norway: '...tired of a play-off with people sitting at Grünerløkka[45] and having an opinion about what we here in the North should do for living'.[46]

The following expression is common amongst Northerners, 'please stay in the South and let us do what we know is best'. It seems to them that the environmental organizations have parked themselves in central Oslo and that they lack, at least, active support from the North and, probably more important, they lack legitimacy. The lack of a foothold along the coast is probably connected to their earlier disapproval of seal and whale hunting, which directly influenced the lives of people and their economic basis, and that different people see different values in different regions.

The anxiety of an oil spill

The inhabitants of Northern Norway are full of expectations for the coming oil and gas development. They demand benefits, which is closely connected to the risk involved with such activity. The main argument and expressed anxiety are over the fear of an oil spill. The risk is said to be low,[47] but people living along the coast have experienced loss of fishing boats, and in the 1980s, some Russian ships being towed ended their journey on the seashore. The Northerners, therefore, have several experiences of accidents on their coast.

The risk can be lowered by precautions, such as a transport corridor, more extensive use of pilots and by an improved state of readiness. The distribution between benefits and burdens is complicated, since the stakeholders involved in this debate are very diverse. *Nature and Youth* is of the opinion that landing oil represents an additional threat to the environment, whilst the coastal politicians, as the previous chapter described, want the economic benefits connected with a landing. Bellona is sceptical about the state of readiness. Even though modern technology can reduce discharge of oil and chemicals from petroleum activity, there cannot be any guarantees against accidents in connection with oil exploration, oil extraction and transport of oil along the coast. People in Northern Norway have anxiously followed the development in the oil transportation from Russia, and the lack of an acute state of readiness has been particularly disquieting to them. Their argument is also reiterated by the Lofot Council: 'There is an immediate need to have in place a tugboat capacity with sufficient strength, increase the competence against contamination in the coastal area, and upgrade the oil protection equipment in several coastal municipalities'. Yet governmental institutions claim that three tug boats and the existing surveillance system are sufficient, and the Coastguard, for instance, has commented that 'the state of readiness is quite good'.[48]

Co-existence is possible

Oil production has been occurring in Norwegian waters for nearly four decades, and without serious contamination other than the Bravo blow out in the 1970s.

The strong resistance towards oil exploration has been based on the threat to the fishing industry in the event of an oil spill. Norwegian fish have the image of being the cleanest fish in the world, and the possibility of this reputation being harmed is discussed.

The mayors in Northern Norway express it like this: 'The Barents Sea fish resources, the inhabitants of the coast and the nation of Norway cannot afford an environmental catastrophe happening due to lack of professional competence and insufficient technical equipment'.[49] The coastal seashore is experiencing a greater variety of economic interests; although traditionally the fishermen used the sea, the competition these days is growing more intense and petroleum companies, as well as fish farms, leisure activities, tourism providers and military activities are among the commercial users of this area. Gradually the fishermen have no longer seen these as competitors, and experience has shown their possible co-existence with other industries. The Norwegian Fishermen Association and the Norwegian Coastal Fishermen Union are stakeholders that have knowledge and experience, and the former has expressed this viewpoint: 'Coexistence is possible; there is room for everybody. But, there is a need for common rules, which take care of the environment and what areas that can be explored'.[50]

The demands from the fishermen are based on four basic principles. First, all activity must be based on zero discharge. Second, the petroleum industry must not occupy any important fishing areas. Third, the government must have a close dialogue with the fishing associations when they consider new areas for oil and gas explorations. And fourth, the state of readiness must be increased to an acceptable level.[51] In the official debate, there is much more tension between environmental organizations and petroleum interests on the conditions for the fisheries than that raised by the fishermen's organizations. The Norwegian Pollution Authority and the Institute of Marine Research have indicated what fields ought to have severe restrictions, while the fishermen limit their claims to: 'We need to respect that some areas are more vulnerable than others'.[52]

The fishermen's associations continue to claim that the map data in the area are not good enough to point out areas of conflict, and the fishermen demand more focus and more money to do this work. It is particularly important for the fishing industry to keep the oil and gas transportation at longer distances from the coast and outside the fishing grounds, which will allow more time and opportunities to handle potential oil spills.

The fishing industry is afraid of losing employees to the petroleum industry. They want to focus on educating more people who can serve both sectors, to avoid strong competition between the two. The Norwegian Fishing Vessel Owners' Association puts it this way:

> It is clear that the oil and gas industry is the biggest competitor when it comes to the labour market. The fishing industry is totally free of subsidies, but we have to compete against an industry that is subsidized. We need equal conditions.[53]

It's our right

This section is about the Sámi people (Textbox 9.4) and their reaction to the coming petroleum industry. Northern Norway has experienced a tremendous upheaval among the Sámi people during the last decade, and understanding this development is essential for understanding today's demand that 'it's our right'. It is also essential to understand that, even though they are said to be one people, the Sámi do not represent one homogenous opinion, which is also true for the question of petroleum development. Some Sámi will claim they have been suppressed and are still colonized, while others regard themselves as Norwegians with a Sámi ethnic culture.

The Finnmark Act has faced massive criticism, and the main critique from an indigenous perspective concerns the lack of proper identification and recognition of Sámi rights to their lands at the individual, community and collective levels. The importance of the international standard, ILO Convention No. 169, is stressed, especially in relation to Articles 14 and 15. The Sámi Council emphasizes that the Finnmark Act is not a victory for the Sámi people, but it is a compromise with which the Sámi people can live. They think the Act should have differentiated between Sámi people and local people when it comes to rights, and that it should be more compatible with ILO 169.[54] When evaluating the impact that activities such as drilling for oil and gas have on Sámi culture, one should note that CERD[55] has already underscored the important role of traditional Sámi

---

Textbox 9.4 The Sámi people

The Sámi are the northernmost indigenous people in Europe, and the only ones in the Nordic region. The Sámi people span the borders of Sweden, Norway, Finland and Russia. Approximately 40,000 of the estimated 85,000 Sámi live in Norway. In Norway, the right to keep reindeer also covers subsidiary rights, such as fishing and hunting rights (the reindeer husbandry rights are compounded of many rights of use). A section of the Sámi population in Norway and Russia are coastal: Sámi people who from time immemorial have been fishing along the Norwegian and Russian coast and at sea. Both lawyers and the main Sámi organizations have asserted that certain coastal rights exist for the coastal Sámi population in Norway. These rights, however, have not been confirmed in legislation. The Sámi have been given more legal rights, and there are still internal processes ongoing for establishing cultural, social and political structures, and external processes. So far the Finnmark Act has authority over the land, lakes and rivers of Finnmark, but no jurisdiction of resources below the surface, or in the fjords and outside the coast.

livelihoods, such as reindeer husbandry, hunting and fishing, in preserving and developing the Sámi culture. The Sámi Council's view is:

> Given all people's equal right to maintain and develop their culture, Norway must prevent the non-Sámi society from expanding into Sámi territory in a manner that prevents, or diminishes, the Sámi people's chances of maintaining and developing its culture.[56]

A panel led by Professor Carsten Smith gave their opinion in 2007 about the property rights of the Sámi along the coastal area. The Norwegian Government has said that they are willing to discuss the matter, but are not prepared to make a decision on the case. The Sámi parliament hopes that the Norwegian Government is prepared to accept the resolution from the UN Human Rights Council, which decided that indigenous people's rights shall also include resources in and below the sea. Sámi President, Aili Keskitalo, who belongs to the largest political party in the Sámi parliament, the National Association of Norwegian Sámi (NSR), claims that indigenous people have rights when the nation-state collects tax from resources off the coast of the Sámi's land. She says:

> The international law gives the Sámi people, as an indigenous people, rights to oil and gas resources in our areas. I am not claiming that we have sole rights to the petroleum resources in the northern areas, but the Sámi people do have such rights as an indigenous people.[57]

President Aili Keskitalo's point of departure is that: 'Oil, gas, fish and minerals in the Sámi area are also Sámi resources. Norway is founded on two people, and that people must act accordingly'.[58] The demands from the Sámi parliament specify, first and foremost, that regulations must be issued to ensure that the Sámi are always included and that adequate importance is attached to Sámi views. Secondly, substantive rules must be laid down that recognize and strengthen Sámi rights and access to resources. Thirdly, regulations must be adopted that guarantee Sámi self- and co-determination with regard to resource management. All in all, this 'package' must be adopted within the parameters of indisputable international law. In practical politics, this means that the Sámi parliament should be in dialogue with the government and governmental institutions operating in their land. During the process of making the Integrated Management Plan, the Sámi parliament was regarded only as a receiver of information and not as a body entitled to comment. This changed during the process. President Aili Keskitalo asserted their right to benefit from the petroleum resources in the Barents Sea, and their principal starting point is that the Sámi have a right to the petroleum resources present in Sámi areas, which also includes areas off the coastline. Vice President of the Sámi parliament, Johan Mikkel Sara, supports this view. He says that the Sámi parliament does not in principle oppose petroleum production and development of the Barents Sea, but that certain demands must be fulfilled. Sara claims that Sámi people should have joint decision-making rights in all parts of the process, and that the Sámi have rights to the oil and gas, and he insists that an indigenous dimension must be in place. He also says that the issue has not been sufficiently elaborated, and that the influence of terminals and plants on the reindeer industry, for example, is currently unclear.

But not all stakeholders who have something to say about these matters agree. The Norwegian Pollution Control Authority (SFT) says:

> The Sámi People are not greatly affected by increased oil and gas activity. This is different from other parts of the Arctic, where much of the petroleum activity is onshore and has direct consequences for the indigenous people's traditional activities. ...The people that perhaps are most affected by increased oil and gas activity are the coastal fishermen, including Sámi fishermen. Even then, the Sámi fishermen are only fishing inside the boundary line and not in the open sea.[59]

As the Norwegian Government prepares to open the Arctic areas for oil exploration, some of the country's indigenous people not only want to protect their interests, but they also want to receive their fair share of the riches that oil and gas activity can bring. They seek their share of oil wealth and are trying to team up with other indigenous groups in the Arctic to create a common platform from which to negotiate land rights and a share of the profits from the extracted energy resources. Johan Mikkel Sara has visited communities of indigenous people in Northern Canada to learn how the local groups manage to negotiate deals with the national authorities and the oil companies. He says: 'We can no longer passively stand by and beg for money. We have to learn from the Canadians and start thinking in a new way'.[60] The agreements that Canada's indigenous people have negotiated inspire the Sámi people. The indigenous people of Canada have become a role model when it comes to economic compensation for activity in indigenous areas. Sara continues:

> Why should the Sámi people be left with low-status jobs when industry is unfolding in their traditional settlement areas? It should not be possible! The Sámi people undoubtedly have the right to receive parts of the oil and gas resources as long as it comes from their traditional areas.[61]

But again there are several counter-forces, as reflected by the members of the Energy and Environmental Committee from the Conservative Party:

> All the natural resources in Norway must belong to the whole Norwegian population, whether we are talking about fish or oil. The oil is a non-renewable resource, and cannot belong to a single group. I cannot imagine any geographical area or group of people who should have any special right to the oil. This is the community's property.[62]

Both the Sámi and other Norwegians wish to benefit from the economic stimulus of increased shipping, tourism and oil and gas development in the Northern waters. Many see the potential for economic development, including the director at The Resource Centre for the Rights of Indigenous Peoples, who offers this solution:

> One possibility is to establish a petroleum fund for employment, education, the establishment of business etc for the Sámi people. The Sámi people demand 'benefit sharing' from the profits of oil and gas activity in the North. In southern Norway, they think that the Sámi live on public subsidies. It depends on whose resources these really are; if they are Sámi resources, which are being used for the general public, then perhaps it is the Sámi people who are gradually are sponsoring southern Norway.[63]

This statement is disputable. Many people in Norway consider the Sámi people as part of Norwegian society and, therefore, feel that they should not gain more benefits from the oil and gas activity than other Norwegians. A member of the Social Left Party, who is also a member of the Energy and Environmental Committee in Parliament, stated that she does not understand any Sámi demands, especially rights concerning oil and gas. She said: 'These resources belong the nation as a whole. Assignment of special privileges on an ethnic foundation is completely out of the question'.[64]

The NSR wants adequate participation in the processes that affect the Northern areas. They say:

> All operations in the Barents Sea and coastal areas are in Sámi interests. Therefore, mechanisms must be established that secure considerable value supplied to the Sámi society from existing and future extraction of non-renewable resources in Sámi areas. The Sámi people have on many occasions said that they want part of the money from oil and gas activity off Finnmark to be spent in Sámi areas.[65]

The Sámi hold both a right to self-determination and the right to manage their own natural resources.[66] The right to self-determination means the right to veto for the Sámi parliaments, said Professor Carsten Smith, leader of the expert group that prepared the draft for the convention, which also stated that the convention determines minimum rights and that each national state is at liberty to expand them. Issues of major importance to the Sámi must be subject to negotiations with the Sámi parliament prior to any public authority decision, the convention draft states, adding that such negotiations must be scheduled to give the Sámi parliaments ample time to influence the process and the outcome. Neither Norway, Sweden nor Finland have granted such negotiation rights to the Sámi parliaments. Bjarne Håkon Hanssen, the Norwegian Minister of Labour and Social Affairs, believes that it is politically and formally important that the nation-states and Sámi parliaments agree on the convention before it is implemented. Mr Hanssen said, 'Most of the convention could be implemented without major problems. However, veto rights for the Sámi parliament, as presented by Carsten Smith, would create a debate over fundamental principles in Norway'.[67] President Aili Keskitalo demands that Norway involve the Sámi parliament in all stages of the investigation process. She claims that the Sámi should receive funding from the Foreign Ministry to help them protect their rights, and she hopes that the oil companies will take their own ethical guidelines seriously.

The international dimension

The *US Geological Survey World Petroleum Assessment 2000* announced that 25 per cent of the remaining petroleum deposit was to be found in Arctic waters. This, in combination with political instability in other major oil-producing regions, has put the Norwegian Arctic on the international political agenda. Decisions made by Norway and Russia on petroleum activity in the Barents Sea are followed with great interest in the United States (US) and in Europe. Norway's focus is to ensure political stability and sustainable development through energy delivery. The Norwegian Foreign Minister, Jonas Gahr Støre, explained:

> We will continue to make our contribution to cooperation, stability and predictability in the region. We will position ourselves in the technological and industrial forefront of this historic chapter of oil and gas exploration that is being opened in the Barents Sea. And we will do this in a way that takes due account of the needs of the vulnerable Arctic environment.[68]

Opportunities and responsibilities are what the Foreign Minister emphasizes in the attempt to consolidate a new perspective for Norway and for the whole of Northern Europe. It is about the Norwegian responsibility towards the environment and natural resources, and for the continuation/development of peace and security. The metaphor 'a sea of opportunities' is centred around possibilities for the whole Northern areas, and he further stated, 'The High North needs a sound policy'.

Previous Norwegian Foreign Ministers have travelled to the Barents region, but Gahr Støre went to the High North. It is difficult to find a definition of what the High North is, and yet the term is used more and more both by the Ministry of Foreign Affairs and by the international community. In a Norwegian context, it is often referred to as 'Our High North'. In a 2005 speech about Foreign Policy to the Storting, Foreign Minister Støre said, 'Our High North policy involves our relationship with Russia, the environment and climate issues in vulnerable Arctic areas, the rights of indigenous people, and important aspects of our relationship with our neighbours and partners'.[69] It seems clear that 'Our High North' involves most of the story lines presented in this chapter, and the phrase has become a geographical term. The government's vision for the High North is for it to be a sea of cooperation with clear boundaries, clear rules for economic activity and high environmental standards, where Norway is recognized by its neighbours as a predictable and reliable coastal state that meets its management responsibilities.[70]

Norway needs to combine and unite the requirements of different countries, counties, municipalities and the indigenous people in its policy-making.

The policy must also include the utilization of the fisheries and petroleum in a way that takes into account the vulnerability of the Arctic environment. On the departmental level, the energy issue has overshadowed other issues and changed Norway's perspective. Norwegian authorities have given new emphasis to the High North, making the Barents Sea a vital part of foreign policy priorities. Foreign Minister Støre also said:

> Our main challenge now is to see the High North not as an area with historic security limitations but one with future economic opportunities. Technology makes it possible to strike a balance between industrial activity and environmental concerns.[71]

This petroleum debate is a classic conflict between the interests of industry and environmental groups. In the story line of 'Drilling to help the environment', one finds statements arguing that it is critical to start the exploration as soon as possible, primarily to set an environmental standard that will help to maintain/conserve the environment in the Arctic Ocean. Statements about the environment have traditionally been used by stakeholders who are against drilling in the Arctic, but now supporters of oil and gas activities in the Arctic use the same arguments and are challenging the validity of their opponents' claims.

Environmental concerns about the possibility of a Russian entrance

Together, the Russian and Norwegian authorities govern important fish resources in the Barents Sea. The Russians have discovered some of the largest offshore gas resources in the world in this common sea. Therefore, it is natural for the Norwegian petroleum discourse to debate where environmental issues overlap with the Russian petroleum industry, and the Foreign Minister is using concepts like sustainable or reasonable development in this context. The Norwegian Foreign Minister formulated the cautiousness required by the situation like this in Moscow:

> Both our countries extend into the High North, and we are both responsible for harvesting and managing its resources – fish and energy – in such a way that they will benefit future generations. This is our common future.[72]

One research report explained how the supporters of drilling in the Barents Sea have taken the environmental rhetoric and turned it into an argument in favour of oil exploration.[73] In this way, it could seem as though the environmental movement has lost its own arguments in the battle over the Barents Sea. A representative from the Ministry of Environment says:

> It is a common understanding in the steering committee (of the Integrated Management Plan) that absence of or no activity on the Norwegian part of the Barents Sea would be unfavourable or negative seen from an environmental point of view.[74]

Finnmark County officials share the same opinion: 'It is important to have a close cooperation between Russia and Norway. This, we think will be of benefit to the oil and gas industry, but most of all the environment will benefit from this too'.[75] These arguments are repeated often, and because they have never been directly challenged, they are accepted as true. The Norwegian Pollution Control Authority is sceptical about these arguments, which suggest that drilling will help the environment. They say: 'We have nothing to teach the Russians. The Russian have stricter regulations on discharge than Norway. They demand absolutely zero discharge'.[76] The Social Left Party opposes the direction of this debate and maintains:

> The argument 'Drilling to help the environment' is nonsense. Stakeholders that want to increase the activity in the North always bring this argument forward. This is about Norwegian stakeholders trying to get positioned to the Shtokman field.[77]

Russia is the second largest petroleum producer in the world and, although Russia has great experience, Russian petroleum production has mainly been onshore. The resources in the Barents Sea are at the bottom of a rough, cold sea, so petroleum activity in this kind of environment is relatively new to Russia. In his Moscow speech, the Norwegian Foreign Minister said:

> No one can match the Norwegian experience of petroleum production in rough northern waters under extreme conditions. This is why it makes such good sense for us to work together with Russian partners in the Barents Sea. The combined experience and know-how of Norwegian and Russian oil companies and authorities will create the best possible conditions for efficient development of petroleum resources in these northern waters.[78]

The petroleum industry in the Norwegian Barents Sea must meet the highest environmental and safety standards in the world. As the Norwegian Conservative Party has said:

> It is very important with respect to the environment that we participate together with Russia. Through cooperation we can influence and introduce technological solutions and set standards. The Russians are far behind us when it become to handling such questions (co-existence between oil, environment, indigenous people, etc). It is important that we set the international standards for how the companies can carry out their activities.[79]

Norway considers that it has something to offer and is prepared to transfer their state-of-the-art knowledge and technology within all areas of offshore exploration and production in Northern waters. This is probably also why two Norwegian companies, Statoil and Hydro, were on Gazprom's short list of possible foreign partners in the Shtokman field, 'and later got access to as the merged company StatoilHydro': They are willing to trade their offshore expertise and technology with Russian partners with regard to access in the gas field.

Using more energy to help the developing countries

The Norwegian oil company, Hydro, claims that we have to find more oil and gas and produce more energy to solve the energy crises and to help the poor countries to develop.[80] However, both the Minister of the Environment and several NGOs, like the WWF, claim that these assumptions are wrong and that it would be the wrong method, either for helping the world's poorest or for tackling pollution. According to the secretary-general of the WWF, the argument about greater use of oil and gas is a 'recipe for how to go to hell by first class'.[81] The former Petroleum and Energy Minister, Torhild Widvey, has said that Norway has a global responsibility to provide the world with more energy, which will contribute to increased wealth in the developing countries:

> In the next 25 years the poorest countries will be next in line to want their energy needs satisfied. With what justice can we say that once we have satisfied our energy requirements, then it is time to shut down, and not let other countries take part in a similar wealth development.[82]

The International Energy Agency (IEA) assumes that the world's energy needs will rise by 60 per cent until 2030. Eighty per cent of this increase will come from petroleum. Today 20 per cent of the world's populations live in the rich Organisation for Economic Co-operation and Development (OECD) countries, which use 50 per cent of the world's energy resources. Torhild Widvey says that it cannot continue like this and has further said:

> The developing countries must get a larger proportion of the energy resources, and Norway possesses a large part of the undiscovered natural resources in the world. It is the Barents Sea and the Lofoten area, along with the Middle East, that will be the future energy regions. We have to use our competence to explore and produce the petroleum resources in an environmentally friendly way. Therefore, an increased activity in the North will contribute to a fairer world.[83]

This statement is considered as irresponsible, disloyal and selfish, according to Hallgeir Langeland, the Social Left Party's spokesman on environmental issues. Langeland underlines that climate change has to be considered when talking about energy production and further petroleum activity in the North.

The world's need for energy supply

The IEA is urging the Norwegian government to consider opening up currently closed areas in the Lofoten area in Northern Norway for oil and gas exploration.

The issue of exploration activities in the Lofoten area, an important area for the Norwegian fishing industry, is a contentious one in Norway. One viewpoint is that it is ill-suited to the environmental agenda in Norway and the political framework of the current coalition government, as formulated in the Soria Moria Declaration. Claude Mandil, IEA Director, said in 2005 that the IEA's view is that exploration activities can live side by side with fishing activities: 'We think it is absolutely compatible to explore and produce oil and gas without harming the environment and fisheries'[84], Mandil stated, adding that the IEA is eager to ensure there is enough exploration and production worldwide, as constraints have recently increased oil prices and deliveries do not seem secure. From this, it appeared that the IEA was asking the Norwegian government to consider opening up Lofoten (which in practical terms means the Nordland VI block) for exploration activity.

The Norwegian Petroleum and Energy Minister, Odd Roger Enoksen, said that there are three key words for the government in the context of policies for the North: activity, presence and knowledge. The Norwegian Arctic is a strategic, main interest for Norway. The Northern strategy is also part of an international policy. The Energy Minister has also stated that Norway's political goal is to be a good and responsible actor or stakeholder in the North, which also means on the international stage. International politics involves regulation, management of values and exchange of viewpoints, which will all influence our national interests and access to resources. Norway will discuss the Barents Sea not only with Russia, but also with everyone who is affected. Talks have already been initiated with France, Germany, the European Commission and Finland, and there will also be talks with the US and Canada.[85]

The global demand for oil and gas products will increase, even if the price of crude oil doubles. China's increasing oil consumption and the US's inability to stabilize their consumption are important driving forces behind this demand (GEO Year Book, 2006).

> As global energy demand increases, so does pressure to explore and develop undiscovered energy resources that may reside in the Arctic. The lifting of an embargo on offshore hydrocarbon exploration in the Norwegian Barents Sea in 2004 has renewed activity there. Regulation of exploration is an important political issue. Debate in 2005 focused on environmental protection and establishing areas free of oil development. According to the International Energy Agency (IEA), if existing energy policies continue, the world's energy needs will be almost 60 per cent higher in 2030 than in 2004. Arguably, this increase in demand could be met from present known fossil fuel reserves.
>
> (IEA, 2006a)[86]

Oil and gas will probably continue to dominate the global energy mix for the foreseeable future, unless alternatives to fossil fuels are found. Alternatively, major changes in global energy patterns may be driven by concerns about energy security, access and the negative impact of current patterns of energy use – particularly on climate change and the health issues associated with air pollution. Already, there is some action in this direction. Norway's Minister of Petroleum and Energy, Odd Roger Enoksen, has said that there are many joint challenges needed to be identified and acted upon, the two most important elements being security of energy supply, and the need to further increase sustainable and environmentally sound oil and gas activities. This will contribute to an improvement of the political and public perception of the oil and gas sector. He has also expressed a concern for establishing a sustainable energy agenda for the North, as this region may prove to become a new petroleum province in Europe and thus become important in ensuring the security of the petroleum supply.

The European Union (EU) and Russia have agreed to strengthen the links between them. This is of major importance since Russia is the largest exporter of gas to the EU. But several of the member states have reminded Norway, as the second largest gas exporter to the EU, that it also plays an important role. One of the points highlighted by the EU is that the development of the Barents Sea, on both the Norwegian and Russian sides, is important for future energy supplies, because there is an increasing dependence on imported energy in the EU.

Russia is the most important country for Norway to have a dialogue with on opportunities and responsibilities in the North. However, as the government recommended in 2007, Norway must also engage our European and American partners in a close dialogue. This process has already started. The countries have responded positively to the Norwegian invitation to participate in what Norway calls 'The High North dialogues', most likely because they realize that energy security gives new substance to the concept of geopolitics. An industrialized country that is unable to secure stable energy deliveries is heading for big trouble. On this matter, the Norwegian Foreign Minister, Jonas Gahr Støre, said:

> Now it is the energy issue that is overshadowing other issues and changing our perspective, not only here in Norway and in neighbouring Russia, but in all countries that are concerned with energy production, supply security and climate and environmental challenges.[87]

It's all contradictions – it's about values

This chapter has shown the major discourses concerning petroleum development activities in Northern Norway. Some stakeholders are more prominent and some have more urgency than others. Urgency is connected to power, and money rules. The economic considerations are an important part in the debates for regional development and for the Sámi's claims. The politicians in Northern Norway are working hard to defend the region from just being a raw materials supplier. The petroleum activity is known to include risk and, therefore, benefits are demanded. But still the most active stakeholder in several of the discourses are the environmental organizations, and what they lack in urgency they have in legitimacy; at minimum, they have support among the technical elite and in the Southern part of Norway.

The discourses are basically about people, values and the facts. But facts and values cannot be separated, and there are several uncertainties, including about ethics. At stake are, among other things, multiple values and a diversity of perceptions, including landscape value, cultural value, economic value, ecological values and political legitimacy.

Regarding ecological values, is there any reason to believe that Norwegians share a common perception in how they value different parts of the environment? For instance, concerning visual pollution, do the locals define oil rigs on the horizon as having a negative impact, or is it their concern for tourism that counts the most? From earlier environmental conflicts, we know there are dividing lines between urban and rural, between left- and right-wing political views, between the educated and the less educated, and so on. To elaborate on that observation, our research did not find one single instance of Northern Norwegian protests based on ecological values. Their major concern is about the conditions for the fishermen, whereas the numbers of Northern Norwegian fishermen and the economic value are not of severe importance. The whole debate surrounding fish is really a debate about cultural values.

Their identity, the feeling of being a Northerner, is strongly tied to being a fisherman, yet today only 6 per cent of them are employed in the marine sector in Finnmark. The numbers have fallen dramatically over the last 20 years, and first reduced their urgency and then their legitimacy as an important stakeholder. Therefore, their voice is not often heard, and the Fishermen's Association even tend to be positive. This needs further investigation, but clearly the cultural transformation, based on the earlier economic transformation, is now paving the way for the petroleum industry to open and explore the High North.

Above or below these community transformation processes are the geopolitical realities. In the government's Soria Moria Declaration,[88] the High North is defined as the 'most important strategic target area in the years to come'.[89] The semantic change in how the region is referred to is mainly caused by the optimism in the petroleum industry, but it is naive not to include the geopolitical dimensions of this. The Norwegian Arctic is being called 'a land of opportunity', and the government have started to use the High North as a synonym for the northernmost part of Norway. The government has strongly entered the debate about oil and gas activities in the Norwegian Arctic, and the Minister of Foreign Affairs, Jonas Gahr Støre, has especially emphasized his department's interest in the province. He has tried to form a coalition across all the discourses brought forwards in the Arctic oil and gas debate. In his speech at the EPC[90] Policy Briefing in Brussels, he said:

> The High North is not only the High North of Norway – and Russia. It is the High North of Europe. Norway's policy in this region – which covers fisheries, the environment, transport, indigenous peoples – and, of course, energy – is at the same time a key component of the Norwegian regional policy and Norwegian European Policy.[91]

Oil and gas activity in the Norwegian Arctic is of great importance: It is about energy, regional policy, the environment, employment and economic growth. In the long term, it is about safety and sovereignty in an area where the world sees renewable and non-renewable resources and opportunities. The Minister of Foreign Affairs continued his speech in Brussels with these words:

> Through cooperation with Russia, Norway is seeking a comprehensive and coherent development of the Barents Sea as a petroleum province. ... It is here that modern Norway is addressing the sustainable development of living resources, including some of the world's most precious fish stock and the impact of climate change – as the polar ice melts. And it is here that we are seeking to develop Europe's youngest energy region side by side and in cooperation with our Russian neighbour.[92]

The political landscape contains supporters, sceptics and opponents. Primarily the more to the political right a group is, the more positive to oil and gas activity they are. The social leftists and opponents against oil and gas activity are strongly supported by the environmental organizations. These organizations all agree that it is unsafe to open up the Lofoten area (Nordland VI/VII), and they are also sceptical of further development of the Barents Sea. They are especially against oil and gas activity in the North Barents Sea. Politicians from Northern Norway support further petroleum development, which they believe will lead to more employment and industrial modernization.

The inhabitants of Northern Norway are mostly positive about increased oil and gas activity, a view that also criss-crosses the political parties. Local people, communities and municipalities in the Northern areas support increased oil and gas activity, but at the same time, they speak for the value of the fisheries. This means we have discourse coalitions that shift depending on the stories. Research and education are important factors for all the stakeholders. If the oil companies are to succeed, they must have a qualified labour force and adequate solutions for development under harsh conditions. The challenges in the Arctic need new methods for carrying out that development. The oil and gas branch is, of course, an important participant in the debate, and there is a constant demand by companies for larger fields for exploration. The Sámi people have not been a prominent stakeholder so far, but lately they have made statements about their indigenous rights.

As the indigenous people in Northern Norway, the Sámi have gained increased power and legitimacy over the land during the last ten years. They have politicians, and after the Snøhvit project was accepted, they realized how influenced they were by petroleum development. Any type of increased business development causes harm to a grassland that already is too small. They are learning step by step; for instance, the Canadian agreements are often mentioned as an example for them to follow, and they are raising their voices. The Finnmark Act was implemented in 2006, and a detailed statement concerning the Sea Sámi's civil rights is expected. In an international setting, they have a high degree of legitimacy, but still their urgency is limited.

The Norwegian Petroleum Directorate (2007) presented their scenarios for petroleum development through to 2046. Their presentation forecast four different scenarios for developing Norwegian petroleum activity. The scenarios range from big discoveries and activities in all sectors of the NCS, to 'Sorry, we're closed'. The NPD's basic concept is that all relevant areas, except the area of overlapping claims, Jan Mayen and Antarctica, will be open for exploration and production. The four-development path goes from A to D. A (full gAs) assumes low oil and gas prices, but the whole NCS will be open and large discoveries will be made. B (TechnolaB) has lasting high prices, the whole NCS is open, and we will find more oil and gas. C ('Sorry, we're closed') has high prices, but the exploration result is so bad that the activity is closed. The last one is D (BlooD, sweat and tears) has high prices, but too few discoveries, and part of the NCS must be closed down owing to environmentalists' demands.

The presentation of these scenarios has heated the debate among stakeholders in the Norwegian discourses. In the news it was said that 'Norway is exhausting the oil and gas resources and coming generations will curse the greed of our generation'.[93] The environmental organizations claim that we must pay more attention to the climate. They say that we all know that production will decline. The question is: how we can prepare for it? The environmentalists say that the best thing for Norway and the climate is to prepare for phasing out of the oil and gas activity, and not grant new licences for exploration areas. If no new discoveries are made in the Norwegian or Barents Seas, and Norway wants to maintain its income from the oil and gas activity, there is no doubt that it must open up areas around Lofoten for exploration. NPD believes that, without new discoveries, production will drop by 60 per cent over the next 20 years. A sudden drop in income will be unacceptable for the majority of the Norwegian people; therefore, the government will be forced to change the Integrated Management Plan in 2010 and open up the temporary closed areas. But to do so, the government also need to satisfy climate targets and invest in new and improved clean technology. One way to keep production high and improve the recovery rates is to invest in new technology in existing fields. The Norwegian Oil Industry Association has estimated that the potential from implementing intelligent oilfields or integrated operations is 250 billion NOK. The use of new technology and changes in organization will give the oil companies opportunities to make better, faster and more accurate decisions. But this development will also mean that it is possible to control both existing and future oil and gas fields remotely, which means that this may not generate many local jobs, or have large spillover effects and benefits to the local inhabitants.

According to the IEA, global energy demands will increase by 53 per cent by 2030. All EU countries, including Great Britain, are today net importers of oil and gas. The EU is importing more than 50 per cent of its present energy needs, a figure that is estimated to increase to 70 per cent by 2030. And like other regions, this increase is mainly due to a steadily rising need to import oil and gas. Norway has been exporting most of its oil and almost all of its gas to Europe since the 1970s, when oil was first produced from the Ekofisk field. Decreased production at the NCS will result in the EU looking for other ways to import energy. Some energy may come from renewable sources, but renewable energy cannot substitute for a drop in production of oil and gas in Norway. The proponents of the Arctic oil and gas will therefore argue that is it both necessary to develop a petroleum industry in the Norwegian Arctic, and that it is possible to do so in harmony with the environment. 'Going North' is therefore not only a new petroleum province of Norway, but also a new petroleum province of Europe. Most likely, this also includes oil and gas activities in the areas around Lofoten.

Notes

1 Traditionally, the parties where positioned on the left–right axis like this: Socialist Left Party, Labour Party, Liberal Party, Christian People's Party, Centre Party, Conservative Party, Progress Party. The Centre Party, however, has moved to a position close to the Socialist Left and Labour. As seen above, however, the energy axis is quite different.
2 http://www.dep.no/fin/english/topics/p10001617/
3 Owing to our different purposes, this chapter and the Arctic Human Development Report (AHDR: Arctic Council, 2004) operate with different demarcations.
4 Ministry of the Environment [Miljøverndepartementet] (2001) Report No. 12, (2001–2002), *Protecting the Riches of the Seas*.
5 Ministry of the Environment [Miljøverndepartementet] (2001) Report No. 12, (2001–2002), *Protecting the Riches of the Seas*.
6 WWF (2003) *The Barents Sea Ecoregion: A Biodiversity Assessment*.
7 Informant interview at The Norwegian Pollution Control Authority (SFT) May 23, 2006.
8 A chronicle in the Norwegian newspaper, *Aftenposten* October 11, 2005 by Gray, Ugland, Aagard and Bjørlykke.
9 A chronicle in the Norwegian newspaper, *Aftenposten* October 15, 2005 by research director Ole Arve Misund at the Institute of Marine Research.
10 Informant interview, Norwegian Petroleum Directorate October 5, 2005.
11 Presentation given by Statoil at UN Symposium on Sustainable Development Doha, February 6-8, 2006.
12 Informant interview at ENI Norge, June 28, 2006.
13 Informant interview at BP, June 26, 2006.
14 Marion Cheatle, department director in the UN programme to http://www.dn.no, February 8, 2006.
15 Referred to in http://www.greenpeace.no, February 9, 2006.
16 At Norsk Hydro seminar, November 7, 2002: http://www.hydro.com.
17 Bellona fact sheet, September 2, 2004.
18 Press release from OLF February 6, 2006; http://www.olf.no/nyheter/ntb/2006/02/?30179.
19 News from Friends of the Earth, March 31, 2006; http://www.naturvern.no/cgi-bin/naturvern/imaker?id=83098.
20 Conclusion of the GEO Year Book 2006; http://www.unep.org/geo/yearbook/yb2006/052.asp.
21 The Bellona Conference, September 16, 2004.
22 NHO is the Norwegian abbreviation for Confederation of Norwegian Enterprise.
23 The Soria Moria Declaration (2005) is the platform of the red–green government, which consist of The Socialist Left Party, The Labour Party and The Center Party.

Going North 237

24 Storting Proposition No. 35, 2001–2002.
25 Finnmark County (2005) *Petroleum Strategy for Finnmark County 2006–2009*, 2005: 19.
26 Thesen and Leknes (2005).
27 Aasjord (2005).
28 Grimsrud (2004), Barlindhaug (2005a, b).
29 Sandersen et al. (2002) and Brastad et al. (2004).
30 Martin Henriksen, president of the Norwegian Labour Youth (AUF), to Radio Nordkapp, January 24, 2006. Forthcoming leader of the AUF.
31 Barlindhaug (2005b).
32 County Council decision, December 8-9, 2005, Vadsø.
33 Prime Minister, Jens Stoltenberg, to NRK Troms and Finnmark, March 7, 2006.
34 Chairman of the Board of North Norway's largest consulting firm, Barlindhaug AS, of the University Hospital in North Norway, and very active in debates concerning the future economic development of North Norway.
35 Chairman of the Board of Petro Arctic.
36 Spoken by the ENI board leader, Sverre Bore, at the dialogue meeting between the Storting, the petroleum industry, and the County Troms stakeholders, September 8, 2006.
37 Proposal to the Storting no 100 2001–2002.
38 Statement to the Norwegian newspaper, *Finnmarken*, June 15, 2006.
39 Statement by the ENI Board Leader, Sverre Bore, at the Dialogue Meeting, September 7, 2006 in Tromsø.
40 Statement by County Council Synnøve Søndergaard at the OLF Dialogue Meeting, September 7, 2006 in Tromsø.
41 Nordland Labour Party Leader, Gunnar Skjellvik.
42 Informant interview, Finnmark County, April 7, 2006.
43 Informant interview, The Executive Committee for Northern Norway (ECNN), February 20, 2006.
44 Chairman of the Board of Petro Arctic, Arvid Jensen, in the television debate, March 30, 2006.
45 A fashionable district of Oslo.
46 Chairman of the Board of Petro Arctic, Arvid Jensen, in the television debate, March 30, 2006.
47 Integrated Management Plan at DNV.
48 Åge-Leif Godø, Norwegian Coastguard.
49 Coastal Mayors' meeting in Tromsø, September 23, 2003.
50 Statement by the Norwegian Fishermen Association at the OLF stakeholder meeting in Stokmarknes, April 28, 2006.
51 Statement by the Norwegian Fishermen Association at the OLF stakeholder meeting in Stokmarknes, April 28, 2006.
52 Statement by the Norwegian Fishermen Association at the OLF stakeholder meeting in Stokmarknes, April 28, 2006.
53 Statement by the Norwegian Fishing Vessel Owners' Association at the OLF stakeholder meeting in Stokmarknes, April 28, 2006.
54 Informant interview the Sámi Council, May 4, 2006.
55 Committee on the Elimination of Racial Discrimination.
56 Letter to the Committee on the Convention on the Elimination of all forms of Racial Discrimination from Nathalie Prouvez, Secretary of the CERD Committee Office for the High Commissioner for Human Rights (OHCHR).
57 The Sámi president Aili Keskitalo to http://www.nordlys.no, August 15, 2006.
58 The Sámi president Aili Keskitalo to Norwegian newspaper, *Aftenposten*, February 1, 2006.
59 Informant interview, SFT, May 23, 2006.
60 Johan Mikkel Sara to http://www.aftenposten.no, May 12, 2006.
61 Johan Mikkel Sara to http://www.aftenposten.no, May 12, 2006.

62 Informant interview, the Conservative Party, April 6, 2006.
63 Informant interview, The Resource Centre for the Rights of Indigenous Peoples, April 5, 2006.
64 Informant interview, the Social Left Party, June 9, 2006.
65 Part of the statement from the NSR party conference, June 11, 2006.
66 The Sámi President, Aili Keskitalo, to http://www.samiradio.org, November 25, 2005.
67 Bjarne Håkon Hanssen to http://www.samiradio.org, November 25, 2005.
68 Speech at London School of Economics by Foreign Minister Jonas Gahr Støre, October 26, 2005.
69 Foreign Policy Address to the Storting by Foreign Minister Jonas Gahr Støre, February 8, 2006.
70 Foreign Policy Address to the Storting, February 8, 2006, by Foreign Minister Jonas Gahr Støre.
71 Foreign Policy Address to the Storting by Foreign Minister Jonas Gahr Støre, February 8, 2006.
72 Speech at Moscow State University by Foreign Minister Jonas Gahr Støre, February 17, 2006.
73 Jensen (2006).
74 Informant interview, The Ministry of Environment, April 20, 2006.
75 Informant interview, Finnmark County, April 7, 2006.
76 Informant interview, SFT, May 23, 2006.
77 Informant interview, The Socialist Left Party, June 9, 2006.
78 Speech at Moscow State University by Foreign Minister Jonas Gahr Støre, February 17, 2006.
79 Informant interview, The Conservative Party, April 6, 2006.
80 Said by Director Eivind Reiten, Hydro, at the The Norwegian Polytechnic Society on November 29, 2005.
81 Secretary General Rasmus Hansson to http://www.offshore.no on November 30, 2005.
82 The former Minister of Petroleum and Energy, Torhild Widvey, to *Aftenposten*, October 13, 2005.
83 The former Minister of Petroleum and Energy, Torhild Widvey, to *Aftenposten*, October 13, 2005.
84 Director IEA, Claude Mandil, to offshore247.com, November 29, 2005.
85 Foreign Policy Address to the Storting by Foreign Minister Jan Pettersen, February 15, 2005.
86 GEO Year Book (2006) *An Overview of Our Changing Environment*, UNEP.
87 Speech at University of Tromsø by Foreign Minister Jonas Gahr Støre, November 10, 2005.
88 The Soria Moria Declaration is the platform of the red–green government, which was formed after the 2005 parliamentary election.
89 Chapter 2 in the Soria Moria Declaration.
90 European Policy Center.
91 Speech by Jonas Gahr Støre, Minister of Foreign Affairs, at an EPC Policy Briefing in Brussels, October 10, 2006.
92 Speech by Jonas Gahr Støre, Minister of Foreign Affairs, at an EPC Policy Briefing in Brussels, October 10, 2006.
93 Statement in the news, May 29, 2006.

References

Aardal, B. (1993) *Energi og miljø. Nye stridsspørsmål i møte med gamle strukturer*, Rapport nr. 15, Oslo: Institutt for samfunnsforskning.
Aasjord, B. (2005) 'Splitt og hersk i Gulf Barents', Chronicle in *Nordlys*, February 10, 2006.

Agenda (2003) *Utredning av helaarig petroleumsvirksomhet i omraadet Lofoten Barentshavet. Beskrivelse av samfunnsmessige forhold.* Temarapport 9A ULB. Oslo.
Arctic Council (2004) *Arctic Human Development Report*, Akureyri, Iceland: Stefansson Arctic Institute.
Barlindhaug (2005a) *Petroleumsrettet næringsutvikling Nord Norge.* En forstudie: Tromsø.
Barlindhaug (2005b) *Petroleumsvirksomhet i Barentshavet. Utbyggingsperspektiver og ringvirkninger.*
Bay Larsen, I. (2005) *Fossile og fornybare ressurser – en komparativ analyse,* Arbeidsnotat 1012:2005, Bodø: Nordlandsforskning.
Brastad, B. et al. (2004) *Holdninger til olje- og gassutvinning utenfor Lofoten. En studie blant befolkningen i Lofoten og Sandnessjøen,* Bodø: NF rapport 02/2004.
Convention No. 169: *Indigenous and Tribal Peoples in Independent Countries, Adopted on 27 June 1989 by the General Conference of the International Labour Organisation at its seventy-sixth session, Entry into force 5 September 1991.*
Finnmark County (2005) *Petroleum Strategy for Finnmark County 2006–2009.*
Jensen, L.C. (2006) *Drilling for the Environment,* Oslo: FNI report 2:2006.
GEO Year Book (2006) *An Overview of Our Changing Environment,* United Nations Environmental Programme.
Grimsrud, B. (2004) *Verdiskapning, sameksistens og miljø,* Oslo: FAFO rapport 462.
Gvozdic, M. (2001) *Analyse av ringvirkninger av olje- og gassvirksomheten i Norskehavet.* Molde: Møre og Romsdals fylkeskommune.
HT (2007) http://www.ht.no/debatt/kommentarer/article81144.ece.
Knutsen, O. (1997). 'From old politics to new politics: environmentalism as a party cleavage', in Strøm, K. and Svåsand, L. (eds) *Challenges to Political Parties. The Case of Norway,* Ann Arbor: The University of Michigan Press.
Kotte, P. (1997) *Tjeldbergodden. Et midtnorsk industrieventyr,* Kristiandsund: KOM forlag.
Letter from the Sámi Council to the Committee on the Convention on the Elimination of all Forms of Racial Discrimination. UTSJOKI 19 February 2004 Dnr 3/2004 Ark. 902.
Miljøverndepartementet (2001) *Protecting the Riches of the Seas,* Report No 12 2001–2002.
Norwegian Petroleum Directorate (2007) *The Resource Report. The Petroleum Resources on the Norwegian Continental Shelf.* Stavanger: Norwegian Petroleum Directorate.
Olsen, K. H. (1988) *Vertskommune for storindustri. Om Tysværs erfaringer med Kårstøutbyggingen,* Stavanger: Rogalandsforskning rapport 118:88.
Sandersen, H.T. et al. (2002) *Den første olje – En intervjuundersøkelse i Lofoten om holdninger til oljeutvinning,* Bodø: Arbeidsnotat Nordlandsforskning1016:2002.
Statistics Norway (2006) http://www.ssb.no/emner/06/01/regsys/
Thesen, G. and Leknes, F. (2005) *IRIS kommentar til Nordlandsforskningsnotatet 1012:2005,* Stavanger: IRIS.
Willoch, K. (1996) *En ny miljøpolitikk,* Oslo: Gyldendal Norsk Forlag.
World Commission on the Ethics of Scientific Knowledge and Technology (COMEST; 2005) *The Precautionary Principle,* UNESCO.
WWF (2003). *The Barents Sea Ecoregion. A Biodiversity Assessment.* Online. Available HTTP: http://assets.panda.org/downloads/barentsseaecoregionreport.pdf (Accessed 18 November 2007).

10 The Russian model

Merging profit and sustainability

Elena N. Andreyeva and Valery A. Kryukov[1]

Introduction

This chapter is devoted to big changes that are taking place in Russia. These changes encompass administrative and resource policy, the interaction between business and federal authorities, regional opportunities to use money from resource development and changes in attitude to indigenous peoples in their struggle for rights to land and natural resources. The processes of change in Russia are similar to those that have taken place in many Western countries over the last two centuries. For Russia, the changes are compressed in time. This is the case for both changes within the government and legislative body, and those within communities and business. People and institutions have to adapt very quickly to new rules and requirements, and simultaneously, have to build new relationships and responses to change. This transformation process is not always smooth, and contradictions and conflicts often arise.

The most active part of society is business groups that have to carry out their activities in a very complex and tense social environment. This applies mainly to companies involved in business spheres connected with strategic resources, which play an extremely important role in modern Russia. The activity of these companies is under the close attention of different circles of society, as well as their cooperation with foreign partners.

The process of legislation on key questions of resource use is not yet complete and creates uncertainties for stakeholders. This is the reason why this situation is described in detail in this chapter as well as the main developmental trends in the area.

Issues of sustainable development are steadily being included in governmental policy and becoming part of the regional authorities' programs of social and economic development. Unfortunately, contradictory situations are very common during transformation periods, which present large obstacles to implementing a sustainable approach to resource development, or in regional policy regarding aboriginal peoples and their vital interests.

The chapter should help to the widen the circles of business and other stakeholders inside Russia. It should also provide a better understanding of the reality of the current situation and some developmental trends in resource regions of the Arctic in forthcoming decades.

Unlike other countries, Russia classifies its proven oil reserves as a state secret, thereby causing considerable damage to its investment image. Estimates of Russia's proven oil reserves vary between 10 and 20 billion tons – enough to keep producing at the current level for 22–45 years.[2] The opening of new offshore fields in the Arctic is expected to cover the world's growing demand for oil and gas.

Contemporary Russian oil and gas discourse has been dominated by the search for a new model of development. Despite the novelty of this 'Russian model', some of the conflict lines are quite old; most have been inherited from the collapse of the Soviet system. Among the key stakeholders, the Russian state emerges as the most powerful voice. The regions of the Russian Arctic, which experienced major difficulties during the transformation period of the 1990s, now expect to ride the petroleum wave, and fix their social and economic problems. New oil and gas projects are expected to prevent the decline of business activities in industrial areas, and reduce social tension caused by high unemployment rates, population migration and a poor quality of life.

However, major oil and gas development will contribute to a growing anthropogenic pressure on social systems of indigenous peoples of the Arctic. Issues of aboriginal rights on lands and resources are neither legislatively nor administratively solved, which causes conflicts between groups of indigenous peoples and industrial companies. At the same time, the governmental structures encourage implementation of sustainable development principles regarding exploration of hydrocarbon and renewable resources in the Arctic. In the following chapter, we attempt to discuss these issues from the viewpoint of the various stakeholders.

The Russian Arctic today: the post-Soviet transition

Despite its huge natural and intellectual resources, Russia is encountering many difficulties in the transformation period towards a market economy. The disbanding of the Soviet Union in 1991 required transitions in three key areas: the economy, political institutions and center–periphery relations. The economic transition from a centrally planned to a market economy resulted in privatization of state property and opening up to international markets. The political transition was from a one-party state to a democracy. Russia also had to go through the process of changing from a unitary state to a federalist system. After the collapse of the USSR, the 89 units, or 'subjects of the federation', that make up Russia demanded more autonomy and more power for local and regional governments. These widespread changes resulted in a lot of uncertainty during the 1990s. Later Putin's team started administrative reforms oriented towards concentrating power at the federal level, and clear and distinctive distribution of power between the federal, regional (provincial) and municipal levels. The regions of the Russian North have experienced the most difficulty in adapting to the new requirements. During the times of a planned economy, the Arctic regions received strong support from the federal budget.

The Northern territories of the country (including the Murmansk and Arkhangelsk regions, the Nenets and Yamal-Nenets Autonomous Okrugs) face complex economic, social and environmental problems. Their productive and

242 *E. N. Andreyeva and V. A. Kryukov*

economic infrastructure was created in the framework of centralized planning and management of the former Soviet Union. The industrial potential of these territories was developed to serve national goals. In the case of the Murmansk area, the aim was to maintain a strong military–industrial presence at the strategically important Northern region.

Now, under the more decentralized market system of the Russian Federation, these settlements cannot be maintained without development of new industries (i.e., oil and gas). According to statistical data, the regions of the Russian North lost more than 800,000 people to other regions during 1989–2002 (Zayonchkovskaya, 2003). The Northern regions are losing their population first and foremost due to the critical economic situation. The first wave of migration from the North took place in the first perestroika years of 1989–1993. In this context, any growth in economic activity is accepted and welcomed in the hope it will provide regeneration and future economic well-being.

Oil and gas as Russia's new strategic assets

There are significant hydrocarbon resources in Arctic Russia – both offshore and on land (Map 10.1). According to the estimates of Russia's leading geological institutes and agencies,[3] the majority of the country's liquid hydrocarbon resources are located in the Western Siberia, 14 per cent are in the region of Volga–Ural, 13 per cent in Eastern Siberia (including the Republic of Sakha), 6.6 per cent in the European North and about 11 per cent in the Arctic and Far East offshore areas. The distribution of oil and gas, categorized by reserves and resources, is given in Table 10.1 and 10.2.

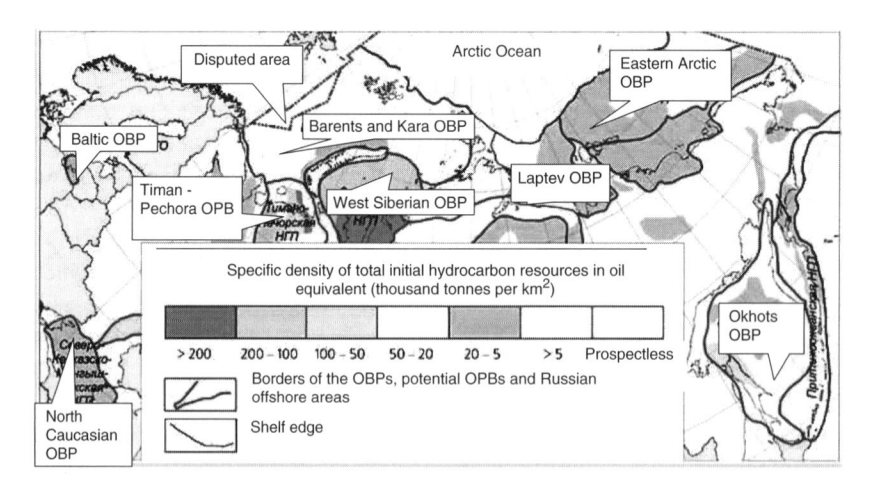

Map 10.1 Prospective oil-bearing provinces (OBPs) with specific density of initial general reserves. (Source: Dmitrievsky, A. and Belonin, M. (2004) 'Russian shelf development perspectives', *Nature*, N9.)

Table 10.1 Distribution of oil and gas reserves and resources in oil- and gas-bearing provinces, as a percentage of Russia's total

Province	Proved current reserves of categories		Preliminary estimated reserves		Prospective and predicted resources	
	Oil	Gas	Oil	Gas	Oil	Gas
Timan-Pechora	7.8	1.5	10.8	1.1	7.7	5.0
West Siberian	69.0	75.6	70.0	51.5	48.0	27.0
East Siberian	3.6	5.8	9.0	17.0	19.8	23.0
Northern seas	-	6.8	1.0	21.0	12.8	34.0
Far Eastern seas	0.99	1.5	3.0	1.7	4.0	5.0

Sources: *Mineral resources of Russia. Economics and Management*, 2002 (4): 12–20; *Bulletin of the Tyumen Regional Duma/Tyumen*, 2002 (9).

The distinctive feature of developing hydrocarbon fields in the Russian Arctic is the high cost of production and transportation of the hydrocarbons produced. However, over the past 10–15 years, new construction technologies for oil and gas production (both on land and offshore), new drilling methods (particularly, horizontal drilling) and methods of prospecting and exploration have noticeably changed ideas regarding the economic expediency of developing fields previously considered inefficient: (see Textbox 10.1 below).

Textbox 10.1 Oil and gas history of the Russian North

Active development of the oil and gas sector in the Russian North is closely connected to the beginning of exploration for oil and gas resources in Western Siberia. The oil and gas fields of the Komi Republic began to play a significant role after the opening of Vyktyl gas condensate field in the 1970s with subsequent large-scale development of the Timan-Pechora fields. Currently, the Northern regions of Russia are the leading producers of oil and gas (see Figure 10.1).

Although Russia has been producing oil on a commercial scale for more than a century now, most of its crude oil has been coming from onshore fields. For many years, Russian oil producers could not afford to develop offshore resources. Russia's huge territory contained many oil fields less costly to develop than offshore ones. But, with low-cost production declining, the time has come to start the exploration of enormous offshore areas. Three-quarters of onshore oil and gas fields are now in development, with the average resource depletion rate close to 50 per cent. Russia's continental shelf area is the largest in the world, accounting for more than 6.2 million km² of which 4 million km² has oil- and gas-bearing potential, as compared with a similar onshore area of about 6 million km².

> Offshore developments are capital-intensive, large-scale and long-life projects, which require good organization and planning, timely risk identification and precise schedule implementation. Failing to meet these requirements will strongly impact both the environment and the economics of the project.
>
> Russian oil producers require new technology to develop the offshore oil fields in the Arctic. Ice-resistant production platforms are needed, and these are currently in production at the Zvyozdochka shipyard by a manufacturer of nuclear-powered submarines.
>
> Ryashin, V. (2005) 'The benefits of mutually rewarding partnership', *Oil of Russia*, (3).

A significant proportion of the Russian Arctic projects are located in the coastal or offshore zone. Consequently, these projects (especially those producing liquid hydrocarbons) potentially have a greater degree of freedom in terms of choosing the direction of hydrocarbon delivery for oil and gas companies.[4]

In recent years, world interest in the production of hydrocarbon resources located in the Arctic has grown significantly, and Russia is no exception. There are several reasons for this increased attention. In short, these can be divided into two main categories: external (international) and internal (domestic).

External (international)

An unprecedented growth of energy consumption in the world as a whole has driven the demand for oil and gas to bring new fields into economic production.

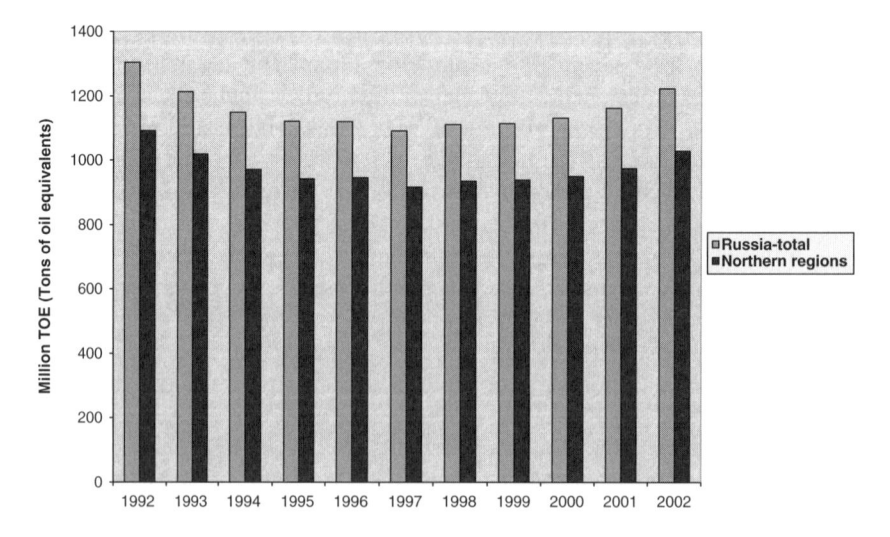

Figure 10.1 Dynamics of oil and gas output in the Northern regions of Russia compared to the total output for the country.

The oil and gas fields situated on the shelf of the North and Norwegian seas were easiest and economically effective to develop, but have now matured and are on the decline. At the same time, the leading companies working in this area–among them Norwegian Statoil and Norsk Hydro (now StatoilHydro)–have accumulated enormous technological expertise and operational experience on the shelf of the Northern seas. Logically, an expansion to the shelf of the Barents, Pechora and other Northern Russian seas would be a natural next step.

In terms of global security and risk handling, the shelf of the Northern Russian seas is located far from the ongoing, emerging or potential zones of armed and diplomatic conflicts. Consequently, the Russian Arctic has a geopolitical advantage for delivering reliable, uninterrupted supplies of oil and gas to the leading markets of energy consumption in the United States (US) and Western Europe.

Internal (domestic)

The increased attention on the Russian Arctic also has domestic influences. Currently, all the known reserve areas of oil and gas production on land are on the decline in Russia. The main production area of natural gas in the polar regions is the Yamal-Nenets Autonomous Okrug, namely the Nadym-pur-Tazovsky region, which is located in the North of Western Siberia. At present, most of the gas (92 per cent) of the JSC Gazprom is produced in this area, and this region will play a key role in the maintenance of gas production until 2010 and in future prospects. At the same time, the country's major gas fields, such as Urengoi, Yamburg and Medvezhye, which at their mature stage produced almost 500 billion m^3 of gas per year, have now come to the stage of declining production, and by 2030 the level of production is predicted to be about 130 billion m^3 per year.[5]

Together with declining gas production in the major fields, other problems have started to emerge. First of all, employment opportunities for the local population are limited to the petroleum industry. As V. Kovalchuk, the mayor of Nadym, pointed out, 'the future of the city can only be connected to the further development of oil and gas projects. The city is to be transformed into the base settlement for development of the fields of the Yamal Peninsula'.[6]

The existing situation puts a strain on the regional budget of the Yamal-Nenets Autonomous Okrug, which most likely will face a severe reduction in the level of incomes in the near future Table 10.1. This, in turn, will considerably curtail the ability of local authorities to solve the social problems in the area.

For more than ten years, the amount of oil produced has not been replaced with new additional reserves. The same is generally true in the gas industry. In 2005, however, the leading Russian gas company, JSC Gazprom, reported a net gain (in available reserves) after reassessing the potential of the Shtokman. Oil and gas resources are strategically important assets for Russia, both for reasons of domestic economical growth and for the country's own position in the international division of labor among other developed industrial countries.

In the political sphere, the emerging Russian state system is drifting towards a rigidly vertical power and away from the federative model declared at the beginning of the 1990s. The federative model was based on regional and municipal rights and powers, including the federative units' right to manage the natural resources on their own territories. The social political sphere is handicapped by the complexity of the transition to new civil society institutions and the search for an efficient model of their functioning. In the economic sphere, new economic institutions are still to be formed; and foremost, procedures for dealing with the after-effects of the earlier system are still to be invented. It is rather a complicated task as:

> To form a new system, the countries implementing transition should not only eliminate the old system and replace it with a new one, but they should also correct the consequences of functioning in the framework of the old system during a long period of time.[7]

Such consequences include both technological infrastructure (above all, the inter-regional one) and the network of settlements created within the system of centralized planning and management.

The Russian model

Several factors have contributed to determining the strategy of oil and gas development in the Arctic, both offshore and within the bounds of the Northern territories of Russia. These are crucial for understanding the principles and assumptions behind what is already referred to as 'the Russian model'.

First of all, the political processes of the emerging Russian state system emphasize the accumulation of all decisive functions and powers at the federal level, including those relating to hydrocarbon resources development. Additionally, the corporative and bureaucratic role of governmental structures' procedures is strengthening, owing to Russia's weak juridical system and lack of supervision. Finally, the need for corrective measures to dealing with the after-effects of the planned economy is still apparent; in particular, there can be no rapid change to the system of settling people in the North.

Consequently, a distinctive Russian model of hydrocarbon resources development has emerged over the past ten years. Its distinguishing features are centralism, corporatism and paternalism. Centralism refers to a growing concentration of decisive and legislative powers at the federal level (including oil and gas revenue flows; see Table 10.2).

Corporatism refers to the existence and dominance of one large state company or a quasi-state corporation taking over the development of significant areas of the Northern sea shelf, as well as territories on land. Paternalism refers to a top-down process of providing solutions to social problems in the regions according to the priorities and directions as defined by the state government bodies and corporations (the Arctic territories and their inhabitants are increasingly taking on the role of passive 'recipients').

Table 10.2 Russian oil and gas revenue flows (federal/regional as of 2006): primary distribution of revenues

Tax base assignment	Federal	Regional
1. Extraction tax		
Oil	95%	5%
Gas	100%	
Oil and gas shelf and offshore	100%	
2. Export duties		
Oil and gas	100%	
3. Excise tax (petrol, diesel)	40%	60%
4. Corporate income tax	27%	73%
5. Royalties SPA		
Gas	100%	
Oil	95%	5%
6. Corporate income tax SPA	20%	80%

Source: Art. 50 ff. Budget Code RF.

The state as a powerful stakeholder

In addition to stakeholders such as the dominating corporation owners (mostly financial institutions acting as representatives for private individuals' shares), the Russian state is the most powerful stakeholder. Oil has been produced in Russia for over 100 years and, during the entire existence of the Soviet Union, the oil and gas industry, like all other industries, belonged to the state. The present Russian authorities see the oil and gas sector of the economy as the basis of the Russian state's power within the world economy. Many states that produce oil and gas for sale to other countries declare their energy resources as national property, and take control of the income from exports. This is done not only by taxing excessive income from private companies, but by taxing the production capacity itself. A similar process can be observed in Russia.[8]

This is why Russia has used various methods – ranging from taking Yuganskneftegaz away from YUKOS for its tax debts to buying Sibneft from its shareholders at market price – to raise the state share in oil and gas production to one-third, and simultaneously regained the controlling share in JSC Gazprom, which produces 90 per cent of Russia's natural gas. The state's long-term plans most likely include increasing control over oil production to half of the country's production volume.

Watch out, the pipes are closing

All active major oil pipelines in Russia, except the KTK Pipeline (Caspian Pipeline Consortium; the state control over it was legally lifted), are owned by Transneft. In 2002, YUKOS, LUKOIL, Sibneft (now – Gazprom Neft), TNK (now – TNK-BP) and Rosneft planned to build a private major oil pipeline from Western Siberia to Murmansk for supplying oil to the US. The Prime Minister at the time, Mikhail Kasyanov, had said earlier it was legally allowed, but added that a private pipeline is out of place in Russia. He explained that the principle of equal access for all oil companies to any pipe in Russia could not be broken. His successor, Mikhail Fradkov, was more precise in 2004: private pipelines cannot exist in Russia at all.

However, private major pipelines in Russia do exist, including the Shell pipeline within the Sakhalin-2 project, and the TNK-BP pipeline in the Kovykta deposit. The law does not demand the nationalization of these pipelines; however, every new major pipeline can only be state-run (Butrin, Vesloguzov and Rebrov, 2006). Recently, Shell announced that they will turn over the controlling share in Sakhalin-2, the largest oil and gas project on the Russian continental shelf, to JSC Gazprom. Analysts consider that the company gave in to pressure from Russian authorities.

Striving for monopoly over new deposits

In November 2006, Russian Cabinet of Ministers made a forecast for natural gas production in Russia by 2015. They predict that independent producers would have only a 17-18 per cent share in gas production, despite owning 24 per cent of the resource base. Gas production experts say this forecast will be correct if JSC Gazprom keeps acquiring major gas deposits of Russia, including those already belonging to private companies. According to its energy strategy, Russia will be extracting between 742 and 754 billion m^3 of natural gas by 2015. This increase will be mostly due to developing the Yamal Peninsula and Shtokman deposits, those in the Irkutsk region, including the Kovyktinsk gas-condensate field, and a number of deposits in the Far East (Grib, 2006).

In December 2006, there was an unexpected announcement that offers from Western partners concerning participation in the Shtokman field development could be considered again. First Deputy Prime Minister Dmitriy Medvedev declared at the Davos Economic Forum that Arctic resource development would be a joint activity with various countries.[9]

> During negotiations with potential partners, Gazprom was aiming at the access to end users of natural gas in Europe and in the United States in exchange for participation in development of this Europe's biggest field. But the topic hasn't been finally closed yet. It could be raised again if some interesting offers are received from foreign partners.[10]

On the same day, Putin's aide Igor Shuvalov in Washington and the Industry and Energy ex-Minister, Viktor Khristenko, in Moscow spoke about possible and desirable involvement of foreign companies in the Shtokman project 'as suppliers of equipment' or 'in some other form'. Russia will independently develop the first stage and provisions for joining the project have remained the same. Of the five Western bidders that reached the final stage of the Shtokman tender in June of

2006, spokesmen for three companies (Total, ConocoPhilips and Statoil) confirmed they had noticed nothing new in the attitude of the Russian authorities (The Energy Shtokman Therapy, 2006).

The licenses were revoked by the Russian Agency of Sub-soil Resources (Rosnedra) because of alleged mismanagement of the license conditions. Experts believed federal authorities might transfer the licenses to the state-owned Rosneft company. The licenses might be in the field of interest for the companies Rosneft and BP, which recently signed a cooperation deal committing companies to cooperate in the Arctic.[11]

The principal decision was made at a meeting in the presence of Mr V. Putin: rights to develop all fields within the Russian continental shelf should be allotted to state-owned companies – JSC Gazprom and Rosneft.[12]

Legislative and institutional challenges for Arctic projects in Russia

The European North of Russia plays an important role in the 'Energy Strategy for Russia towards 2020' that was approved by the Russian Government in May 2003. The strategy document estimates that the Arctic shelf of Russia and Yamal Peninsula will become the most important regions for Russia's gas production strategically. However, the document also recognizes that, in order for this to happen, a stable investment framework and foreign expertise will be required.

Several oil and gas fields have been discovered and a lot of prospects have been mapped seismically in the Russian part of the Barents and Pechora seas (Map 10.2). However, the area is large, and it is also considered an undeveloped exploration province. The key success factors for exploration in this area will include the size and concentration of reserves, the reservoir parameters with high well productivity and predictions of oil (light versus heavy) as opposed to gas. The area is also undeveloped with respect to production. The first projects, 'opening projects', will also need to develop the necessary infrastructure in order to sustain the development of large fields with high profitability. This will in turn benefit other projects and will inspire exploration activity on a long-term basis. Prirazlomnoe and Shtockman will be such 'opening projects' for oil and gas in the Arctic offshore region. Experts from JSC Gazprom and their partners assume that several small discoveries along with known structures individually are too insignificant to be stand-alone projects. Area planning supported by the authorities along with cooperation between license holders will help to make many of these fields commercially viable.

Mineral legislation – how to make it stable and predictable?

A distinctive feature of the development of oil and gas resources in the Russian Arctic is that the main participants are large companies – the oil companies JSC LUKOIL and JSC Rosneft, as well as the gas company JSC Gazprom. In 2006,

Map 10.2 Oil and gas fields and hydrocarbon-bearing structures of the Russian Arctic shelf.

another leading Russian oil company, JSC Surgutneftegas, also declared its interest in the work offshore. It signed an agreement with the Norwegian company Statoil on cooperation in the sphere of development of oil and gas resources offshore.[13] In this way, the company will significantly expand its participation in the Northern projects, in addition to the ongoing development of the fields in the North of Eastern Siberia (Talakanskoye and four leases in the territory of Republic of Sakha).[14] The principal legislation covering the approaches to the development and use of hydrocarbon resources in the Russian Federation (RF) comprises: the RF Constitution (1993); the RF Federal Subsoil Law (1992); the RF Federal Law on the RF's Continental Shelf (1995) and the RF Federal Law on the Production Sharing Agreement.

The legal and commercial framework is very important for determining the level of activity in the region, owing to the large investment required, the risks involved and the slow return on investment. During the 1990s, in the Russian mineral legislative sphere, an attempt was made to combine public law and civil law approaches. The Federal Subsoil Law was based upon public law principles (i.e. dominance of the state in the decision-making process, as in determining license conditions). Such an approach protects the company against any changes in tax or conditions. A more stable and predictable approach is based upon civil law principles (i.e. agreement between the state and the company) – a production sharing agreement – where any changes have to be discussed by both sides before they can be approved. This is why nearly all Russian oil companies formed with foreign investors lobbied for the civil law approach. The process of formation started in 1993; it was not until 1994–1995 that Russia and its oil and gas sector experienced a great investment shortage.

Thomas Walde, Professor at the Center for Energy, Petroleum and Natural Resources Law at the University of Dundee, observed the functioning of Production Sharing Contracts (PSCs) in the initial stages of the oil and gas development in the new market economy conditions:

> We had lately several references to the dissatisfaction in Russia with the PSCs concluded with several major oil companies. From my own, minor, role in Russian oil–gas law reform I have some observations. First, this history started with the 'Houston' project in the 1990s; I considered this a largely illusionary project by the US oil community to 'take over' the Russian oil industry. I remember a very weird tax law professor calling me in 1991 at night urging me to make him visiting professor in Dundee so he could learn how to run the Russian oil industry. Second, at one time the idea emerged that PSCs were the right instrument for Russia to organise foreign investments. PSCs, one has to remember, have been used exclusively in developing countries; their particular feature is that they are very opaque on who runs and controls the operation. They can be something like a service contract under the control of the state company, but they can also in effect simply continue foreign control like a traditional concession/license contract, but give the impression the host state now owns and controls the operation. That opacity was the first main advantage and reason for PSCs, the other one was, and is, that it operates an in-built tax stabilisation mechanism by way of the cost recovery system (and that mechanism–diluted in Russia–is I understand now a major issue in Russia). Cost overruns lead simply to faster and more extensive cost recovery and that seems currently a major criticism of the Sakhalin contracts (notwithstanding that the costs in the oil industry have seriously inflated everywhere–as is normal in any industry experiencing a sudden ups and downs) and we have a prior, low-oil price induced shrinkage of capacities (facilities, human resources).[15]

Unfortunately, the development of a production sharing agreement (PSA) regime makes it more uncertain and difficult to obtain PSA terms for projects. This is problematic for both Russian companies and their foreign partners. Suggestions for simplifying the procedure have been submitted for consideration to the state Duma and it is hoped that the regime position can be revised.[16] Another issue is the pending overhaul of the Subsoil Law. The final edition failed to gain government approval, which makes the outcome of the important changes unclear. Nevertheless, the great potential of the hydrocarbon reserves (the Shtokman field alone is considered to be larger than all the Norwegian gas fields discovered during the last 30 years) and the expected long-term production (more than 50 years) attract foreign companies to participate in this project.

The 'two keys' principle

In compliance with the RF Constitution (Article 72, Item 1) in Russia, the issues of oil and gas resources located onshore are to be solved under the joint jurisdiction of the Federation and its constituent entities (see Textbox 10.2), which implies

---

Textbox 10.2 Legal framework

The principle of joint jurisdiction (Article 76 of the RF Constitution)

The principle of joint jurisdiction presumed actual participation of both the regions and the federal center in granting the rights to use subsurface mineral resources. However, practical implementation of this principle was not clearly formulated. This led to a number of serious problems in interrelations between the companies – users of subsurface resources and regional authorities. Finally, in 2004 the principle was suspended. Several amendments revoking the principle were introduced to the Underground Resources Law.

Underground Resources Law (1992)

The fundamental legislation on use of subsurface resources includes the following:

- subsurface resources are a state property and are granted to companies–land users for a certain period of time for the purpose of survey, exploration and development of mineral deposits;
- entitlement for the use of subsurface resources should be carried out on a payment basis;
- subsurface resources are jointly owned by the federation and constituent entities of the federation, apart from the fields located on the shelf and in the closed seas that are under the federal jurisdiction.

One of the main problems is execution of the law, which contains several reference rules. A new edition was, therefore, submitted for revision in June 2006 by the Minister of Natural Resources. In addition to clarifying the role of the Federal Center in regional subsurface management, this edition attempts to integrate the civil law propositions into the resource management frame.

> **Production sharing agreement (or contract) (PSA; 1995)**
>
> An alternative legislation, in the form of production sharing agreements, was a response reaction of Russian and Western companies to investment instability in new development projects. Procedures accompanying implementation of PSA are extremely complicated: about 28 different coordination and agreement stages have to be passed through.

they have to reach a certain consensus. The issues include the possession, use and management of land, subsurface, water and other natural resources, as well as management decisions concerning nature, preservation of the environment and environmental safety control and use of natural resources within specially protected natural territories (Moe and Kryukov, 1998).

The principle of joint jurisdiction presumed (Article 76 of the RF Constitution), that:

> on the issues of joint jurisdiction of the Russian Federation and constituent entities of the Russian Federation: Federal laws, as well as RF constituent entities laws, and other statutory legal acts are issued, and accepted in compliance with the Federal ones.

The 'two keys' (joint jurisdiction) principle arose as a result of the political processes that prevailed in Russia in the beginning of the 1990s. The principle reflected the aspirations of the territories to solve the social problems they inherited after the collapse of the old system, and to meet the social challenges of the future. Initially this principle was based on the Subsoil Law (1992) and later became one of the cornerstones of the new Constitution (1995). Among the key stakeholders affected by the formation of the principle of joint jurisdiction are the population of the oil and gas territories (including the native peoples of the North), oil and gas companies, state institutions and other bodies at both the federal and regional level. During this period, the oil and gas companies – as their formation was at the initial stage – were not active participants in the discussions over grants for the rights to use subsurface resources. The main concern was on the issues of ownership of the companies' assets, as well as access to export infrastructure.

The process of the companies' formation was basically completed by the end of the 1990s. As this process was coming to an end, the position of the major companies on the issues of subsurface management began to change. This was mostly due to the reappearance of the real proprietors, who were interested in receiving payment for their assets purchased during privatization. The companies supported the cancellation of this principle, as the presence of two participants on the part of the state, in their opinion, created additional complexities.

However, this was not without a controversy. Victor Orlov, Chairman of the Committee on natural resources and preservation of the environment under the

Council of Federation of the RF (in the 1990s, Minister of Natural Resources), pointed out that 'the regions deprived of any interest in subsurface management, may become opponents and not supporters of the reforms being carried out'.[17] In Orlov's opinion, in Russia's emerging democracy, the 'two keys' principle provided stability both to the companies' economic activity and life in the regions.[18] The subsurface management agencies, which were created in the resource regions and were made responsible for 'the second key', appeared to be more consistent and stable. Up until 2002, the regions actively participated in investment of the mineral resources base. Moreover, investments in geological prospecting exceeded the volume of federal investments by 2-3 times.

The conflict of Vladimir Butov, Governor of the Nenets Autonomous Okrug (NAO), with the 'oilmen' became one of the catalysts for facilitating changes to the Underground Resources Law. From December 2001 for a period of two years, the head of NAO neglected to sign 43 permissive documents on the work of the affiliated company PC LUKOIL. As a result, the company was declared to have infringed license agreements.[19] The Archangelsk regional assembly directed amendments of the Underground Resources Law to the state Duma, prepared by experts of the company. The essence of these amendments was to deprive the head of the regional authority of the right to dispose of the natural resources in the region. However, the NAO governor acted not so much in the interests of the local population, but rather in his own interests – he created 'the petroleum company of regional development' – the Nenets petroleum company. Therefore, the position of the company JSC Arkhangelskgeoldobycha (an affiliated branch of the JSC 'LUKOIL') was based on the assumption that the law in its 1992 edition impeded the development of oil and gas complexes, as the general director of the company Oleg Moldovanov noted. The governor of another oil-producing region – the Tomsk region – Victor Kress stated 'it is much safer for the state when "the key" belongs to two, rather than to one owner'.[20]

Nevertheless, on August 3, 2004 the deputies of the state Duma ratified amendments to the RF Ministry of Natural Resources to the Underground Resources Law in the second reading and actually deprived the governors of the right to 'the second key'.

The Underground Resources Law

The management of subsurface resources of the Russian Federation continental shelf was assigned to the RF authorities for regulation (Article 3, the RF URL – Underground Resources Law). There are several fundamental points that provide the framework for legislation on subsurface resources. Thus, the subsurface resources are considered to be state property, and are granted to companies/land users for a certain period of time for the purpose of geological survey, exploration and development of mineral deposits. Furthermore, entitlement for the use of subsurface resources should be carried out on a payment basis. Also, the subsurface resources are jointly owned by the Federation and its constituent entities (apart from the fields located on the shelf and in the closed seas, which fall under federal jurisdiction).

By the time the administrative reform began in 2001, a number of conceptual changes were introduced to the Underground Resources Law. Many of those directly affected the interests of the regions. Hence, some of the amendments accepted in 2001 fundamentally changed the system of payment for land use. Instead of royalties and deductions for mineral replacement, a tax on production was introduced. Distribution of oil and gas production tax between the budgets has changed drastically since that time. From 2007 onwards, it has been totally directed to the federal budget.

The balance and distribution of power between the federal and regional level of the state system changed fundamentally with respect to control and management of subsurface resources in 2004. The changes were introduced by the Federal Law:

> On introduction of changes and amendments to the RF Underground Resources Law of August 22, 2004, # 122-FL (Federal Law). These amendments reflected the need to bring the RF Underground Resources Law in compliance with the Federal Law.[21]

'On general principles of legislative (representative) and executive public authorities organization in the Russian Federation' of July 4, 2003, # 95-FL.

Proxies of the federal body of the state subsurface resources fund control significantly has been expanded; the earlier existing principle of the 'two keys' – joint decision-making at the federal and regional levels on the issues of use and management of subsurface resources – was replaced. From that time on, all the important decisions concerning oil and gas projects were supposed to be made at the federal level.

In our opinion, the 'two keys' principle most profoundly ensured accommodation of both the State's own interests and the interests of the regions where the leases and land users are located. The principle was directed at smoothing economic, environmental and social conflicts connected to geological survey, exploration and development of subsurface resources. It also created incentives for regional participation and encouraged the regional bodies to create a socially oriented system of natural resources management.

During the 1990s, the Underground Resources Law contributed to obtaining significant financial and social–economic advantages from the implementation of subsurface resources management projects in the producing regions, including the Arctic. Thus, rental income received by the regional budgets from hydrocarbon production at that period, enabled significant increases in the living standard of the population in the corresponding constituent entities of the Russian Federation.

More importantly, the Underground Resources Law allowed investment flows that were directed at implementing large infrastructure projects at regional level. The construction of bridges and roads of inter-regional and national significance in Yamal-Nenets Autonomous Okrug and Khanty-Mansi Autonomous Okrug may be an illustration. These projects were the first positive experience among structural reforms to be implemented in practice at the regional level.

In June 2006, the Minister of Natural Resources, Yuri Trutnev, submitted a new edition of the Underground Resources Law. At present time the new version has not been approved yet. The main changes apply to rules of foreign access and investment to resources that are to be developed. The fields with reserves that are referred to as 'strategic assets', may be developed only by organizations that are under Russian companies' control (for gas, reserves of more than 50 billion m^3; for oil, more than 70 million tones).[22]

These limitations, however, are not referred to geological reconnaissance. If the foreign company discovers a field with such 'strategic reserves', it should form a joint enterprise with a Russian company, which would have working control of the stock share. Although the new edition is not accepted at governmental and parliament level yet, the President, Vladimir Putin, welcomed these new changes: 'Now, these new rules make the resource use procedure more transparent, and these limitations serve the interests of national economic security'.[23]

One of the key problems of implementating major oil and gas projects in Russia lies in the fact that the current tax system is directed at withdrawal of super profits resulting from a favorable price climate. Moreover, this tax system does not take into account the recurring need for large capital investments. The fiscal regime is not supportive of new capital-intensive projects, such as offshore development of the Northern seas shelf, or the on-land fields located at high latitudes.

Operation on the basis of production sharing agreements remains one of the most attractive forms of investment. But the procedures behind the preparation and implementation of PSAs are extremely complicated. Before the implementation can begin, about 28 different coordination and agreement stages have to be passed through. This, among others, includes statutory acts, auctions in order to obtain the rights for continental shelf development, adoption to and correspondence with two federal laws and about a dozen regulations at the level of the RF government, not to mention executive orders (decrees, directions and conclusions on bills). In addition, about 15 sessions of the PSA preparation commission, 15 rounds of negotiations with the investors, over 20 coordination meetings with ministries and departments, as well as about five expert examinations (environmental, the Central Committee for Development, the Advisory Council, the Institute of Legislation and Comparative Jurisprudence and the Ministry of Economic Development and Trade) are needed. As a result, the procedure of closing a PSA drags on for two to three years.

All the projects related to development of the oil and gas field in the Russian Arctic – both onshore and offshore – are export-oriented. Proximity to potential markets for produced hydrocarbons is useful for several reasons: not just because of its relative closeness, but also because it is possible to avoid the use of JSC Transneft's pipelines. Transneft occupies a monopolistic position in the service rendering and transportation market.

Another aspect that makes the hydrocarbon production in the North so special is the significant remoteness of the production areas from the consumption areas. Therefore, the transportation factor plays a critical role in providing an economically expedient and acceptable level of production. The main patterns of hydrocarbon transportation have traditionally included pipeline transportation, and sea transport of liquid hydrocarbons. Additionally, conversion of gas hydrocarbons to the liquid state represents another potent alternative. In addition to purely technical functions, hydrocarbon transportation also plays a significant economic role. It provides new job opportunities and creates several socio-economic impacts in the Northern regions. However, most of the impacts are of a short-term nature, mostly during the period of construction.

The case of the Shtokman and Prirazlomnoye fields illustrates how the decisions made at the governmental level as well as the extent of state involvement in the leading Russian companies (JSC Gazprom and JSC Rosneft) affect the implementation of large oil and gas projects in the Russian Arctic.

The original plans in the mid-1990s that were based on active regional participation failed to follow through. To a growing extent, a unitary (or paternalistic) model of the project implementation is about to be created. The real active participants are the federal level and the companies owned by the state, while the constituent entities of the federation and municipalities are to a greater extent assigned a role of passive recipients of certain indirect benefits from the project implementation.

Sustainable development in the Russian context: providing social guarantees

Before sustainable development can be considered, it is imperative for Russia to emerge from a state of economic and structural crisis. It is the most important task at hand. However, it is also imperative to take into account the restrictions imposed by sustainable development defined by scientifically supported ecological requirements. Sustainable ecological development is the first restriction. Should the restrictions be accepted, they should prove to be profitable both socially and economically (Kondratjev, 2003).

In the mid-1990s, Russia attempted to incorporate the principles of sustainable development into the process of economic reforms. In 1996, the President's Decree 'On the concept of the Russian Federation's transition towards sustainable development'[24] was published. It was pointed out that the 'transition to sustainable development should ensure a long-term balanced solution of social-economic development and preservation of favorable environment and natural-resource potential, satisfaction of needs of the present and future generations'.[25]

The basic principles and the main directions of development of the Arctic Russian regions were determined in 2004 by the Federal Law 'On the fundamentals of the State regulation of social–economic development of the Russian Federation's North'[26] and in other normative documents. The prime tasks set out in the documents were focused on providing a solution to the socio-economic problems of the Arctic regions by modernizing the basic branches of economy, enhancing their competitiveness and developing transport infrastructure. Furthermore, these tasks included restricting economic activity in environmentally sensitive territories, maintaining traditional subsistence and conditions for traditional modes of life and nature management of the native minorities of the North. Additionally, the need for stimulating scientific and innovative activity in the Arctic, creating international cooperation in the sphere of ecology, science and engineering was emphasized.

In February 2006, the draft of the 'The concept of sustainable development applied to the RF Arctic' was submitted to the meeting of the enlarged Collegium of the RF Ministry of Regional Development. It was emphasized that, by the beginning of the 1990s, the old model of Arctic development based on the principles of planned economy, faced serious problems. The RF's transition to the market economy made it impossible to continue financing and planning all types of economic activity and all forms of social support for the population of the Arctic regions.

The collapse of the Soviet system eventually led to a drastic decline and deterioration of the quality of life and living standard of the population in a number of sub-Arctic regions. A low average level of income prevented from carrying out of people's different demands, including that for self-dependent migration. At the same time, the financing the federal migration programs only in recent years has been cut twice in comparable prices.

The infrastructure that developed during the Soviet era failed to function in an effective and competitive way. In order to keep maintenance costs to a minimum, a number of settlements were closed down, and the traffic flows connecting these settlements, including air transport, dried up. The traffic flow along the Northern Sea Route has fallen by a factor of 2.5 over the past 15 years and, in the Eastern part of the Route, the traffic flow is 30 times lower. Dikson and Anderma, two of many ports along the Arctic cost, have practically stopped functioning; the port of Tiksi is struggling to survive.

Thus, the most critical issue of sustainable development approach in the Arctic Russia is providing a long-term solution for socio-economic problems. In the Russian discourse, this is often referred to as 'providing social guarantees' for inhabitants of the Arctic. In several researchers' view,[27] this task requires active participation of the state, including implementation of large national state projects and programs, and dealing with the after-effects of the previous model of planned development for the Arctic regions. All this undoubtedly includes implementation of major oil and gas projects. The question is to what extent the current solutions and approaches to development of oil and gas resources in the Russian Arctic can facilitate changes towards the sustainability in the future.

As Tables 10.3 and 10.4 demonstrate, the economic indicators of volumes and dynamics of industrial production and investments, budget expenses, and the level of personal incomes exceed the country's average in all Arctic regions dominated by the petroleum industry. Three oil and gas regions – Nenets, Yamal-Nenets and Khanty-Mansi – are at the top three list of best salary level. In these regions, the average wages are 3-4 times higher than in other remote regions of the country, while the level of federal budget supply per capita is 4-6 times higher than the Russian average. This allows regional authorities to provide stable systems of social security and participate as co-investors in regional infrastructure projects.

Table 10.3 Development indicators for the key oil and gas regions of the Russian North

Region	Oil and gas production per capita (tons of standard fuel)		Volume of industrial products per capita (thousands of rubles)		Gross regional product per capita (thousands of rubles)		Investments in fixed capital per capita (thousands of rubles)	
	2003	2004	2003	2004	2003	2004	2003	2004
Russian Federation	9.15	9.67	58.94	78.12	80.74	No data	15.124	18.981
Yamal-Nenets AO	1346.52	1361.97	366.4	569.43	633.58		323.211	302.15
Khanty-Mansi AO	248.18	269.57	419.97	617.12	522.21		112.786	123.499
Republic of Komi	18.09	19.24	70.8	87.84	112.33		25.148	30.171
Nenets AO	263.57	372.86	456.12	803.54	600.93		411.043	363.64
Republic of Sakha	2.49	2.53	102.59	140.43	140.33		32.091	30.836

AO, Autonomous Okrug.

Table 10.4 Development indicators for the key oil and gas regions of the Russian North

Region	Tax proceeds per capita (thousands of rubles)		Budget proceeds per capita (thousands of rubles)		Budget expenditure per capita (thousands of rubles)		Average per capita monetary proceeds of population (per month; rubles)	
	2003	2004	2003	2004	2003	2004	2003	2004
Yamal-Nenets AO	230.97	352.89	96.63	102.87	96.58	98.17	15,962	18,868
Khanty-Mansi AO	203.66	363.16	66.52	97.89	66.37	80.09	12,892	14,972
Republic of Komi	32.05	40.98	16.26	20.52	17.19	20.53	7477	9301
Nenets AO	169.02	328.66	70.89	110.62	71.05	88.47	14,197	20,122
Republic of Sakha	25.18	30.36	42.05	45.09	43.79	48.73	8240	9633

Note: Calculated by *Regions of Russia. Social and Economic Indicators, 2005: Statistical transactions*, Moscow: Rosstat Publishers, 2006; *Russian Statistical Annual 2005: Statistical transactions*, Moscow: Rosstat Publishers, 2005.
AO, Autonomous Okrug.

Petroleum challenges to the environment

There are several social and ecological problems connected with restructuring the energy sector and the exploitation of oil in the Russian Arctic. In 1995, a member of the Russian Socio-Ecological Union, Alexei Grigoriev, pointed out the following issues that were considered top priority for stakeholders to resolve (Grigoriev, 1995), most of which remain unresolved to this day:

1. *Lack of public information and debate*. The World Bank Country Assistance Strategy (CAS), including the section on restructuring the energy sector, is not publicly available and is virtually unknown to the Russian public. It has never been publicly discussed, nor has it been debated or approved by the Russian Parliament. The Russian public has had no opportunity to discuss whether or not it is necessary to exploit the oil fields. Publicly available information about the project and the agreements and contracts, related to its development, is very limited.
2. *Implementation of unsustainable strategies*. There has been no debate or discussion of the environmental consequences of the strategy or the implementation of the energy sector plans. Most projects are not based on an environmentally sustainable development strategy, but rather on a sharp increase in Russian oil extraction for export. According to some estimates, Russian oil reserves will be exhausted in 20 years at the present rate of extraction. If the attempts to restore lost levels of production succeed, this will happen even sooner.
3. *Increase in greenhouse gas emissions*. Ironically, oil and gas projects mean that donor nations and their taxpaying citizens will invest money to promote greenhouse gas emissions. This direct investment in global warming will be much larger than the investments by the same donors and taxpayers to prevent global warming, for example, in support of Global Environmental Facility (GEF) projects also managed by the World Bank.
4. *Adverse impact on the development of renewable energy sources*. An increase in Russian oil supply to the world market will decrease motivation for industrialized countries, mainly in Europe, to reduce their energy consumption and develop renewable sources of energy.
5. *Environmental destruction*. One of the main reasons why most of the Arctic oil fields have remained unexploited thus far is because of environmental considerations related to the implementation of projects in a very vulnerable environment. Any large oil spill will most likely affect downstream areas of this important region. There is also an obvious risk that oil pollution will impact the sea and shores of the whole Arctic.

From 2004 to 2006, Russia presided in the Arctic Council. Sustainable development of the Arctic is one of the most important issues of this supranational structure. The practical ways of implementing the principles that lie behind the sustainable development in each country are constantly under discussion during meetings of the Council. In March 2006, at a meeting in Salekhard (Yamal-Nenets Autonomous Okrug), the participants delivered the following statement:

> Economic basis of existence in the Arctic Region is not only mineral but biological resources. Today, it is necessary to advance and preserve the traditional resources, which fulfill major functions in reproduction of the system of cultural-economic skills, social stabilization and ethno-saving significance for the indigenous peoples. The projects that are directed at development of production facilities of reindeer breeding, fishery, and fur breeding by the means of modern technology are necessary.[28]

The assistant to the Minister of Economics and Trade of RF, Boris Morgunov, said the following: 'There is no sense to accumulate economic possibilities of countries at the expense of health of the population and their future gene pool. Growth of the population is a major indicator of steady promoting Arctic communities to prosperity'.[29] In turn, the vice-governor of Yamal-Nenets Autonomous Okrug noted:

> Natural resources of our region are great, and environmental themes together with indigenous people issues like traditional economy and cultural heritage are of high priority to us. Our okrug may be the model region for pilot projects for many current themes stated at the Arctic Council meeting.[30]

These statements of top officials give some hope that social and environmental problems will be considered as the first priority in the twenty-first century. Unfortunately, the real practice in the resource regions of the Russian Arctic reveals a different picture. Economic goals and corporate interests dominate, while the federal government and legislative structures are still discussing the policies of sustainable development in the Russian Arctic. As a result, much of the designed mechanisms for implementation of sustainable development remain declarative and not executed.

However, oil and gas development in the Arctic also represent a major threat to the fragile Northern environment. The potential environmental impacts can be enormous and not always predictable.

Thus, according to Turuntaev (2006), the offshore oil and gas production in the Barents Sea region may lead to purely technogenic earthquakes and also bring about an increase in natural seismicity and the intensity of deformation processes. In addition, field facilities themselves may be exposed to seismic events and geological deformations. Therefore, a greater attention should be given to passive seismic monitoring. The production of hydrocarbons induces weak seismic events, which can only be recorded using a highly sensitive seismometer network.[31] Among the regions already struggling with major environmental problems, is the Kola Peninsula around the regional center of Murmansk and its bordering areas.

Arctic offshore projects as the economic savior of the bordering regions

Over the past decades, Russia has witnessed a most interesting evolution of the arguments in favor of the implementation of oil and gas projects in the Arctic zone. The debates started at the time of drastic political and economic changes at the beginning of 1990s. There were several factors that contributed to keeping this lively discussion going. First of all, the production and technical potential that many of the Arctic regions possessed in the times of the planned economy was suddenly not applicable any more. This was especially critical for the military and industrial complex in the Murmansk and Arkhangelsk regions.

Following the collapse of the Soviet Union, several regions strived for sovereignty. Struggling with chronic budget deficits and shortages in payment, the regions of the Arctic aspired to implement their own export-oriented projects, such as production of hydrocarbons and their subsequent delivery for export. And, as mentioned earlier, the principle of joint jurisdiction really worked until 2002.

Using offshore projects to rejuvenate the regional economy in the Murmansk and Arkhangelsk regions

In 1992, the state 'Russian company for shelf development' (Rosshelf) was founded with the participation of enterprises from the military–industrial complex. In compliance with the RF President's decree of November 30, 1992, Rosshelf obtained the license for developing the hydrocarbon fields on the Arctic shelf – Shtokman and Prirazlomnoye.[32] The issue of keeping incremental oil and gas production was not as acute then as it is now. It was assumed that execution of the projects would contribute to stabilizing and then improving the economic situation in the Northwestern region of Russia.

Restructuring military defense enterprises to suit the needs of the growing petroleum industry was intended to establish an industrial framework for the development of the Arctic shelf fields, which would create about 120,000 jobs. Furthermore, construction of a fixed offshore platform would provide ongoing permanent employment for about 10,000 people in the city of Severodvinsk.[33] Implementation of the project would also result in the retention of the scientific and technological potential of the state enterprises PA Sevmashpredpriyatiye and SE Zvyozdochka.

Overall budget receipts at various levels during the implementation of the Prirazlomnoye project alone were estimated at US$1.9 billion. The Murmansk and Arkhangelsk regions were considered as potent suppliers of materials and machinery for drilling operations, field development, transportation of gas and gas-condensate, construction works, servicing the fleet and social welfare. Moreover, implementation of the projects was supposed to carry out gasification of the Murmansk and Arkhangelsk regions, as well as the Republic of Karelia.

It was assumed by JSC Gazprom authorities that the realization of the technological solutions accepted in the projects of the Shtokman and Prirazlomnoye fields would give the Russian participants an opportunity to take part in new development projects on the foreign shelf.

Starting in 1992, a general policy towards survey, exploration and development of hydrocarbon fields on the shelf started. It was characterized by heavy involvement of the state in obtaining the rights for using subsurface resources. Preferences are given to the companies owned by the state, such as JSC Gazprom – at that moment the Russian Joint Stock Company.

On May 23, 1996, the RF President issued a decree (N765): 'On creation of the industrial production base for development of hydrocarbon fields on the continental Arctic shelf'.[34] This document not only issued licenses for the Prirazlomnoye and Shtokman fields to Rosshelf, but simultaneously appointed the PA Sevmashpredpriyatiye as the general contractor for construction of ice-resistant platforms. Thus, on the one hand, the requirements for 70 per cent participation of Russian contractors stated by the PSA Law were fulfilled, but, on the other hand, choosing a contractor without any competition placed implementation of the shelf projects 'outside the market'. The general contractors were now able to dictate their conditions to the customer.

The basic centers for Shtokman field development are supposed to be located in Murmansk and Arhangelsk. The coastal zone of the Murmansk region facilitates main infrastructure objects, meanwhile the industrial enterprises of the Arkhangelsk region become involved in production of the equipment. The famous plants of the Soviet defense industry (Sevmash, Zhvezhdochka) now have been partly adjusted to civil production, and their productive and intellectual potential will play a crucial role in the supply of heavy metal equipment. With a help of the Norwegian company Statoil, a new center, Sozvezhdie, was opened in Arhangelsk in the year 2005. Its main purpose is to create a network of equipment and service supplies for Shtokman's needs. Moreover, Statoil is planning to invest US$3 million in socio-economic development programs in the Arkhangelsk region.[35]

Participation of the Murmansk region in realization of this project will contribute to the diversification of the regional economy, and its further social and economic development. An arrangement between Gazprom and the Murmansk area authorities in 2006 was already in place at the first stage of the project's implementation, and it is assumed that the Murmansk regions will receive annually 1 billion m^3 out of the total 20 billion m^3 of gas produced. All of this will enable gas to be supplied to several large enterprises and private households in the area.

Prospective plans for a factory producing liquefied gas on the Kola Peninsula estimate that about US$4.5 billion will be invested in the Murmansk region, along with creating new job opportunities, as well as rejuvenation of the local and regional budgets. According to the PSA Law, no less than 70 per cent of all the equipment and materials orders should be placed with Russian national enterprises during the development of the Shtokman field.

It is also important to mention that future large-scale industrial activity requires a well-advanced system of power engineering. This suits the Murmansk region well, as its atomic power station produces excess power and delivers electric power to the integrated power grid for other regions of the RF, and for export to Finland and Norway.[36]

In Russia, taking into account the importance and strategic nature of hydrocarbon resources in the Arctic, only the spokesmen and representatives of the state interests are being granted the rights for using subsoil resources. However, in our opinion, the attempts undertaken in the 1990s to create domestic companies concentrating in their hands the entire process of preparation and implementation of such petroleum projects and the implementation of the state programs on the development of the Arctic shelf appeared unsuccessful.

This can be explained by several factors. Foremost, cooperation with the world's leading oil and gas companies on implementing similar projects was badly underestimated. Severodvinsk and Murmansk have succeeded in facilitating unique sea defense complexes, which certainly showcases their potential. However, this cannot be directly applied to the implementation of commercial projects as development of oil and gas resources in difficult climatic conditions and in a fundamentally different field of activity. Therefore, one cannot but agree with the opinion of Academician I. S. Gramberg that 'neither Gazprom nor any other domestic gas or oil companies are in position to start development of the Arctic fields independently. The rigid protectionism in the absence of real investment opportunities is simply ridiculous'.[37]

The case of the Yamal peninsula

Yamal-Nenets Autonomous okrug is the main natural gas and oil production area of Russia (Map 10.3). It includes a very large part of the Northern areas of Western Siberia. About 80 per cent of the Russian oil and gas reserves are located here. Exploration in the region began in the 1960s, and today the industry faces the need to develop the most difficult sub-region – the Yamal Peninsula. This area is characterized by extremely severe natural conditions, high costs and financial risks, the need for innovative technology, a tense social situation and extremely high sensitivity and vulnerability of the ecosystems.

The traditional raw material base in this region – the Nadym-Pur-Taz gas area with its unique fields, Medvezhje, Urengoyskoe and Yamburgskoe – are close to entering the stage of falling extraction. By 2020, the fields currently being exploited will produce no more than 30 per cent of the volume demanded. Taking into account that consumption of natural gas is still growing in the world and in Russia (where the share of natural gas in the fuel–energy balance was more than 52 per cent in 2005), the main producing company of Russia, JSC Gazprom, has determined that development on the Yamal Peninsula is essential. Despite the very complex conditions of development and extremely high costs, the resource potential of the area will help to solve the growing deficit in gas production of old gas areas.

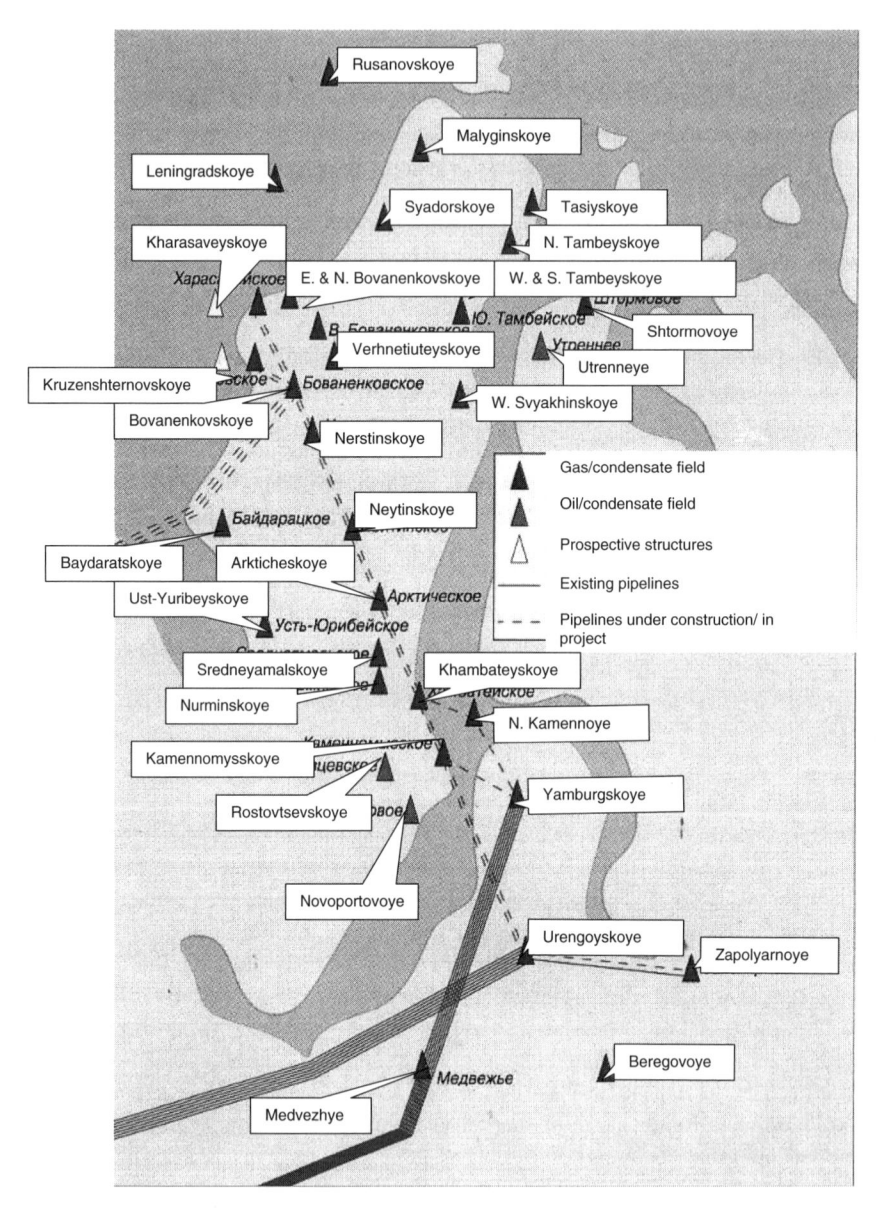

Map 10.3 Location of gas fields and pipelines in the Yamal-Nenets Autonomous Okrug.

Yamal's only proved reserves of 11 gas and 15 gas-condensate fields ready for development, exceed 10.4 billion m^3 of gas and 228 million tones condensate. The licenses for extraction belong to NadymGazprom, a company 100 per cent owned by Gazprom. In 2002, Gazprom teamed up with the government of the Yamal-Nenets Autonomous Okrug District to draft a sweeping development program for the oil and gas fields of Yamal and its offshore areas, which is now being reviewed by the Russian government. Gazprom is currently priming the Yamal fields for drilling. The company holds mining licenses for the largest Yamal fields – Bovanenkovskoye, Kharasaveyskoye and Novoportovskoye – which have ultimate reserves estimates of 5.8 trillion m^3 of natural gas and 100 million t of condensate. Owing to very complex natural conditions, gas will cost much more to extract in Yamal compared to existing gas fields. Solid government backing will, therefore, be needed to tap into the Yamal reserves.

The experts noted that Gazprom is not in a hurry with realization of these plans and continues to prepare different documents such as background for investments, environmental assessment of the project and analysis of possible transport for delivering gas to consumers.[38] Compensating for decreased production volume, Gazprom decided to start the development of small field-satellites, located close to existing infrastructure that had originally been built for field-giants (see Figure 10.2): As for the fields off the Yamal Peninsula, Ob-Taz Bays, they are considered as a major strategic asset for the gas industry.

Several options have been proposed in order to bring stability and prosperity to the region.[39] The structure of the Oil and gas sector should most likely be changed towards the emergence of small- and medium-sized enterprises operating roads, terminals and transport infrastructure, as well as pipelines, oil separation units and tank farms. This will hopefully facilitate investments in small oil and gas projects, which in turn requires simplified licensing procedures and fast issuing

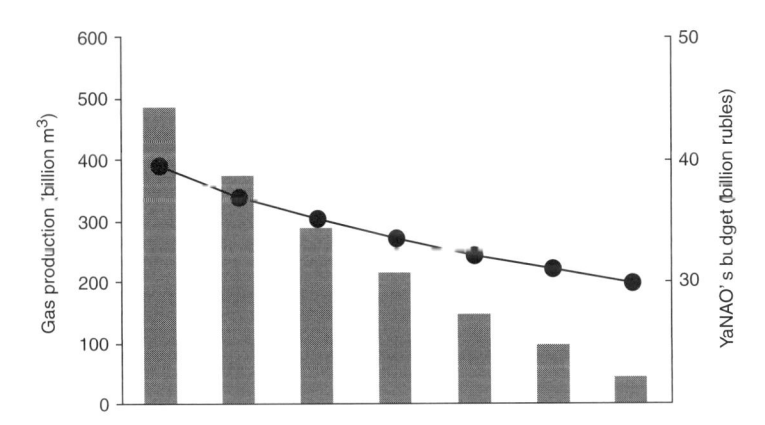

Figure 10.2 Dynamics of receipts of the budget of the Yamal-Nenets Autonomous Okrug (YaNAO) v. gas production dynamics.

of various permission documents. At a larger scale, there is a certain need for creating a market for delivering service products to the oil and gas industry (drilling, geophysical, etc.).

As a whole, the oil and gas sector of the Russian economy needs diversification. Some examples are already to be found in: the Gubkinsky petroleum refinery producing non-polluting motor fuels; the Novourengoisky petrochemical complex processing gas and condensate; the Gubkinsky and Muravlenkovsky gas-processing plants processing associated petroleum gas and directly in the fields of the Purovsky area with their output of liquefied gases and diesel fuel; as well as the construction of two new gas-processing plants in the Tazovsky and Purovsky areas.

In the opinion of Y. Neelov, Governor of Yamal-Nenets Okrug, the potential of both the Yamal Peninsula and the Nadym-Pur-Tas area should dramatically increase not only owing to modernization and commissioning of hydrocarbon fields, but also owing to the creation of new sectors of economy, such as the advancement of production of solid minerals and the creation of a mining industrial region in the Polar Urals zone, petrochemical works and, in prospect, electric power industry on the basis of low-pressure gas.[39]

Experts have also called for establishing 'inter-regional shift teams' from the cities of Nadym, Urengoi, Yamburg and Salekhard, which all are located close to the new projects' territories.[40] This will remove the problem of 'superfluous' population. The population of Novy Urengoi is 106,200; however, including those working on a rotational basis and temporarily registered in the area, the total number of the population of municipal unit approximates to 120,000. In recent years, the rotational team method has been used for field development on a large scale. At the same time, the main staff for the rotational teams is recruited outside the Yamal-Nenets Autonomous Okrug. These teams usually consist not only of qualified workers from other regions, but also less qualified staff. Such workers are available in the cities of Nadym, Novy Urengoi and Labytnangi, which are located much closer to the fields.

In the opinion of A. Kim, Deputy Governor of the Yamal-Nenets Autonomous Okrug, the inter-regional rotational teams formed at 'a long distance' should include two further components. 'Medium-distance' rotational teams for gas fields should be formed of highly skilled experts living in the industrial cities of the Ural and Siberia located further south. 'Near-distance' rotational teams should be formed in the cities located nearby the developed fields, and include manpower from the cities of Nadym, Novy Urengoi, etc. With such organization of rotational teams, unemployment rate in the cities of the Yamal-Nenetz Autonomous Okrug can be appreciably reduced.

Since the mid-1990s, a program of resettlement from the North has been implemented. In the territory of Nadym-Pur-Tazovsky, resettlement was initially carried out at JSC Gazprom' expense, and later was paid for by the local and federal government. For example, from Nadym and the Nadym area, approximately 8000 people (about 2500 families) were moved to other areas of the country during 1994–2003. One in ten inhabitants left, but then many of them came back. The main reasons for this were high wage levels and more stable social conditions in the Yamal-Nenets Autonomous Okrug than in other regions of the country.

Arctic oil and gas projects can certainly contribute much to the development of all regions located in the immediate proximity – Murmansk, Arkhangelsk regions, Nenets and Yamal-Nenets Autonomous Okrugs. However, turning these major opportunities into reality requires not only excellent subsurface resources management, but also needs to take into account the interests of the local population. Only in this case it is reasonable to speak about a model of development that is close to sustainable.

Being indigenous in the petroleum age

Starting in 1999, three federal laws dealing with the rights of indigenous peoples have been adopted, namely 'On guarantees of rights of indigenous numerically small peoples of the Russian Federation', 'On general principles of organization of communities of indigenous numerically small peoples of the North, Siberia and the Far East of the Russian Federation', and 'On territories of traditional nature use of indigenous numerically small peoples of the North, Siberia and the Far East of the Russian Federation'. These laws guarantee the protection of the primordial habitat and traditional lifestyle of indigenous numerically small peoples (see Textbox 10.3), the right to organize communities with tax benefits for traditional nature use, gratuitous use of land of traditional habitation and economic activities, participation of indigenous numerically small peoples in co-governance of natural resources on territories of their traditional habitation and economic activities. However, the practical implementation of the RF government's

---

Textbox 10.3 Indigenous peoples of the Russian North

Aboriginal peoples of the Russian North, Siberia and the Far East are represented by 28 minorities, of which around ten minorities belong to the Arctic group. The biggest groups are Sámi, Nenets, Enets, Nganasans, Evenks, Evens, Dolgans, Chukchee and Eskimo. The majority live in small villages close to their subsistence areas, where they pursue traditional occupations like reindeer herding, hunting and fishing. But the reality these people face today is anything but an idyllic carryover from the past.

Since the colonization of the North, large expanses have gradually been converted into areas for alien settlement, transportation routes, industry, forestry, mining and oil production, and have been devastated by pollution, irresponsibly managed oil and mineral prospecting and military activity.

In tandem with the environmental disaster went the social decay of the indigenous societies since the early Soviet era, with collectivization of subsistence activities, forced relocations, spiritual oppression and destruction of traditional social patterns and values. The result was the well-known

> minority syndrome marked by loss of ethnic identity, but minorities were involved in work at collective farms, received free medical care and education for their children. Their traditional economic activity was donated from federal budget considerably and it became the radical contrast with current situation.
>
> The recent socio-economic crises of Russia, which came along with the transition to a market economy, have led to a breakdown of most of the supply and transportation systems in the remote areas of the North. Having been incorporated into the alien Soviet economic system, made dependent on modern infrastructure and product distribution, the people now find themselves left alone without supplies, medical care, rising mortality and the economic means and sufficient legal expertise to deal with the situation. The desperate road back to the old ways of life has tempted many, but is often hampered by the degradation or destruction of the natural environment.
>
> Reindeer herding is the fundamental, subsistence-related occupation of many Northern peoples. It is not necessarily the most typical native occupation, but the most characteristic one that still has economic significance. Furthermore, it is not just an economic occupation, but has developed into a way of life closely connected with ethnic identity. There are large-scale, extensive herding cultures like those of the Nenets, Chukchee and Koryaks, and small-scale breeding mainly for draught and riding animals as a subsidiary occupation for many taiga people. Reindeer herding, however, is very sensitive to environmental changes. Modern development has created a severe threat to reindeer herding and its related cultures.
>
> Environment, health, legal issues and the economy are today on the agenda of the indigenous associations. Russian Association of Indigenous Peoples of the North (RAIPON) and associated organizations are working hard towards the Russian authorities concerning the emplacement of a satisfactory legal basis for indigenous rights. So-called ethnic communities are formed, where the native population executes a sort of self-determination in terms of traditional subsistence. Environmental violations have been brought to trial. Health-related development projects are being initiated. Native communities are trying to go back to their traditional social clan structure and to revive the old ways of life in order to survive the present socio-economic crisis.
>
> Source: Dallman, V. K. (2001) *Indigenous People of the Russian North*, Tromsø: ANSPIRA/ Norwegian Polar Institute.

policy prevailing since 2001 has clearly shown that the RF government has failed to execute these federal laws adopted in 1999–2001 (Bogoyavlenskiy and Murashko, 2004).

According to the information collected by regional associations of the Russian Association of Indigenous Peoples of the North (RAIPON), 246 communities have been legally registered during the past three years of the established federal law, although there are more than 700 villages with concentrated indigenous populations in Russia. In some administrative units of the RF, there is not a single registered community, while in others there are several, although their majority still remains unregistered according to the existing legislation. It has not yet been determined which state body should be responsible for the registration of communities or the issuing of law-making standards with regard to their concessionary taxation. The consequent legal instruments concerning communities are intentionally intricate. Hence, in some regions it is assumed that communities should be registered in state legal bodies, while in others it is believed to be a matter for the tax inspectorates. In some regions, communities are exempt from taxation and free from charges, while other communities are facing exorbitant claims in connection with taxes, and communities are forced to go into liquidation.

During the first three years since the federal law on Territories of Traditional Nature Use (TTNU; adopted in May 2001) went into effect, not a single TTNU under federal administration had been formed, while the majority of land incorporating TTNUs is land of federal subordination. All the applications to establish TTNUs have been met with the RF government's refusal, first of all because the criteria for delimitation of TTNU were not precisely described in the Federal Law. In some regions, long before this law, regional governments formed TTNUs under regional administration. For example, in the 1990s there were about 500 in the Khanty Mansi Autonomous Okrug. Traditionally they are still called 'lineage-based kinship areas' covering about 26 per cent of the Okrug's territory, but more than 40 per cent of these lineage-based kinship areas have already been leased to oil companies on long-term contracts. In Nenets Okrug, large-scale reindeer breeding units were formed on the basis of defunct sovkhozes (Soviet state-operated farms), whose grazing areas covered about 60 per cent of the Okrug's territory. According to Bogoyavlenskiy and Murashko (2004), these TTNUs are threatened by a gradual leasing handover to oil companies. The RF government has already sold licenses for oil and gas production in sectors included in the TTNUs. The RF government holds that the establishment of regional TTNUs is illegal before the adoption of the law and without clear criteria for that. The tax assessment authority demands TTNU rental payment from communities. On one hand, it is legally fair in accordance with the new RF Land Code, but on the other, it contradicts the federal law 'On payment for land', in accordance with which Northern indigenous peoples are relieved from payment for land. The intentional confusion in the laws creates uncertainty among indigenous peoples about their future, and leads to closing down of their communities.

The ecological environment of indigenous peoples' habitation is systematically disturbed. Offshore oil production operations have already adversely affected the environment in traditional settlement areas of indigenous peoples, the quality of marine bioresources known to be the indigenous peoples' main food. RAIPON has repeatedly approached the RF government with letters about the unacceptability of worsening the ecological situation in the seas of the Far East. RAIPON has become one of the claimants mounting a lawsuit against the RF government defending the habitat of grey whales in the Sea of Okhotsk. At present, RAIPON is receiving information concerning the full-scale realization of the state program of offshore oil and gas production in the Far East – a program which has not been submitted to public evaluation until now – as well as the pipeline construction project from Sakhalin all across the Khabarovskiy and Primorskiy territories and over the border, and about oil and gas production projects and oil pipeline construction in Chukotka and Buryatia. These projects will have and already have an inevitable impact on the territories of traditional habitation and economic activities of several indigenous numerically small peoples of the Far East. Nonetheless, these projects have not been discussed with indigenous representatives, and the opinion of the local population and indigenous inhabitants has not been taken into account during the projects' realization.

In March 2003, a letter came from the president of the Sakhalin Association of Indigenous Peoples concerning the beginning of prospecting operations in Piltunskiy Bay, which is an area where traditional fishing takes place. RAIPON reacted to this letter by sending an inquiry to the Ministry of Natural Resources (MNR). It has become known from MNR's answer that the oil company, a branch of Exxon, carried out prospecting operations not only without coming to any agreement with the local indigenous peoples, but even prior to obtaining a positive conclusion from an environmental expert evaluation. Nonetheless, MNR justifies the company's actions and informs RAIPON that prospecting has been carried out without any infringements. RAIPON has repeatedly called the attention of the RF president, RF government and the RF Federal Assembly to these violations, and suggested ways to solve the problems by setting up a federal body focused on indigenous peoples and adopting necessary changes in and supplements to the existing legislation.[41] All the proposals made by RAIPON, despite the favorable disposition of the RF president and RF Federal Assembly, have been blocked by the RF government as economically inexpedient.[42]

Oil and the worsening economic state of indigenous peoples of the Arctic

Small- and medium-sized regional oil and gas companies were created in order to provide social and economic support to the native peoples of the region. Hence, native peoples of the North received a part of the shareholdings of privatized oil and gas companies located in the territory of their habitation. Approximately 5 per cent of shares of all the oil and gas enterprises (at initial privatization) were allotted for these purposes. Several petroleum companies (like Evikhon and YugraNeft) received the status of:

> regional development companies … in the interests of minority peoples of Khanty and Mansi. Formally, several investment and unit funds were supposed to focus on maintenance of long-term economic interests of native minorities, e.g. The Investment company of social protection and development of minority peoples of the North.[43]

However, previous experience with similar structures has been far from successful.

In 1992, the Decree of the Russian Federation Government[44] stated that it is necessary:

> to agree upon foundation of the joint-stock petroleum company Evikhon as the regional development company ... for geological survey, development and production of oil in the Upper and West Salym fields in the interests of the native minorities Khanty and Mansi, living in the territory of these fields.[45]

Sadly, the further activity of this company was connected only to a very small degree to the interests of the native minorities. In 1995, the joint-stock company Evikhon imported duty-free cigarettes from abroad for US$112,262 thousand which accounted for 31 per cent of the market share for imported cigarettes in Russia at the time. These sales in no way contributed to solving the problems of the native minorities of the North.[46]

Further 'traces of presence' of any native minorities in the property of the company Evikhon were finally lost. In 1993, the Shell company, or its affiliated structure Shell Salym Development (SSD), won the rights to the subsurface resources of the Salym fields. In 1996, Evikhon and SSD created conditions of equal shareholding in the joint venture Salym Petroleum Development (SPD), which was supposed to begin development of the Salym fields. In 1998, the license on the Salym field was transferred from Evikhon to SPD. At the end of 1998, the English company Siberia–Energy (SE) purchased 20 per cent of Evikhon's share in SPD (the bargain was carried out on behalf of private investors, and at roughly US$30 million) and, at the end of 1999, an additional 55 per cent. As a result, the share of SE at the end of 1999 already accounted for about 82 per cent in Evikhon, and 78 per cent in Yugraneft. Hence, at the beginning of 2007, the key owners of SPD are Shell and Gazpromneft companies. Shareholdings in the oil and gas companies transferred to various investment and voucher funds 'in the interests of the native minorities' are no less confusing. Therefore, it is apparent that the interests of the minorities have not been safeguarded. Not only were the forms of public control absent, but also those of the state.

Development of hydrocarbons production has sharply aggravated the already existing problems connected to the traditional subsistence of the native minorities. In the territory of the Purovsky area, there are 4000 representatives of different native minorities. Basically all the economically active population is engaged in traditional occupations, and is incorporated into agricultural cooperatives and national communities. All the national communities are concentrated in the south of the Purovsky area, which is the area of intensive oil development. In the opinion of Vladimir Marbik, the Vice-President of the Committee of Native Minorities of the Purovsky area:

> expansion of activity of the oil-and-gas companies leads to dramatic reduction of reindeer breeding. Traditionally, the nomadic population of the region lived an isolated life. After introducing oil-and-gas projects to the region, the territories of traditional economic activities are surrounded with communication

networks, oil-gas-condensate pipelines, motorways, pumping units and drilling rigs. This dramatically reduces the native peoples' opportunities to roam freely and pursue the traditional nomadic way of living.

Combining oil and traditional subsistence in Yamal

The first attempt to develop the giant Yamal oil fields – Bovanenkovo-Kharasavey in 1988–1989 – was considered unsuccessful owing to the environmental and social problems in the region. The technical feasibility report of 1989 found numerous deficiencies, and the project was closed for a while, but a comprehensive program of research was initiated in several directions. The program was headed by Gazprom in cooperation with a large number of scientific research institutes. The perestroika period of the 1990s brought about changes in the financial and managerial structure of the development, and the operational opening of Yamal was eventually postponed.

In the 1960s and 1970s, environmental control over industrial activity was very poor, as elsewhere in the world. Only in the late 1980s, the negative environmental impact the oil and gas industry has on ecosystems as well as the local population's health came to the attention of the Russian public. The 25 years of gas development in the Yamal-Nenets Autonomous Okrug has had a large effect on the local population, its habitats, natural resources and the environment. Owing to the lack of legislation and control over the Northern territories, the industrial enterprises occupied 10-30 times more land than would have been permitted according to existing standards. Pasture lands have been destroyed by fire and heavy vehicle traffic. As a result, more than 1 million hectares of pasture has been removed from the population in the rural areas. Several rivers and lakes were spoiled, which has affected the balance of the natural water systems of the region. However, these negative impacts were eclipsed by the great success of the gas industry development, the appearance of new urban settlements and a fast-growing population.

The Nenets people of Yamal are not only the most numerous of all the peoples of the Russian Arctic, but are also considered both politically and socially aware of the significant role the indigenous peoples of the Russian North represent in the oil and gas development. Indigenous peoples (33,500) represent 19 per cent of the total population of the okrug (172,600), but the three sub-regions impacted by industrial activity – Tazovsky, Nadymsky and Yamalsky – do not have an equal distribution of these people. Most (more than 10,000) live in the Yamalsky sub-region, 6683 live in Tazovsky and 2048 in Nadymsky.

The initial stages of the oil and gas development in the 1970s and 1980s took place mainly in the Nadym sub-region, where the indigenous population was low. The first phase of industrial operations resulted in severe miscalculations regarding the vulnerable ecosystems of the Polar zone. However, since these activities affected only small numbers of indigenous peoples, the degradation of pastures that took place in the sub-region did not have the crucial negative impact that it is now expected in the Yamalsky and Tazovsky sub-regions.

Reindeer breeding is the main traditional subsistence activity of the Nenets people, and the people of the region found a way to save and even increase the number of herds during the economic crises of the reformation period. Unlike some other regions, the local authorities of Yamal-Nenets Okrug supported the indigenous peoples and let them increase the number of private reindeer herds while simultaneously staying in collective farms. The state unions rendered assistance to the reindeer breeders who wanted to keep their large herds and to process the products of reindeer breeding. Today, Yamal has the largest herd in the Russian North, with around 600,000 animals, almost two-thirds of which are privately owned. Several other regions of the Russian Arctic, where reindeer breeding was an important part of the indigenous peoples' economy, have lost considerable numbers of reindeer, sometimes to a catastrophic level. This applies to the Chukotsky Okrug, where reindeer herds were reduced to a fifth of their original size. The Yamal-Nenets Okrug continues to increase the size of its herds even though the Yamal pastures are overexploited. Today, the Yamal and Tazov sub-regions already have 70 per cent of the reindeer herds while possessing only 31 per cent of the pasture resources. Currently, this problem is very tense in the Yamal sub-region, where the actual number of animals exceeds the intended capacity of the pastures by 87 per cent. The Nenets people even have to sell reindeer to the neighboring Khanty-Mansi, Nenets and Evenk Okrugs, but the problem of pasture resources is still very acute. Hence, the local people consider the forthcoming expansion of the gas companies on the Yamal Peninsula as a serious threat to their lifestyle. It will almost certainly result in withdrawal of land for industrial construction, drilling pipelines and contamination of rivers and lakes.

However, despite the threat to their present living standards, most people do not wish to leave their native places of residence. Even whilst having large pastures available, the Nenets refuse to change the ways of reindeer migration and habitats. Each family or tribe have ethnic ties to specific areas. If the move is forced, the results can be devastating. Surgut district, which is located in the Khanty–Mansi Autonomous Okrug, the main oil-producing area of Russia, can serve as an example. Several families lost their habitats when most of their tribal lands were taken over by the industrial development. In exchange, the aboriginals were moved to a settlement specially designed for their needs. Without any opportunity to carry out their traditional lifestyle and living on welfare, people simply lost interest in life. The suicide rate and level of alcohol abuse soared. The Governor of the Okrug at the special meeting of the local authorities and scientists in June (2001) had to recognize publicly that this experience was not socially sustainable.

The Russian ethnographer, K. B. Klokov (2001),[47] while analyzing the complex and difficult situation of recent changes in the indigenous communities, comes to a rather specific conclusion. The dominating society perceives reindeer breeding of the Arctic peoples as somewhat foreign to the economy. Thus, it tends to assimilate reindeer breeding into the farming pattern, which for the dominating society is considered a customary, reasonable and effective means of economic activity. As a whole, during the last decades, the subsistence of the Northern peoples has been under a rigid control of the dominating society. Prior to the reformation period, efforts were directed at raising the productivity of the Northern peoples as a supplier of raw meat and fish. Now the purpose is changed; the prior goal of indigenous policies today is to make reindeer breeding a maximum commodity in the exchange economy and to open it up for commercial market.

In response to this, reindeer breeders can adopt one of three strategies. They can cooperate with the dominating society (passive accommodation), isolate themselves from the dominating society (escape) or attempt to change the interaction environment with the dominating society. During the Soviet times, the first two strategies were often applied. These, however, had different consequences for peoples' living conditions. Nomadic families seldom wanted to follow the official requirements, and therefore did not receive any support from the state. Passive adaptation led to the steady abolition of reindeer breeding as a rural activity.

It is very indicative that those Northern regions that had a well-organized Soviet-plan economy also suffered the most during the transfer to market conditions. In contrast, those who did not receive strong support from the federal budget can adapt more easily to the new conditions and expand their traditional subsistence. Currently, indigenous peoples are trying to create new relationships with the dominating society, searching for partnerships with other stakeholders that are interested in their lands. While the movement of large industrial companies into the region has brought many negative impacts to the indigenous population, there are some positive aspects as well. Despite losing some pasture lands, the indigenous peoples now have access to the market for traditional products and, therefore, receive some financial support from the companies as compensation for withdrawn lands. According to the conclusions drawn up in the ethnographical research mentioned above, a moderate industrial development and contact with the local population is preferable to either an industrial boom or a complete absence of industrial activity in the territory.

However, despite the existence of several federal laws regarding protection of indigenous rights, the interaction between industrial companies and indigenous peoples has many difficulties. The main challenge to the functioning of these laws is determining the exact territorial borders of the indigenous people's lands, an extensive task that has not been performed until recently.

So each time there is a need for indigenous peoples' permission for operations on their lands, it may proceed in different ways. Typically, industrial companies sign an agreement concerning social support of the indigenous group and the local community. The format of the agreement has not been standardized. Therefore, this is usually handled by corporate lawyers in the companies' favor, and without any actual assessment of the benefits and disadvantages for local peoples. Sometimes, aboriginal associations invite scientists and lawyers from Moscow, St Petersburg or Tyumen in order for these agreements to be more just. One such successful agreement was signed between Lukoil and an indigenous agricultural production community known as Erv. The main idea of the agreement was not just to compensate financially for withdrawn lands, but also to support the traditional economic activity, such as by purchasing new equipment for reindeer breeders, supplying equipment for maintaining reindeers' health, and constructing dwelling houses for aboriginals. The 'oil people' agreed to purchase traditional meat and fish products, while the 'tundra people' undertook to provide high-quality products.

On the other hand, many of the indigenous peoples' claims and requirements from the oil and gas industry are not clear-cut or well-worked out. Naturally, it is difficult to determine the extent of real damage caused by oil operations. It is also unclear how to assess the losses connected to the withdrawal of pasture lands and wildlife areas. The main point is that these cannot be viewed purely in economic terms. The loss of lands for indigenous peoples does not mean a loss of money but a loss of the possibility to exist. Such uncertainties allow representatives of industrial companies to diminish the real value of compensation during negotiations with indigenous peoples, and can obtain permission for oil operations almost for free. A good example of this is shown in the results of an ethno-ecological study conducted by a group of scientists – anthropologists, biologists and lawyers – for the company Gazflot. The report assessed the possibilities for a geological survey and drilling in the area where the Ob Bay and the Taz Bay merge, a very rich gas field with initial total reserves of 7.5 trillion m^3 (2001). The main concern to the local population was the danger of losing the most important fish habitat in Western Siberia. The merging area of these two bays is famous for its rich diversity of the most valuable kinds of white Siberian fish. For the locals, fishery is one of the main sources of employment along with reindeer breeding. During the time of economic transition, many people turned back to pursuing traditional subsistence in order to provide their families with cheap local products. The results of the ethno-ecological study confirmed that the probability of hydrocarbon pollution is high, and that this will bring about changes in the natural ecosystems and fish populations. The final conclusion of the expert commission required any activity to stop in this vulnerable and biologically valuable area. Despite the discussion that followed at the local parliament hearings, the industrial company managed to convince the local society that many of the requirements highlighted in the report could be rejected because of the uncertainty connected to the impact assessment methods. Besides, the company promised to carry out all the necessary precautionary measures to decrease the possible negative impacts. As a result, the company got permission to start their operations without any formally formulated compensation amount or statement of losses and damages to the locals.

Nevertheless, it is important to mention that the Nenets people have been strongly supported by the regional authorities. The latter managed to realize a number of social programs for indigenous peoples aimed at education, health, employment and investments in the local economy. All these programs were possible to implement owing to revenues the regional budget received from the gas companies operating in the area.

Aboriginals' demands and ideas for the future development of the region are often expressed at the corporate meetings of oil and gas companies, and in the State Duma. Khotyako Ezengi, the former chairman of a collective farm in Yar-Sale, has been appointed by its present employer, the oil company NadymGazprom, to liaise with the native peoples regarding the company's decisions. At the international conference 'Oil and Gas of the Arctic' (June 27, 2006) he stated:

> We don't want to tell what is more important for the Yamal area – reindeer or natural gas. Both are in great demand by the people. During the last 30 years a reindeer breeder has never said that he is against industrial development. But we do have one modest requirement – will you respect our vital interests and protect our habitats?[48]

The leaders of RAIPON, Sergei Haryuchi and Anatoly Todyshev, regularly submit drafts and take legislative initiatives to protect the rights of the aboriginals in the region, but unfortunately deputies of the State Duma do not pay adequate attention to such actual issues of indigenous peoples at their committees.

Oil and gas – a new threat to the Murmansk region?

The Murmansk region belongs to the North-West Federal Region of Russia and is one of the most urbanized areas of the country. Several industrial centers were created in the region during the Soviet era of planned economic development. The share of urban population accounts 798,400 people (or 92 per cent of region's total) (*Murmansk Annual Statistic Book*, 2005).

The environmental problems of the Kola Peninsula are connected to the lack of satisfactory technological procedures for waste utilization in large industrial enterprises. Hard and liquid wastes have been polluting the natural environment for several decades. Table 10.5 presents figures from the Federal Program 'World Ocean' and a state document 'Concept of sustainable development in the Arctic zone of the RF'.

The situation improved slightly during the perestroika period when most industrial enterprises were in a state of deep economic crisis and had to cut their production.

Table 10.5 Complex ecological assessment in the Arctic regions of RF, 2003

	Arkhangelsk region	Murmansk region	Nenets region
Discharge of sewage in surface aqueous objects (million m³)	794.6	1893.3	1.3
Discharged polluted wastes in atmosphere (thousands of tons)	278.4	278.4	17.7
Formed toxic wastes (thousands of tons)	293.5	468.9	28.6
Consumed and neutralized from total volume of toxic wastes (%)	20	59.1	16.7
Duration of period with positive average day temperature of air in AZ (days)	150–200	140–180	110–160
Period of freezing	165–185	170–190	200–220
Recurrence of dangerous hydrometeorological phenomena in AZ (ph. per year)	5–25	0–25	5–35

Now, the region has its own sustainable development program overseen by an international team and regional managerial structures.

The main positions of this program found their way into the regional budget, which provides hope that the situation will change for the better. Since 2005, the Ministry of Economics and Trade, together with the scientific centers of the Murmansk region started a pilot program of integrated resource management in the coastal areas entitled 'World Ocean'.

Today, the major environmental problems in the Murmansk regions concern the Kola Bay waters and coastline, with several industrial settlements and military bases located along the coast. The river flow of the Kola Peninsula to the Barents Sea is relatively insignificant (26.3 km^3), so the contamination by polluted waters in the shelf and offshore area is not very apparent. The present state of the Kola Bay waters has been assessed, according to the complex scale of estimation, as 'polluted' and is in the V class of quality. Permissible concentrations of oil products are exceeded 20–30 times and phenols by 5–6. The waters of the Kola Bay are also contaminated with ammonium nitrogen, fluidized and organic matters, chlorine organic pesticides and radionuclides (Denisov, 2002).

The expansion of the oil and gas industry in the Murmansk area will most likely bring further environmental problems to the region. Therefore, a very strict standard should be imposed prior to the exploitation of new facilities and improvements to the pollution from other sources set up.

The coastal zone of the Barents Sea close to the Murmansk region is the only all-year ice-free area in the Arctic Ocean. It has significant economic potential for both the Northern part of the federal region and the whole country. Today, the main focus of the environmental discourse is pointed towards the issues of future development of the Barents Sea shelf, along with the rejuvenation of the Northern Sea Route, and the increase in the shipping traffic to the West with export goods, including oil and liquid natural gas.

The development of hydrocarbon fields in the Barents Sea and the delivery of raw oil and liquid gas by tankers and other vessels to consumers in Russia, European countries and the United States represent a real danger to the natural environment of the Kola Peninsula, its adjacent waters and the offshore marine ecosystems. This issue also seems to concern the companies currently performing in the Barents Sea – Morneftegasproject, Gasflot and Gazprom. According to Borisov, Osetrova, Ponomarenko et al. (2001), a new program preventing marine pollution is currently under development in Russia. The program is built upon the latest achievements of the Russian specialists in this area, and uses the results of several scientific studies conducted in the Barents Sea on vulnerability of biological resources and possible scenarios of dealing with oil spills (Borisov, Osetrova, Ponomarenko et al., 2001).

Nevertheless, the Murmansk region is not a homogenous region in term of resources or socio-economic and environment aspects. These differences determine the diversity of attitudes towards the new petroleum development among different social groups. The economic crisis of the reformation period resulted in mass population migration, growth of unemployment, and bankruptcy of small- and medium-sized businesses. The industrial centers of the Murmansk region were the first to feel the impact of the changed market conditions. The highly urbanized area of the Kola Bay, which was stricken by the after-effects of the system collapse, is now hoping to be the center of the new development that will bring job opportunities, high wages and infrastructure improvement. The harbor facilities are expected to receive new investments in order to facilitate the future technological expansion. According to the prospects of this new development, the products of offshore oil and gas fields will be processed in the Murmansk harbor for further delivery to European and American consumers.

The first draft for the marine pipeline from the Shtokman platform to the coast of Teriberka settlement was changed in favor of Vidyaevo, a famous naval base with available labor resources and a more attractive geographical position in direct proximity to Murmansk. Teriberka was a center of fishery and fish processing, one of many that existed on the Eastern coast but at which activity has now ceased. There is little doubt that the already weak position of the Russian fisheries in the settlements along the coast would be even more disturbed by the oil activity. Nevertheless, after long discussions and protests of local people the decision was made by JSC Gasprom in favor of Teriberka.

In 2005, the regional authorities of Murmansk county and JSC Gazprom signed an agreement (on November 9, 2005) stating that the region will receive around one billion cubic meters of natural gas for its consumer needs. Additionally, a new plant for liquid gas production will be built, along with a terminal for the tanker gas transportation. As one option, tankers could deliver liquid gas (LG) (around 15–18 million tons) to the US market. As other projects have demonstrated, the oil and gas activity has been proven to lead the remote regions into a new level of economic development. No wonder that the authorities of the Murmansk oblast and the population of the Kola Bay region are interested in accelerating exploration of the Shtokman reserves. A similar view is expressed by the population of the Western parts of the region, situated close to the Norwegian border and Varanger-fjoiord. Another plan for the new coastal terminal for oil and gas condensate in Linahamaari, the Pechenga Bay is under discussion. Moreover, some naval bases located on the Western coast are now aiming at performing civilian activities connected with oil and gas.

Industrial development and the Russian Sámi

The Eastern part of the Kola Peninsula (the Lovozero region) differs considerably from the urban Western part of the area. It has a rather large territory and a small aboriginal Sámi population (around 14,000 people) that have lived on the Kola Peninsula since the beginning of the first millennium. The share of indigenous peoples on the Kola Peninsula now is so negligible, less than 1 per cent, that hardly any information or statistical data can be found about them and their traditional economy in official documents. Nevertheless, Sámi people have maintained a rather stable population with very slow growth since 1959 from 1700 to 2000 in 2005. Their main occupations are reindeer breeding, hunting and fishing.

These people depend on the state of biological resources, free access to use them and the quality of environment. They are the first to feel climate changes and impacts of anthropogenic activity. Fortunately, the main industrial centers and the mining industry are located in other parts of the peninsula. However, the Eastern part of the peninsula is rich in infrequent and rare earth metals that are being processed at the plant in Revda settlement. The plant is managed by the closed joint-stock company Lovozersky GOK.

The Lovozersky area is the only location of the Eastern branch of the Scandinavian Sámi in Russia. During the Soviet period, they suffered significant hardships – some of them (around 40 per cent) had been assimilated into the majority population, speak the Russian language and live and work in the urban settlements. The rest, who kept the traditional economic activities, have been hustled away from their habitats to the Eastern part of the Murmansk region. Their main activity is domestic reindeer breeding along with fishing that gives them a substantial addition to their traditional diet. In 2000, the Sámi had more than 60,000 reindeers in their possession. Now, this figure is going down rather rapidly for different reasons.

It is rather indicative that the Sámi's reindeer herding suffers from the effects of current climate change. Seasonal migration patterns have been strongly affected by warmer autumns and early springs. Together with other aboriginal peoples, the Komi-Izhems, the Sámi work within two cooperative unions (transformed kolkhoz) – Tundra (Lovozero) and Olenevod (Krasnoshelje settlement). The traditional Sámi economy is strongly influenced by the current administrative and legislative policy in the region. They lose their reindeers owing to the growing incidence of poaching and cannot gain appropriate protection from the local authorities. Hunters from the industrial centers kill reindeers; while the aboriginals are prohibited from carrying a rifle from March until August.

Biological resources and the land used by Sámi for centuries are now attracting the commercial attention of different groups both from the Murmansk region and from abroad. Lately, a new long-term leasing agreement (15 years) between American and Finnish businessmen and the regional authorities was signed on the largest river of the Kola Peninsula – the Ponoy. The river is rich in Atlantic salmon that comes to spawn to the river from the sea. Now, the American–Finnish company, G. Lumis Outdoor Adventure, brings tourists from all over the world to fish salmon, while the native peoples have to purchase a license prior to fishing in the river where they earlier could catch about 40-80 tons annually.

The accommodation of foreign tourists on the Sámi territory became very popular owing to the unspoiled nature and wildlife diversity of the area. However, the distribution of tourism income does not reach the Sámi people. Moreover, when a few Sámi enthusiasts decided to open their own small businesses for tourist accommodation, they could not outperform the larger companies and were, therefore, forced to stop their activity. In addition, the construction of roads and industrial facilities is causing great concern, as there is a danger that the reindeers will lose their pastures and migration ways will be disturbed. The Sámi say: 'if the reindeer disappear, Sámi people and their culture will disappear as well'.[49]

The problems discussed above became the starting point of a new project called 'Civil Society of the Lovozero Region' in March 2005. The participants of the project are represented by the Public Organization of the Murmansk Sámi region, Association of the Kola Sámi, the Sámi tribes, Lovozero local authorities, Worls Wildlife Fund (WWF) (Moscow Division), the Kola Center for Protection of Wildlife (Apatity), the Murman Fishery Agency, along with the Forestry and Hunting Management structures. The project includes conducting seminars to help the stakeholders understand the situation and prepare suggestions for joint implementation of the civil actions.

In their struggle for survival, the Sámi people try to cooperate with other indigenous peoples and international organizations. Since the end of 1980s, the Russian Sámi communicate closely with the Sámi of the Scandinavian countries and the Canadian Inuits. Researchers from the Arctic Institute of North America and Calgary University provide the information on the experiences of indigenous peoples of other Arctic countries in their struggle for land rights. They also have support of some public organizations of the Murmansk region that constitute a so-called 'Northern Coalition'. The coalition includes Geya, Bellona-Murmansk, the Kola Center for Protection of Environment and Wild Nature, Barents Office of WWF–Russia and Nature and Youth. The main goal of the Northern Coalition is to minimize the ecological risks and damages during the industrial development of oil and gas resources, and to optimize the profits for sustainable development of the Northern territories. Since 2003, the Coalition has facilitated annual conferences with delegates from different sectors of Russian society: regional government, oil companies, indigenous peoples and ecologists. The issues of oil and gas development in the Arctic were considered a top priority during the discussions.

Oil and gas versus fisheries

The competition for the resources in the Murmansk region is a rather contentious issue, which will become more urgent in the years to come. In the Barents Sea, fisheries and the oil and gas industry compete for the same commercial zones and resources. Oil and gas expansion in the Barents Sea will affect these areas in the forthcoming decade, and it is obvious that an assessment of the outcome scenarios should be started right away.

The most reasonable idea is to conduct a complex ecological and economical assessment of the consequences of both types of activities on all levels from international to local. The next essential measure is to create oil-free zones to protect the most ecologically valuable areas. Such oil-free areas should receive an official status along with a license to precede preservation work, and should include special requirements for the companies working in the field and certain operational limitations. The upper management of companies and government alike must recognize that oil and gas alone cannot satisfy all the needs of the population and that bioresources are of high and long-term value both to the local people and the whole society.

In terms of biological efficiency, the Barents Sea yields only to the seas of the Far East. Stocks of bio-resources of the Barents Sea are in the lead among all seas of the Arctic Ocean, and the extraction of fish in the Western sector accounts for almost 85 per cent of that from all Northern basins. The use of these biological resources plays a significant role in the national economy, and the preservation of marine ecosystems is the responsibility of all adjacent countries, especially Norway and Russia. The Southern part of the Barents Sea is the most biologically valuable, with diverse fish resources presented by cod, Atlantic salmon, capelin, flatfish, haddock, herring, pollack, etc. However, the fish resources have been overexploited for many decades, which has brought about a considerable reduction in extraction. For some years, a critical and even a catastrophic situation with cod and capelin populations were registered (Skjoldal, 1991; Hansen, Hansson and Norris, 1996; Matishov and Rodin, 1996; Alekseev, 1999). Maintenance of biodiversity, particularly for the commercial and valuable fish populations, is a common and very persistent problem for all countries of the Arctic region.

In addition to the conservation challenges, Russia is currently struggling with regeneration of its fishing industry. The Murmansk region possesses a sufficient resource base, and a large proportion of the local population is involved in the industry. Among the factors hampering the regeneration of fisheries are the ineffective national system of licenses and permissions, along with ministerial mistakes that put a spoke in small fishing units' wheels. The laws that have been passed into the legislative system during the last 3–4 years have created such unprofitable conditions for the fishermen that they are forced into searching for ways to get around the law. In turn, fishermen that have been using the same fishing spots for decades currently suffer from poaching on a massive scale. Usually, poachers come from urban settlements or military bases and, in contrast to the locals, are stocked with modern technological equipment, including helicopters and off-road vehicles.

The main conflict in the use of local resources is expected to take place between the oil and gas industry and the branches of traditional subsistence using renewable resources, such as fishing, hunting and trapping. There are reasons to assume that the oil and gas development together with the acceleration of shipping along the coasts and offshore, as well as the construction of industrial infrastructure, will cause many negative impacts on the biological system of the Barents Sea. First, the existing experience of hydrocarbon extraction in other areas of the world, including the Russian North, shows that transport by pipelines or tankers brings about the contamination of marine ecosystems, causes different mechanical disturbances and changes habitats and ways of migration of marine biological populations. Second, the analysis of oil spills confirms that their occurrence is very high due to severe natural conditions: low temperatures, storms, ice movements and strong winds (The precautionary approach to north sea fisheries management, 1997).

It is important to mention that oceanographic research on movement of marine waters along the coast has shown that the direction of flows goes from West to East, and that any contamination will eventually move from Varanger fiord to St Nose.

Because of the decrease in the water depth, contaminants are accumulating on the East side of the Murmansk coast and are not penetrating far into the sea. Third, and most importantly, the industry lacks experience of working under such complex natural conditions. This experience is lacking not only in Russia but also in other countries, even those that already perform offshore, and the conditions in the Russian Arctic are more severe, more complex and unpredictable.

Concluding remarks

The issues of regional socio-economic development dominate the contemporary discourse surrounding the sustainable development of the Arctic oil and gas resources in the Russian context. Undoubtedly, the federal level is the dominant 'player' among the stakeholders working on development of the Northern fields. The presence of Russia in the Arctic North is considered to be one of the major priorities of the state. Thus, the nature of present political orientation often dominates the discourse. While all the decisive and legislative power of the oil and gas resources development in the Russian Arctic is concentrated at the federal level, the participation at the regional level is limited to conditional participation. Regions can decide independently on the usage of land resources in laying pipelines and construction of terminals, but the principal decisions of 'go/no-go' character are centralized at the top. Representatives of constituent entities of the federation are members of the corresponding commissions in accordance with general practice. At the same time, the federal government has transferred a significant part of its decisive power to the largest Russian oil and gas companies, such as the JSC Gazprom and the JSC Rosneft. The lack of state rules and regulations concerning the conduct of oil and gas operations in this situation seems to be compensated by the corporate rules and regulations.

Thus, both the paternalistic and the corporate approach are being realized in the development of oil and gas resources of the Russian Arctic. Starting in 2007, the regional budgets of the constituent entities of the Russian Federation were deprived of the opportunity to receive a tax deduction on the lease of the land to the oil and gas companies. This means that the only role the regions will have in the development and implementation of oil and gas projects will primarily consist of receiving indirect benefits from the activity of oil and gas companies, service center enterprises and the manufacturers of equipment for the industry.

According to this scenario, the territories of the Russian Arctic will not be able to receive tax breaks and payments needed for diversification of the regional economy through the usage of their own resources (e.g. rehabilitation of the fisheries). In quite the same manner, the decisive power of the municipal level and the organizations expressing interests of the native minorities of the North is reduced to simply receiving certain indemnifications for possible damages to the environment or the living conditions of the population.

As a whole, it is important to emphasize that there are more problems than solutions to the issues of the hydrocarbon resources development in the Arctic Russia. Several regulation systems on usage of the subsurface resources are lacking;

opinions and interests of various groups of the population are rarely heard, and technical competencies for sustainable implementation of the projects are in short supply.

Hence, the approach to sustainable development that is currently being implemented in Russia is hardly an exhaustive one in the economic, ecological or social terms. There are several fragments of the system, which should be further integrated into the ultimate framework of an integral sustainable approach.

Notes

1. The authors express their thanks to Yelena A. Borkova, Candidate of Economic Sciences (Novosibirsk), for her assistance in collecting and preparing the source data for this chapter.
2. Yershov, Y. (2006) 'Russia's oil and gas sector in a globalizing world', *Oil of Russia*, (4).
3. All-Russia Oil Research Geological Exploration Institute, Institute of Oil and Gas Geology under the Siberian Branch of the Russian Academy of Sciences, All-Russia Scientific-Research Institute of Natural Gases and Gas Technologies and others (see e.g. *Oil of Russia*, 2004 (4): 8, and (7): 20; *Mineral Resources of Russia. Economics and Management*, 2002 (4): 12–20).
4. Ragner, C. L. (2000) *The 21st Century – Turning Point for The Northern Sea Route?*, Dordrecht: Kluwer Academic Publishers.
5. In the 1960s and 1970s, these giant fields were designed individually. With a more complex approach, both the arrangement of the cities, and the infrastructure of the fields would have been different, more rational. Within the framework of the projects, the city of Nadym and the settlement of Pangody were founded close to the Medvezhye field's surface construction. The city of Novy Urengoi was also built close to the Urengoiskoye field. As a result, JSC Gazprom is still financing the maintenance of these settlements and their related social infrastructure. Some of these expenses can be transferred to regional and municipal budgets, but the local and regional authorities are not ready to bear this burden because of a shortage of means.
6. Kovalchuk, V. (2003) *Social–Economic Partnership of the Local Authorities and Natural Monopolies for the Sake of Sustainable Development of the Region – Problems and Prospects of the Usage of the Low Pressure Gas*, Materials of the scientific-pratical conference 10-13 March, 2003, Nadym, JSC Gazprom-Russian Academy of Sciences.
7. Hill, F. and Gaddy, C. (2003) *The Siberian Curse. How Communist Planners Left Russia Out in the Cold*, Washington, DC: Brookings Institution Press.
8. *Kommersant*, December, 2006.
9. Grivach, A. (2007) 'And welcome again', *Vremya Novostey*, January 29.
10. Spoken by Russia's President Vladimir Putin when interviewed by Mexican publisher, Mario Vazquez Rana.
11. http://www.barentsobserver.com/index.php?id=388437&cat=16285&xforceredir=1&noredir=1
12. Bekker, A., Surgenko, V. and Derbilova, E. (2007) 'Shelf for two persons', *Vedomosti*, 22 January.
13. Derbilova, Y. and Borisov, N. (2005) '"Surgutneftegas" will occupy itself with the shelf. It intends to develop it together with "Statoil"', *Vedomosti*, December 5.
14. Surzhenko, V. (2006) '"Surgutneftegas" stops saving up. The company will spend its savings on development of East Siberian fields', *Vedomosti*, June 15.
15. Smolyakova, T. (2006) 'Nefterazdel – Russia loses hundreds of billions of dollars in case of old PSA contracts', *Rossiakaya Gazeta*, October 21.
16. Gorshkova A. (2006) 'Transparent indications', *Vremya Novostey*, October 21.

17 Orlov, V. (2002) 'Federalism and mineral resource use', *Mineral Resources of Russia: Economics and Management*, (5).
18 As above.
19 Pfilippov, V. (2002) 'Second key be taken away from Nenets Governor', *Izvestia*, July 17.
20 Kruglov, A. (2004) 'Siberian accord going to stop oil leakage for the sake of federal center', *Kommersant-Sibir*, April 22.
21 Walde, T. (2006) *Production-Sharing Contracts in Russia: A Historical Observation*, September 26.
22 Bekker, A. (2006) 'Not for foreigners', *Vedomosti*, May 25.
23 Rebrov, D. (2006) 'Strategic view', *Vremya Novostey*, June 14.
24 The RF President's Decree 'On the concept of the Russian Federation transition to sustainable development' of April 1, 1996, # 440, *Rossiyskaya Gazeta*, April 19, 1996.
25 As above.
26 *Rossiakaya Gazeta*, July 2, 1996.
27 See Lazsentcev, V. (ed.; 2005) *North as the Object of Complex Regional Studies*, Syktyvkar: Russian Academy of Sciences; and Volgin, N. and Alexeev, Ju. (eds; 2004) *Russian North: Problem of Social Development*, Moscow: Dashkov & Co. Publ.
28 www.yamal.org/arctic/index.html#baza.
29 www.neelov.ru/6/2006 3/14/1178.
30 As above.
31 Turuntaev, S. (2006) 'The Arctic awaits', *Oil of Russia*, (4).
32 *Oil and Capital*, (VII-VIII), 2001.
33 Special issue of *Oil and Capital*: 'Russian offshore development', 2006.
34 The RF President's Decree 'On the creation of the industrial production base for development of hydrocarbon fields on the continental Arctic shelf' of May 23, 1996, # 765, *Rossiyskaya Gazeta*, May 30, 1996.
35 *Industrial News*, (3), March, 2006.
36 Yevdokimov, Yu. 'On the brink of the new era of gas production in Russia', *Oil and Capital*, (8), 2005.
37 See, for example, Rybalchenko, I. (2003) 'Gazprom is in want of money', *Kommersant*, May 31.
38 Dokuchaev, D. (2006) 'Gazprom decided to develop Schtokman field alone', *Moscow News*, (39).
39 Kazarinov, V., Lyubovny, V., Chelintzev, J. S., Novy Urengoi, O. (eds; 2005) *Novy Urengoi Heading Toward Sustainable Development*, Yekaterinburg: UralSibpress Publishing House.
40 As above.
41 'Mir korennykh narodov', *Zhivaya Arktika*, (13).
42 'Mir korennykh narodov', *Zhivaya Arktika*, (14).
43 Decree of the Government of Russian Federation 'On formation of oil companies as companies of the Regional Development in Khanty-Mansiisk Autonomus Okrug' of July 21, 1992, # 503.
44 As above.
45 As above.
46 'Are Khanty and Mansi needed for Khanty-Mansiisk?', *Novyie Izvestia*, March 22, 2000.
47 Klokov, K. *Modern State of Circumpolar Reindeer Breeding*. Available at: http://www.jurant.ru/publications/reindeer.
48 Ezingi, Kh. (2006) The problems of indigenous peoples of the North, *Proceedings of the International Scientific Conference 'Oil, Gas of the Arctic', Moscow, June 27-29*.
49 Afanasjeva, N., the President of the Kola Sámi Association, interview 2006. All materials and information in this part of the chapter have been gathered during an expedition to Murmansk region and the Lovozero sub-region in June-July 2006 as part of the research project of International Polar Year 2007–08.

References

Alekseev, A.P. (1999) *Morskoe rybolovstvo Rossii na rubezhe vekov*. World ocean on the eve of XXI century. St.Petersburg: RGO Publishing House.
Bogoyavlenskiy, D. and Murashko, O. (2004) Indigenous peoples of the North: results of the 2002 general census and political situation. Interpretation of 2002 census results, *Indigenous Peoples' World – Living Arctic*, (15), 2003.
Borisov, V., Osetrova, N., Ponomarenko, V., et al. (2001) 'Environmental impact of the offshore oil and gas industry on the Barents Sea living resources', *VNIIRO*, Moscow: Gazprom-Arcticecoshelf.
Butrin, D., Vesloguzov, V. and Rebrov, D. (2006) 'Watch out, the pipes are closing!', *Kommersant*, December 14.
Denisov, V. (2002) 'Eco-geographical principles of sustainable resources exploitation in the shelf seas' (ecological geography of the sea), *Apatity*, Russia.
Grib, N. (2006) 'Gazprom received congratulations on new deposits', *Kommersant*, November 20.
Grigoriev, A. (1995) 'Foreign investments and the future of the Russian oil and gas industry', *Taiga-News*, 13.
Hansen, J. R., Hansson, R. and Norris S. (eds; 1996) *The State of European Arctic Environment*, EEA Environmental Monograph no. 3, Copenhagen: European Environment Agency.
Kondratjev, K. A. (2003) *Global Sustainable Development: Reality and Illusions*, Analytical Report to the Assembly Meeting of RAS, St. Petersburg.
Matishov, G. and Rodin, A. V. (1996) *Atlantic Cod: Biology, Ecology, Fishery*, St.Petersburg, Nauka.
Moe, A. and Kryukov, V. (1998) 'Two keys and many locks: joint management of oil and gas resources in Russia', *Post-Soviet Geography and Economics*, 39 (7).
Skjoldal, H. R. (1991) *Management of Marine Living Resources in Changing Ocean Climate*, Bergen: Institute of Marine Research.
The Energy Shtokman Therapy (2006), *Kommersant*, December 11.
The Precautionary Approach to North Sea Fisheries Management (1997) Oslo. Seminar Report. Fisken og Havet, # 1.
Zayonchkovskaya, Z. A. (2003) *Demographic Future of the Russian Arctic: Drastic Changes*, Paper presented at the Third IISP Inter-Regional Conference, Norilsk.

Part III
Comparisons and managerial implications

11 Human rights and indigenous peoples in the Arctic

What are the implications for the oil and gas industry?

Ketil Fred Hansen and Nigel Bankes

Introduction

There are about 300 million indigenous people in the world today, representing more than 4000 different languages in more than 70 countries (United Nations Human Development Report; UNHDR, 2004: 29).[1] One and a half million of these people live in the Arctic region where indigenous people represent 15 per cent of the total population (compared with 1.5 per cent in the entire world). This is a common trait throughout the circumpolar world, although the proportion of indigenous peoples does vary considerably. In some places, indigenous peoples are in an absolute majority,[2] while in other parts of the Arctic, they represent a small minority.[3] The adoption of the Declaration of the Rights of Indigenous Peoples by the United Nations General Assembly in September 2007, while of obvious global significance, is of particular importance in the Arctic and to the four Arctic states considered in this volume.

This chapter examines the interplay between the human rights of indigenous peoples and the interests of the oil and gas industry in carrying out exploration, production and transportation of oil and gas within the traditional territories of those indigenous peoples. The chapter focuses on the human rights of indigenous peoples in international law, but also tries to assess how these rights in international law form part of the discourse within the domestic legal and political systems of each of the four states: Canada, Norway, Russia and the United States of America (USA). Furthermore, the chapter explores how these discourses influence the commitments of the oil and gas sector to ideas of corporate social responsibility.

Within the field of international human rights law, we focus on the rights of indigenous peoples in relation to lands and resources within their traditional territories. As a starting point, we recognize that there is a close relationship between the right of all peoples (including indigenous peoples) to self-determination as provided for in the two International Covenants,[4] which also recognize the inherent right of all peoples to their natural wealth and resources, and their right not be deprived of their own means of subsistence.[5]

In reviewing the relevant international instruments, it is worth emphasizing that, although some of these instruments have global application (e.g. the International

Covenant on Civil and Political Rights; ICCPR), either because they are widely ratified or because they also encode principles of customary law, others do not bind each of the four states. This may be because a state has not ratified the relevant instrument (e.g. the *Convention 169 Concerning Indigenous and Tribal Peoples in Independent Countries*; hereafter 'ILO 169') or because the norms are regional in nature, for example, the norms of the Organization of American States (OAS) may bind the USA and Canada, but they will not bind Norway or Russia. It is also worth noting that United Nations (UN) declarations once adopted often influence how states deal with relevant issues, even if the declarations are not legally binding.[6] Table 11.1 illustrates the positions, and further discussion follows in the substantive sections of the chapter.

Norway's commitment to human rights instruments directed at the situation of indigenous peoples is notable both within the context of this group of states but also amongst its fellow Nordic countries. Jan Egeland argues that an important part of Norway's reputation and 'power' is founded upon its human rights record (Egeland, 1988). This commitment legitimizes Sámi claims to political inclusion and accommodation within the majoritarian framework of the democratic nation-state (Broderstad, 2001; Oskal, 2001: 235).

The four states observe different practices and rules with respect to the relationship between international law and domestic law. Historically, the legal position in Canada and Norway is very similar with respect to the incorporation of international human rights conventions into the domestic legal system. Both states take a dualist approach in which ratification of an international convention does not automatically result in a change in domestic law. Domestic rules will only change as a result of an enactment of the competent legislature.[7] Furthermore, both states generally took the somewhat self-satisfied view that domestic norms were already compliant or 'in harmony' with the main international

Table 11.1 Ratifications of selected human rights instruments

	Canada	Norway	Russia	US
ILO 169	No	Yes	No	No
International Covenant on Civil and Political Rights (ICCPR)*	Yes	Yes	Yes	Yes
ICCPR, Optional Protocol No. 1*	Yes	Yes	Yes	No
Convention on the Elimination of All Forms of Racial Discrimination	Yes	Yes	Yes	Yes
American Declaration on the Rights† and Duties of Man	Yes	No	No	Yes

*By ratifying Optional Protocol No. 1, a state party to the ICCPR agrees to accept the right of individual petition whereby an individual may complain to the Human Rights Committee that a state has violated his or her protected rights.
†All member states of the Organization of American States are bound by the Declaration by virtue of membership of the Organization.

human rights instruments and that legislated changes to domestic law were, therefore, unnecessary. The judiciary supports this view by applying a presumption to the effect that domestic laws are assumed to comply with the international obligations of the state, and will thus strive to interpret domestic laws consistently with this understanding.

With a view to strengthening the status of human rights in domestic law, Norway departed from this tradition in April 1999 when it adopted the Human Rights Act.[8] With this legislation Norway listed certain conventions as having the force of law in Norway,[9] including the two International Covenants and Optional Protocol No. 1 to the ICCPR. The Act goes on to provide that the provisions of the listed agreements 'shall take precedence over any other legislative provisions that conflict with them'.

US law is certainly more complex with respect to the relationship between international treaty commitments and domestic law, but the Supremacy Clause of the US Constitution (Article VI(2)) stipulates that treaties have the same status as an Act of Congress and thus will trump an inconsistent state law as well as an earlier Congressional enactment. Treaties, however, require Senate approval (Tribe, 2000: 643-656).[10] On the face of it, Russia is the most 'internationalist' of the four states insofar as Article 15 of the 1993 Constitution takes a monist position in which both customary law and treaties to which Russia is a party are automatically a part of the legal system and, in the event of a conflict between a treaty and an ordinary law, it is the treaty provision that will prevail. However, most Russian observers argue that the law and practice in matters concerning human rights and indigenous peoples differ considerably.[11]

Canada, Norway and Russia each offer a degree of constitutional protection to Aboriginal rights. Canada amended its constitution in 1982 to recognize and affirm 'existing aboriginal and treaty rights'. Norway amended its constitution in 1988 to recognize (Article 110a) that the state has an obligation to create the conditions to enable the Sámi to preserve and develop their culture, language and way of life. And the Russian constitution (Article 69) 'guarantees the rights of small indigenous peoples in accordance with the generally accepted principles and standards of international law and treaties of the Russian Federation'. There is no formal constitutional protection for the rights of indigenous people in the USA, although the rules pertaining to the division of powers between the state and federal governments do to some extent insulate indigenous communities from the application of state (but not federal) laws.[12]

The interests of the oil and gas industry

As the oil and gas industry is driven to the frontiers in the search for new resources whether in the Beaufort Sea, the Canadian High Arctic, Greenland, the Barents Sea or elsewhere in the Arctic, it frequently finds itself operating within the traditional territories of indigenous peoples. In most cases, the industry operates on the basis of authorization (permits, licences, etc.) and resource rights (leases, concessions, etc.) secured from a national or sub-national, non-indigenous government. In many cases, the industry will engage in these activities in areas where indigenous peoples are known to have claims to lands and resources, but where these claims are contested by, and have never been settled by, non-indigenous governments.

In some cases, however (e.g. in Alaska and much of northern Canada), the non-indigenous governments have settled by agreement (e.g. the Nunavut Land Claim Agreement) or legislation (e.g. the Alaska Native Claims Settlement Act), at least at some level, the land and resource rights of the indigenous peoples. These settlements may recognize a full ownership claim of indigenous peoples to oil and gas rights and sub-surface rights within at least part of their traditional territories. In these cases, the oil and gas industry will find itself negotiating with an indigenous landowner (e.g. the Inuvialuit people of the Mackenzie Delta) for its oil and gas concession or lease. For the remainder of the lands, however, (and this is the case under all the land claim agreements in northern Canada, including the Inuvialuit Agreement) indigenous peoples may only own the surface rights, in which case the industry will still have to deal with access issues and negotiate impact and benefit agreements. Other aspects of these land claim agreements provide for co-management structures and regulatory authorities that share power and authority between indigenous peoples and the non-indigenous government.[13]

In sum, oil and gas operators working within the traditional territory of indigenous peoples face particular challenges. Where claims have been settled, the ownership position will be clearer, although industry will likely find itself dealing with multiple owners and/or usufruct holders, both the state and indigenous people. Greenfield linear developments (e.g. pipelines such as the Mackenzie Gas Project described in Chapter 8 in this book) that pass through the traditional territories of a number of indigenous peoples will present the greatest challenge. The challenge for the operator will be to observe both the laws of general applications, and the terms of any settlement agreement between governments and indigenous peoples, as well as the extra-legal norms related to corporate social responsibility (CSR) perhaps informed by non-binding international declarations, such as UN Declaration on the Rights of Indigenous Peoples (UNDRIP), and international conventions, such as ILO 169, even if not ratified by a particular state.

Where land claims remain unresolved, an oil and gas operator faces additional and more serious challenges. At a practical level, unresolved claims create uncertainty as to title for operators and their investors. Does the operator have the legal right to proceed? Is it sufficient for the operator to have a set of state authorizations and state-granted property rights, or must it also secure additional authorizations and approvals from the indigenous peoples with respect to the use of their traditional territories? And will governments object if the operator elects to deal directly with the indigenous peoples? An operator who chooses to ignore these legal uncertainties and deal only with the national government faces the possibility not only of civil protest (roadblocks, boycotts, etc.) but also risks the accusation that it is ignoring the human rights of the indigenous peoples. This may lead to legal liability (e.g. liability for trespass on the property of another) or other forms of legal challenge to permits and the right to operate (e.g. the indigenous peoples may argue that authorizations that violate constitutional protections of indigenous rights or that are contrary to international norms that are part of domestic law are simply void or invalid). Finally, some observers argue that an international operator who deals solely with the national government fails to meet certain norms of CSR, thereby damaging the corporation's image and reputation (Fjellheim, 2006).

In some cases, but more prominently in the case of operators active in developing countries rather than in the Arctic, the World Bank or a Regional Development Bank will want to be assured that the rights of indigenous peoples have been fully respected and accommodated before endorsing a resource project, and providing necessary financing or guarantees (see e.g. World Bank, 2006). For example, the International Finance Corporation of the World Bank recommends that extractive industries operating in indigenous (or disputed) territories follow the ILO 169 Convention, even if it is not ratified by the national governments in question, since it is used as a reference point by indigenous peoples themselves (World Bank, 2007). While emphasizing that the legal responsibilities towards indigenous peoples are held by the states, to break with international accepted human rights standards 'can have serious reputational impacts', even if not part of the domestic law of a particular state (World Bank, 2007: 5).

Recent developments in human rights law

Despite important international progress concerning the rights of indigenous peoples over the last few decades, it is an unfortunate reality that there is still significant legal uncertainty in many countries as to the validity and scope of the property and resource rights of indigenous people. At one extreme, this uncertainty manifests itself in the doctrine of terra nullius, the claim that, when European or other settlers arrived, the social institutions of the indigenous peoples were so primitive that there were no indigenous property institutions to recognize. The International Court of Justice has condemned this as a racist doctrine[14] and it no longer informs the domestic law of most countries.[15] But it is still the case in many countries in the Americas, in the South Pacific and in Nordic countries that governments have failed to recognize, demarcate and protect the property rights of indigenous peoples. It is not uncommon for the governments of these countries to pay lip service to the protection of indigenous land rights but then to act as if the state were the full beneficial owner of all traditional territories within the boundaries of the state.[16]

There have been considerable developments in the body of international law that pertains to the rights of indigenous peoples over their traditional territories over the last two decades. It is now possible to assert without fear of contradiction that these issues are not just matters of domestic law and policy (the traditional approach to human rights generally) but also raise questions of international law against which domestic laws and policies may be judged. We can also see in these developments a broadening of the language and content of human rights law to embrace collective rights as well as individual rights. And, finally, we can see in these developments a willingness to explore different avenues for vindicating the rights of indigenous people. One approach emphasizes the need to develop and formulate norms that are specific to the position of indigenous peoples. Another approach argues that it is possible to re-interpret existing human rights protections (e.g. the right to property and the right to equality) so that they protect indigenous peoples as well as non-indigenous peoples.

We will discuss four of those developments here. First, we will discuss ILO Convention 169 and its implementation in Norway. Second, we will see that the recognition and protection afforded to minority rights by Article 27 of the ICCPR has implications for the way in which the non-indigenous governments deal with resources within the traditional territories of indigenous peoples. Third, we discuss the UN Declaration of the Rights of Indigenous Peoples. And finally, we refer to a few examples of the increased use of international fora (international courts and other international investigative bodies) to articulate and vindicate the land and resource rights of indigenous peoples. In discussing these developments we will also note some differences in the law and practice of the four states, Russia, Norway, Canada and the USA. These discourses will give us an impression of the different views and arguments articulated in the debate about oil and gas activities in the Arctic by different indigenous peoples.

The first and third of these developments are examples of efforts to develop new laws that speak specifically to the position of indigenous peoples. The second and last are examples of the use of existing international norms to protect the interests of indigenous people. In their way, each of these developments represents an effort to continue the process of decolonizing the international law of the nineteenth century and to make international law a truly global law rather than a law of European settler societies.

Convention concerning indigenous and tribal peoples in independent countries, ILO 169

In formal legal terms, this Convention represents the most important legal recognition to date of the land and resource rights of indigenous peoples[17] in international law (Thornberry, 2002; Anaya, 2004: 58-61). Before examining its terms, however, it is essential to recognize one important limitation on its application in the present case. Of the four states we are considering in this book, only Norway has ratified the Convention[18] and there is little chance that Canada or the USA will ratify.[19] Russian researchers and indigenous peoples organizations also doubt that Russia will ratify the Convention in the near future.

The ILO 169 Convention (which replaces the earlier ILO Convention 107 of 1957) aims to protect indigenous and tribal peoples and their way of life and culture based upon their own aspirations and priorities (Graver and Ulfstein, 2004: 346). While much of the Convention deals with labour-related issues, land rights are also of central significance. The Convention recognizes that indigenous people have both procedural rights and substantive rights.

Article 6 imposes important procedural obligations on the Contracting Parties. It requires governments to consult whenever considering legislative or administrative measures that may affect indigenous peoples directly. The consultation must be in good faith, appropriate to the circumstances, and 'with the objective of achieving agreement or consent to the proposed measures'.

Articles 14 and 15 deal with the land and resource rights of indigenous peoples and are of central importance here:

Article 14

1. The rights of ownership and possession of the peoples concerned over the lands[20] which they traditionally occupy shall be recognized. In addition, measures shall be taken in appropriate cases to safeguard the right of the peoples concerned to use lands not exclusively occupied by them, but to which they have traditionally had access for their subsistence and traditional activities. Particular attention shall be paid to the situation of nomadic peoples and shifting cultivators in this respect.
2. Governments shall take steps as necessary to identify the lands which the peoples concerned traditionally occupy, and to guarantee effective protection of their rights of ownership and possession.
3. Adequate procedures shall be established within the national legal system to resolve land claims by the peoples concerned.

Paragraph 1 recognizes that indigenous peoples may have rights in relation to two categories of lands, a first category over which they have exclusive rights by virtue of traditional occupation and a second category of lands over which they have used rights. Such use rights might include hunting, fishing and gathering rights, the right to carry on traditional ceremonies and other practices and the right to graze animals such as reindeer. As to the first category of lands, the state must recognize the indigenous ownership and possessory interest and to that end is obliged to identify the lands in question and guarantee their effective protection. Additional measures may be required to protect use rights.

This Article of the Convention and its implementation has been the subject of significant debate in Norway, especially in the context of the Finnmark Estate legislation (discussed below) but also with respect to the offshore. Sámi representatives argue that 'lands' should include both onshore and offshore territories and connected natural resources. Norwegian politicians and the general public normally do not acknowledge any special offshore rights for the Sámi population. This different understanding is important in the context of oil and gas activities in the Barents Sea. Sámi representatives argue that Sámi fishermen have used the sea since time immemorial for their subsistence and that, even if offshore oil activities take place far from land and from places traditionally used by the Sámi, those activities will affect fish and fish habitat. They argue, therefore, that the

Sámi should have a say in offshore drilling and that they have special rights to compensation [see e.g.; Russian Association of Indigenous Peoples of the North (RAIPON), 2002; and different readers' opinions in the daily newspapers *Nordlys* and *Finnmark Dagblad* during the winter of 2006).

For example, the annual conference of Norske Samers Riksforbund (NSR)[21] in 2004 stated that the Sámi people have had property rights and usufruct rights since time immemorial to the marine resources and sea areas of the Sámi territories. The conference emphasized that the term 'lands' as used in ILO 169 includes the concept of territories, which covers 'the total environment of the areas which the peoples concerned occupy or otherwise use'.[22] The large majority of the Sámi, if not all, support this view (see Chapter 9 for more detailed information).

However, the Sámi do not speak with one voice when it comes to whether or not to allow expanded drilling for oil and gas in the Barents Sea. The Sámediggi has no official view on this, and the Sámi president says she is not fiercely against drilling but wants to secure the Sámi right to self-determination on this issue (see Chapter 9 for further elaborations). The question of expanding oil and gas activities is also controversial with the Sámi people. The National Association of Norwegian Sámi has from the very beginning (1968–2004) adopted all resolutions at their national assemblies unanimously. But at its 2005 annual assembly this question divided the party heavily. Many of the delegates voted for opening up for offshore oil and gas activities in the Barents Sea, but a tiny majority voted against.[23]

Article 15 of the Convention deals with rights to resources:

Article 15

1. The rights of the peoples concerned to the natural resources pertaining to their lands shall be specially safeguarded. These rights include the right of these peoples to participate in the use, management and conservation of these resources.
2. In cases in which the State retains the ownership of mineral or subsurface resources or rights to other resources pertaining to lands, governments shall establish or maintain procedures through which they shall consult these peoples, with a view to ascertaining whether and to what degree their interests would be prejudiced, before undertaking or permitting any programmes for the exploration or exploitation of such resources pertaining to their lands. The peoples concerned shall wherever possible participate in the benefits of such activities, and shall receive fair compensation for any damages which they may sustain as a result of such activities.

Paragraph 1 of Article 15 deals with the resource rights that pertain to the lands referred to in Article 14 for which an indigenous people has either exclusive rights or use rights. While Article 14 deals with the private or proprietary aspects

Human rights and indigenous peoples in the Arctic 299

of these resource rights, Article 15(1) deals with the governmental or regulatory sphere and affords indigenous peoples the right to participate in the use, management and conservation of those resources that they either own exclusively or to which they have rights.

Paragraph 2 of Article 15 deals with resource rights which the state reserves to itself. In short, this paragraph recognizes that, in many jurisdictions, mineral resources are the subject of public ownership rather than private ownership and concedes that a state need not recognize an indigenous ownership interest in these resources provided that it does not discriminate against indigenous people as owners. This is the case, for example, for petroleum and minerals in Norway (Graver and Ulfstein, 2004: 367, 370). But while the state is not obliged to recognize private indigenous ownership of these resource rights, it is obliged to put in place a consultation procedure that provides for a prior assessment of the impacts of resource exploration or development on any indigenous exclusive rights or use rights identified under Article 14. Furthermore, indigenous people shall be afforded the opportunity to participate in such activities. However, the consultation language of paragraph 2 falls short of requiring 'prior informed consent', and while the last sentence of the paragraph requires compensation for actual damage, it does not require sharing of economic benefits (e.g. through the payment of a royalty or some other form of revenue sharing).

Perhaps the key question to ask is what difference has ratification of ILO 169 made to the legal position of Sámi in Norway? This is not the sort of question that lends itself to a definitive response (since there are many potentially relevant causal relationships) but it is possible to provide some sort of answer using the recent Finnmark Act as an example. The government introduced the Finnmark Act in April 2003 after numerous official reports (e.g. of the Sámi rights board[24]) and various consultations with indigenous peoples groups. The Finnmark Act was introduced partly to clarify uncertainties surrounding land and water rights in Finnmark, but also to provide a secure basis for Sámi culture (the background to the Act is explored in more detail in this volume in Chapter 9. A key part of the bill involved the transfer of land title in much of the region from the state to the proposed Finnmark Estate, a form of co-management authority where three representatives from the Sámidiggi, together with three representatives from the Finnmark County Council, would be represented on the board responsible for making decisions about land use within Finnmark.

The initial reaction of the Sámediggi to the proposed legislation was negative. The Sámediggi argued that the government had breached its consultation obligations under ILO 169 and that the bill did not afford adequate recognition and protection to existing Sámi rights. As a result, the Storting's Standing Committee on Justice commissioned an independent and extensive legal opinion to assess the conformity of the legislation with Norway's international legal obligations (not only with respect to the ILO 169, but also with respect to the ICCPR and the Convention on the Elimination of All Forms of Racial Discrimination; CERD).[25] In summary, the opinion concluded that the bill did not meet the requirements of Article 14 of ILO 169. In particular, the authors concluded that Norway was obliged to identify the two categories of land to which Sámi had rights, and to afford the Sámi the same legal and factual control over those lands to which the Sámi had exclusive rights as typically accrue to an owner and possessor of property. The bill did not accomplish this. Alternatively, the bill might afford Sámi ownership-like powers and control over all of the lands to be vested in the Finnmark Estate, but it did not accomplish that either given the composition of the Board.

The Act as passed in 2005 contains a number of amendments to better protect the Sámi interest. For example, paragraph 3 makes it clear that the Act is subject to ILO 169 and paragraph 5 clarifies that the Act and the transfer of title to the Finnmark Estate does not prejudice the collective and individual rights of Sámi based on traditional use. A further amendment accords Sámi representatives on the Board of the Finnmark Estate a more important role when it makes decisions with respect to land use in the municipalities where Sámi represent the majority of the population (Karasjok, Kautokeino, Nesseby, Porsanger and Tana). A further result was that the government and the Sámi parliament adopted an agreement on procedures for consultation. Those procedures specifically acknowledge that one of the purposes of the agreement is to 'provide a practical implementation of the Central Government's obligations under international law to consult indigenous peoples'.[26]

Yet, even if the Finnmark Act, as passed in April 2005, was more favourable to the Sámi than the original text proposed two years earlier, Sámi representatives argue that the Act was a compromise they could live with and not a victory. From their perspective, the Act did little more than legalize some Sámi territorial rights that had been rejected by the colonizing Norwegian government for more than a hundred years.[27] In addition, the Finnmark Act was silent with respect to Sámi rights to marine resources and did not recognize the Sámi understanding of 'lands' as including the sea (www.nsr.no).

In conclusion, it seems possible to say that ILO ratification has changed the nature of the debate about Sámi land rights in Norway. First, the debate is now explicitly about the importance of observing international obligations. Second, examination of the proposed legislation in light of the Convention led both to improved protection of existing Sámi rights and to improved consultation procedures between the state and the Sámi parliament.[28] That said, and in the context of oil and gas developments within the Norwegian Arctic, the Sámi argue that the consultation process is more of a formality than an affirmation of the Sámi right to self-determination. For the consultation processes to be authentic, they have to be conducted in 'a spirit of partnership and mutual respect' (United Nations, 2007), 'undertaken in good faith' (ILO 169) and 'with the objective of achieving agreement' (Royal Ministry of Labour and Social Inclusion, 2006).

Taking the Norwegian White paper no. 30 (2004-05) *Opportunities and Challenges in the North* (Norwegian Ministry of Foreign Affairs, 2005) as an example, the Norwegian government argues that they have consulted the Sámi population. However, the Sámi parliament and other Sámi organizations do not accept that they have been consulted in a fair way. Aili Keskitalo, the Sámi president until mid-2007, argues that the Sámi people have not been listened to.[29] For many

Sámi politicians and spokespersons, 'consultation' in the context of the right to self-determination includes a right to veto government proposals. This view has the support of Professor Dr.juris Carsten Smith (the former Chief Justice of Norway) in his elaboration on the draft Nordic Sámi convention.[30] If, say most Sámi organizations and politicians, consultation does not embrace the right to veto, then it can be little more than a play for the gallery. In this context, it is important to note that the newly established guide to the understanding of the agreed consultation procedures between the Sámidiggi and the Norwegian government states that 'the consultation procedures will, to a greater degree, prepare the ground for a partnership between State authorities and the Sami Parliament. However, the State authorities still retain the legal authority and responsibility to make the final decisions in such matters' (Royal Ministry of Labour and Social Inclusion, 2006).

Most indigenous peoples worldwide contend that consultation should be based upon 'free, prior and informed consent' and that this principle implies a right to veto. Andrea Carmen from the International Indian Treaty Council, for example, argues that free, prior and informed consent 'must be applied as the operative human rights framework and standard in conducting all new agreements and arrangements' (Carmen, 2006). Similarly, Erica-Irene Daes, the former Special Rapporteur of the United Nations Working Group on Indigenous Populations [UNWGIP], argues that 'in order to be meaningful this modern concept of self-determination must logically and legally carry with it the essential right of permanent sovereignty over natural resources' (Daes, 2003). And the Sámi Council and the Inuit Circumpolar Conference, in a joint statement at the UNWGIP in 2004 argued that:

> Our organizations have been very concerned with institutions such as the World Bank Group trying to dilute the concept of free, prior and informed consent in recent processes such as the Extractive Industry Review and the review of the Indigenous Peoples Policy, e.g. wanting to transform the right of free, prior and informed <u>consent</u> to free, prior and informed <u>consultations</u>. . . . It is imperative that any third party seeking consensus from an indigenous people, do so only from the body etc. authorized to give consent under the relevant customary legal system of that people. We believe that a thorough understanding of the relevance of customary law is important not only for the safeguarding of indigenous rights. Such an understanding can also mitigate fears sometimes raised that a right to free, prior and informed consent implies a right for any indigenous person to stop every development project.

The act of consultation is also addressed in the proposed new common Sámi convention. The convention states that issues of major importance to the Sámi must be subject to negotiations with the Sámi parliament prior to any public authority decision, and adds that such negotiations must be scheduled to give the Sámi parliament ample time to influence the process and the outcome.[31] Similar views are evident amongst indigenous peoples. As reported in Chapter 7 (Alaska), indigenous people dealing with oil and gas companies say that they have plenty of time: 'We can sit and we can wait'.

Protecting land rights and traditional territory through the recognition of the cultural rights of minorities

While Norway is the only one of the four states that is a party to ILO 169, all four states have ratified the International Covenant on Civil and Political Rights, although only Norway has taken the step of making the Covenant part of domestic law.[32] As we have seen, the ICCPR, in common with its companion Covenant on Economic and Social Rights, begins with the recognition that all peoples (including, therefore, indigenous peoples) have the right of self-determination. But the ICCPR also offers a degree of protection to minorities through its Article 27:

> In those States in which ethnic, religious or linguistic minorities exist, persons belonging to such minorities shall not be denied the right, in community with the other members of their group, to enjoy their own culture, to profess and practise their own religion, or to use their own language. (ICCPR, s 27)

Although Article 27 speaks to the position of minorities, it is clear that, in many if not all cases, indigenous peoples within a settler society will fall within the concept of a minority. Certainly this is true of Indian and Inupiat peoples in Alaska, the First Nations and Inuit of Canada, the Sámi of Norway and the different indigenous peoples of Russia. Furthermore, while the text of Article 27 does not mention rights to land and resources, and seems to speak more obviously to social issues such as language and education, the UN Human Rights Committee, the body responsible for monitoring the implementation of the Convention by the parties (as well the body responsible for considering petitions from aggrieved individuals where the state – Norway or Canada – has ratified the so-called Optional Protocol), has given the Article a broad interpretation. In particular, the Committee has interpreted the Article as imposing upon the state the duty to protect the distinctive cultures of indigenous peoples, and has emphasized that the economic and material aspects of culture may include the traditional territories of indigenous peoples and their special relationship with their territories. The precise contours of this duty are difficult to identify but we can gain a better understanding of the position by examining the Committee's general interpretive comment on Article 27 as well as a set of three related decisions (the Lænsman decisions) that it has rendered in response to petitions from Sámi reindeer herders in Finland.

The Committee's general comment on Article 27 (adopted in 1994)[33] emphasized (at para. 7) that 'culture manifests itself in many forms, including a particular way of life associated with the use of land resources, especially in the case of indigenous peoples'. The Committee went on to say that the 'right may include such traditional activities as fishing or hunting' and that 'the enjoyment of those rights may require positive legal measures of protection and measures to ensure the effective participation of members of minority communities in decisions which affect them' (*ibid.*).[33] The duties imposed on the state therefore include substantive obligations (the duty to protect) and procedural obligations (effective participation in decisions that affect minorities and/or indigenous people).

Perhaps more revealing than the general comment, at least in terms of trying to elucidate the precise content of the substantive duty to protect, are the Committee's three Længsman decisions in 1994, 1996 and 2005.[34] In each case, the substance of the complaint was that the government of Finland, by authorizing resource developments of various forms (either individually or cumulatively) within the traditional reindeer grazing territory of the petitioner had breached the Article 27 rights of the petitioners. But, in each case, the Committee denied the petition. The Committee's reasoning with respect to the most recent such petition is instructive. The Committee noted (at para. 10.1) that while:

> measures whose impact amounts to a denial of the right are incompatible with the obligations under article 27 . . . measures with only a limited impact on the way of life and livelihood of persons belonging to a minority will not necessarily amount to a denial.

The Committee emphasized that it was important to look at the cumulative effect (at para. 10.2) of 'a series of actions or measures taken by a State party over a period of time and in more than one area of the State occupied by that minority'. In the end, and recognizing that the petitioner and the state fundamentally disagreed as to their assessments of the effect of logging operations on herding activities,[35] the Committee concluded (at para. 10.3) that 'the effects of logging carried out . . . . have not been shown to be serious enough as to amount to a denial of the authors' right to enjoy their own culture in community with other members of their group under article 27 of the Covenant'.

In conclusion, the jurisprudence on Article 27 makes it clear that the protection of culture extends to a state obligation to protect the relationship between indigenous peoples and their traditional territory. Consequently, state decisions to authorize resource activities within traditional territories may come under international scrutiny where those activities are so extensive or intensive that they seriously interfere with the ability to practise culturally important activities. While the 'Views' of the Human Rights Committee seem to suggest that the threshold for intervention is high, the Committee has warned that the cumulative effects of a number of developments may breach a state's Article 27 obligations.

The UN Declaration on the Rights of Indigenous Peoples (UNDRIP)

Over the last decade, we have seen one important global effort and at least two regional efforts to develop comprehensive declarations of the rights of indigenous peoples.[36] Here we will focus on the global initiative within the UN to develop a Declaration on the Rights of Indigenous Peoples. This development can be seen as a continuation of the process begun after the Second World War to articulate first a Universal Declaration of Human Rights and later the two international covenants as well as a Declaration on the Rights of Minorities.[37]

Efforts to draft a declaration within the UN began in 1982 with the establishment of the Working Group on Indigenous Populations. The Working Group first adopted a Draft Declaration in 1994 after which the Human Rights Commission formed an open-ended inter-sessional working group to elaborate the draft declaration with the expectation that this work might be completed within the first International Decade of the World's Indigenous People (1995–2005). For some long time it appeared as if the UN process was in serious trouble as key states seemed to be unable to agree on some very basic points (e.g. acknowledgement of the right of indigenous peoples to self-determination) and most of the text remained bracketed. But a breakthrough in 2006 resulted in adoption by the UN Human Rights Council (the successor to the Human Rights Commission) of the UN Declaration on the Rights of Indigenous Peoples on June 29, 2006 by a margin of 30:2 with 12 abstentions. The two dissenting Council members were Canada and Russia. The Declaration was finally adopted by the UN General Assembly on September 13, 2007 against the votes of the USA, Canada, Australia and New Zealand. Russia abstained as did ten other countries.[38]

The text comprises 46 articles. It recognizes the right of indigenous peoples to self-determination. Articles 25-32 deal with various aspects of the rights of indigenous peoples to lands, territories and resources. The key statements in relation to oil and gas development are perhaps Article 26 and 32:

Article 26

1. Indigenous peoples have the right to the lands, territories and resources which they have traditionally owned, occupied or otherwise used or acquired.
2. Indigenous peoples have the right to own, use, develop and control the lands, territories and resources that they possess by reason of traditional ownership or other traditional occupation or use, as well as those which they have otherwise acquired.
3. States shall give legal recognition and protection to these lands, territories and resources. Such recognition shall be conducted with due respect to the customs, traditions and land tenure systems of the indigenous peoples concerned.

Article 32

1. Indigenous peoples have the right to determine and develop priorities and strategies for the development or use of their lands or territories and other resources.
2. States shall consult and cooperate in good faith with the indigenous peoples concerned through their own representative institutions in order

to obtain their free and informed consent prior to the approval of any project affecting their lands or territories and other resources, particularly in connection with the development, utilization or exploitation of mineral, water or other resources.
3. States shall provide effective mechanisms for just and fair redress for any such activities, and appropriate measures shall be taken to mitigate adverse environmental, economic, social, cultural or spiritual impact.

In common with other articles (and indeed the drafting approach of ILO 169), these articles provide not only a statement of rights but also articulate the State's correlative obligations.

John McNee, Canada's spokesperson, referred to the lands and resources articles as a major reason for Canada's negative vote 'the[se] provisions', he said, 'are overly broad, unclear, and capable of a wide variety of interpretations, discounting the need to recognize a range of rights over land and possibly putting into question matters that have been settled by treaty'.[39]

By contrast, the adoption of the Declaration was unanimously applauded by indigenous peoples worldwide. The National Chief of the Assembly of First Nations in Canada, for example, said the 13th of September is an important day in Canada's history: 'It's a day to celebrate, and a day to act'.[40] The Inuit Circumpolar Council and Sámi Council stated that the Declaration 'establish[es] the necessary foundation for sustainable and equitable development in the Arctic and will ensure that Indigenous peoples in the Arctic directly benefit from such resource use'.[41] The UN Permanent Forum also hailed the declaration as a breakthrough:

> It is a key instrument and tool for raising awareness on and monitoring progress of indigenous peoples' situations and the protection, respect and fulfilment of indigenous peoples' rights. It will further enflesh and operationalize the human rights-based approach to development as it applies to Indigenous Peoples.

And the UN Secretary-General called on governments and civil society to urgently advance the work of integrating the rights of indigenous peoples into international human rights and development agendas, as well as policies and programmes at all levels, so as to ensure that the vision behind the Declaration becomes a reality.[42]

It remains to be seen how the Declaration will influence state politics, business behaviour and practical rights for indigenous peoples. Even if the Declaration is not legally binding, the standards and visions emphasized in the text should at least influence those governments that voted for the Declaration. Extractive industries operating within indigenous or contested territories will also need to take the Declaration into account so as not to damage their international reputations.

Fora for articulating and vindicating the land and resource rights of indigenous peoples

At the same time as states and indigenous peoples have been elaborating declarations of rights that pertain specifically to indigenous peoples, some indigenous peoples have also been making efforts to take advantage of existing and general human rights instruments. We have seen some examples of this in the text already, as indigenous peoples seek to use the Human Rights Committee and the 'minorities' provision of the ICCPR.[43] However, this trend has perhaps been most noticeable in the context of the OAS, where indigenous peoples have vindicated their rights before both the Inter American Commission on Human Rights and within the Inter American Court. Resort to these fora has been triggered in each case by conflicts between indigenous peoples and state-authorized resource activities (including oil and gas activities) in traditional territories. These cases include the *Awas Tingni* Case before the Court and the *Maya* Case before the Commission. The former is the more authoritative. While neither of these decisions directly implicates any of the four Arctic states that are the subject of this volume, the decisions do show how general human rights instruments should be relevant considerations when states propose to licence resource development activities within the traditional territories of indigenous peoples.[44]

There are two main human rights instruments within the OAS system: the American Declaration on the Rights and Duties of Man, and the American Convention on Human Rights. The principal distinction between these two instruments for present purposes is that all member states of the OAS are subject to the Declaration by virtue of membership but the Convention binds only those states that have ratified the Convention. Neither the USA nor Canada has ratified the Convention.

In *Awas Tingni*, the community argued that Nicaragua was in breach of two obligations under the Convention, the Article 25 right to an effective remedy and the Article 21 right to property. The trigger for the alleged breach was the decision of the Nicaraguan government to enter into a timber concession within the traditional territory of the indigenous people.[45]

As to the right to an effective remedy, the plaintiffs in *Awas Tingni* noted that Nicaragua's constitution and domestic legislation was supposed to protect the rights of indigenous peoples, but over many years, and after several rounds of litigation, no indigenous people in Nicaragua had ever been able to obtain secure title to their traditional territories. That, in the opinion of the Court, was a breach of the state's obligations and by way of remedy (at para. 138):

> the State must adopt in its domestic law the necessary legislative, administrative, or other measures to create an effective mechanism for delimitation and titling of the property of the members of the Awas Tingni Mayagna Community, in accordance with the customary law, values, customs and mores of that Community.

As to the right to property, the Court began by observing that the right to property protected by Article 21 of the Convention protected not only an individual's right to property but also communal conceptions of property common to indigenous societies. This ought particularly to be the case given the importance that indigenous peoples attach to their relationship to land and traditional territory (at para. 149):

> some specifications are required on the concept of property in indigenous communities. Among indigenous peoples there is a communitarian tradition regarding a communal form of collective property of the land, in the sense that ownership of the land is not centred on an individual but rather on the group and its community. Indigenous groups, by the fact of their very existence, have the right to live freely in their own territory; the close ties of indigenous people with the land must be recognized and understood as the fundamental basis of their cultures, their spiritual life, their integrity, and their economic survival. For indigenous communities, relations to the land are not merely a matter of possession and production but a material and spiritual element which they must fully enjoy, even to preserve their cultural legacy and transmit it to future generations.

As we have just noted, Nicaragua's legal system purported to offer both constitutional and statutory protection to the property rights of indigenous peoples. The difficulty was that the state had failed to implement this protection in a practical and effective way thereby creating (at para. 153) 'a climate of constant uncertainty among the members of the Awas Tingni Community, insofar as they do not know for certain how far their communal property extends geographically and, therefore, they do not know until where they can freely use and enjoy their respective property'. That, in the opinion of the Court, cried out for a remedy. Accordingly, the Court ruled (at para. 153) that:

> the members of the Awas Tingni Community have the right that the State
>
> 1. carry out the delimitation, demarcation, and titling of the territory belonging to the Community; and
> 2. abstain from carrying out, until that delimitation, demarcation, and titling have been done, actions that might lead the agents of the State itself, or third parties acting with its acquiescence or its tolerance, to affect the existence, value, use or enjoyment of the property located in the geographical area where the members of the Community live and carry out their activities.

> .... the Court believes that, in light of Article 21 of the Convention, the State has violated the right of the members of the Mayagna Awas Tingni Community to the use and enjoyment of their property, and that it has granted concessions to third parties to utilize the property and resources located in an area which could correspond, fully or in part, to the lands which must be delimited, demarcated and titled.

The *Maya*[46] decision of the Commission extends the analysis to a different sector (the oil and gas sector) and to different protected rights (the right to equality before the law), and extends the analysis to the Declaration as well as the Convention. But the gravamen of the *Maya* claim was the same as that that of the *Awas Tigni*, namely that the state had alienated resource rights in Maya traditional territory without first demarcating and recognizing the rights of the indigenous inhabitants.

As part of its equality analysis, the Commission made two significant points. The first was that there was a disparity in the protection that the state of Belize offered to different forms of property. State-granted rights were well protected by a state title system. By contrast, the state afforded the customary rights of the Maya no recognition or protection. That, in the opinion of the Commission, amounted to a breach of the Article II right (under the Declaration) to equality before the law. Second, the Commission emphasized that, in the circumstances (and given a prevailing environment in which indigenous land rights had not been protected), formal equality of treatment might not suffice. Hence the state of Belize might be required to take special measures to protect those indigenous property rights.

In conclusion, a growing body of case law in the inter-American system shows how basic human rights protections may be used by indigenous communities to require that states take steps to protect their traditional lands and furthermore require that states take these measures before they grant resource interests to other parties within those traditional territories. This body of law is directly applicable to both Canada and the USA but may also be persuasive in the context of other international human rights instruments.[47]

Discussion and conclusions

In this concluding section, we draw attention to several themes that emerge from the above account. The first theme deals with the competing claims of the state and indigenous peoples to land and resources. The second theme seeks to draw out the importance of legal and political recognition of the colonial context of relationships between settler societies and indigenous peoples, and the third theme explores some of the implications of the material covered in this chapter for the corporate social responsibility policies of oil and gas corporations operating within the traditional territories of indigenous peoples.

Uncertain laws and unresolved land claims: prior informed consent or prior informed consultation

Indigenous peoples regard oil and gas development as one of the greatest threats to their survival as peoples. Indigenous peoples depend on their lands, territories and natural resources for their livelihoods and their cultural survival. As demand for oil and gas increases worldwide, the pressure on indigenous peoples' lands intensifies. According to representatives of indigenous peoples, they have 'legitimate reasons for being deeply concerned about planned oil and gas explorations in their territories as developers' interests normally prevail wherever and whenever indigenous people's interests and rights clash with development projects' (Henriksen, 2005: 655).

This chapter has demonstrated that existing international human rights laws and conventions recognize the right of indigenous peoples to the ownership and control of their lands, territories and natural resources. Yet states, relying on the statist rhetoric of permanent sovereignty,[48] also claim rights over those natural resources including the right to control sub-surface resources to develop them in their national state interest. But state sovereignty over natural resources cannot be absolute. It must be subject to other principles and rules of international law, including human rights norms (Caruso, Colchester, MacKay et al., 2003: 21, Salim, 2004: 59). These competing claims frequently lead to conflict, serving to draw attention to a central issue which lies at the heart of many international debates: is the relevant rule that of prior informed consent before the state authorizes activities within traditional territories of indigenous peoples, or does the relevant norm have a softer content, such as prior informed consultation?

Central in this debate is the concept of time. Indigenous peoples have 'different orientations towards time and space, different positioning within time and space, and different systems of making space and time "real"' (Tuhiwai Smith, [1999] 2005: 55). While politicians think in terms of the next elections (3–6 years), and oil and gas companies plan for the coming 10–20 years, indigenous peoples use a timeframe of many generations. Indigenous peoples have cultural survival as their prime goal, while oil and gas companies aim to maximize shareholder returns. Hence, for indigenous peoples, a fair consultation procedure requires a timeframe that is different from that typically adopted by modern states and the oil and gas industry. It takes time to inform people about specific issues and it takes time to gather different views. And if governments and industry fail to take that time, the process is merely a formality without relevance to indigenous people.

Recognizing the colonial context

Indigenous peoples have suffered under different forms of colonial settlement and conquest for centuries. This is as true in the Arctic as elsewhere, where the colonial powers include: Canada (as the successor to England and France) v. First Nations and Inuit; the USA in Alaska (and as the successor to Russia) v. the Inupiat, Aleut and other peoples; Norway, Sweden, Finland and Russia, v. the Sámi; and Russia v. the Nenets, and the other 40 or so indigenous peoples of Northern and Eastern Russia. As the former UN Special Rapporteur on indigenous peoples observed 'the legacy of colonialism is probably most acute in the area of expropriation of indigenous lands, territories and resources for national economic and development interests' (Daes, 2001: 49–50).

Yet states are frequently reluctant to come to terms with this colonial context and legacy, and may be particularly reluctant to recognize the legal consequences of this legacy. In some cases, we can point to a key event which brings about a transformation. In Norway, for example, we may point to conflict over the development of the Alta river for hydropower in 1981–1982. This conflict resulted in huge protests from the entire Sámi population and was supported by many ethnic Norwegians. It is widely recognized that the revitalization of Sámi politics in Norway began with this event but, as discussed in Chapter 6 (Bankes, this volume), this event also triggered the creation of the Sámi Rights Commission, which led in turn to Norwegian ratification of ILO 169. In Canada one would point to the constitutional protection of Aboriginal rights in 1982.[49]

A key part of recognizing the colonial context involves the recognition of indigenous peoples as 'peoples' within the framework of international law and as 'indigenous'. And while formal instruments (e.g. ILO ratification and constitutional provisions; the UN Declaration) may effect this recognition, it may be more difficult to secure that recognition (and what it may entail in terms of special measures) within the general population. For example, reports in two papers published in Northern Norway, *Nordlys* and *Finnmark Dagblad* during 2006, reveal that many ethnic Norwegians believe that there should be equality of treatment between Sámis and ethnic Norwegians, many of whom have lived in the same area for many generations. Similar views prevail in part of Russia. As the mayor of Archangelsk puts it: 'We are all indigenous'.

Law and regulations: an implementation gap?

According to the Permanent Forum on Indigenous Issues, one of the main challenges that faces indigenous peoples is the implementation of new norms and laws.[50] Most indigenous peoples are governed by governments far away from their localities with limited cultural understanding of indigenous peoples. Most often decisions are taken without involving relevant indigenous peoples. Language issues compound the problem. And while we have certainly seen progress over the last 25 years in the development of constitutional provisions and laws relating to indigenous peoples, there is a significant 'implementation gap' in some cases between the laws themselves and their effective application at the local level. Much too often, powerful economic and political interests actually override the laws themselves, leaving indigenous communities without due protection of their human rights. According to the Special Rapporteur on the situation of human rights and fundamental freedoms of indigenous peoples, Rodolfo Stavenhagen, 'it is urgent that this issue be addressed fully' (Stavenhagen, 2003).

The Russian constitution, for example, apparently offers indigenous peoples important protection, but very few of the indigenous peoples in Russia are recognized as such by the state and, in general, they lack resources and opportunities to express their views satisfactorily (RAIPON, 2002), although groups like the Russian Association of Indigenous Peoples of the North do provide some opportunities. A smaller association, the Indigenous Peoples of the North of Sakhalin Region, had oil and gas development as one of their major themes at their fifth congress in October 2004. The association decided to start defending 'indigenous peoples' constitutional rights via protest actions against oil companies violating indigenous peoples' rights according to the norms and principles of international justice and Russian legislation'. RAIPON has supported this move.[51]

Trans-national companies and corporate social responsibility

What are the implications of all of this for oil and gas companies operating in the traditional territories of indigenous peoples? How should they respond to weak domestic laws or 'the implementation gap'? How should this affect the way those companies develop and apply their own standards of corporate social responsibility (CSR)? Should those standards reflect the principle of prior informed consent or should companies defer to the position of the state?

Rodolfo Stavenhagen, the Special Rapporteur on the situation of human rights and fundamental freedoms of indigenous peoples, has stated:

> Sustainable development is essential for the survival and future of indigenous peoples, whose right to development means the right to determine their own pace of change, consistent with their own vision of development, including their right to say no. Free, prior, informed consent is essential for the human rights of indigenous peoples in relation to major development projects, and this should involve ensuring mutually acceptable benefit sharing, and mutually acceptable independent mechanisms for resolving disputes between the parties involved, including the private sector.
>
> (Stavenhagen, 2001)

Stavenhagen also recommends that 'Governments and business enterprises work closely with indigenous peoples and organizations to seek consensus on development strategies and projects'.[52] This view is supported by the World Bank, yet not from the indigenous peoples rights or respect perspective, but from the perspective of the business corporation itself. The International Finance Corporation of the World Bank argues that to break with international accepted human rights standards or indigenous peoples rights, even if these rights are not part of a states legislature, 'can have serious reputational impacts' and, therefore, recommends businesses to follow ILO 169 (World Bank, 2007: 5). Most operators in the oil and gas industry recognize the importance of setting environmental, social, health and product stewardship policies as part of their CSR policies. They are aware that, if they operate within the traditional territories of indigenous peoples and yet deal solely with the non-indigenous government, they will most likely fail to observe certain norms of CSR, thereby damaging the corporation's image and reputation (see e.g. Fjellheim, 2006).

Most oil and gas companies also argue that they provide direct benefits to the communities in which they are operating. These benefits include education and training, health clinics, transport and communication facilities. They also may provide significant economic rents to host governments. Yet for many indigenous peoples this will not be enough. They are all too aware of the short time preferences of oil and gas companies, and that they will be left with any damages when the oil and companies have long since extracted their resources and departed;

hence the emphasis on prior informed consent and access to a share of the rents to supplement any employment opportunities and other potential social and economic benefits.

Notes

1. No general, internationally accepted, definition of indigenous peoples exists. It is typical of indigenous populations that they do not represent the dominant population in the larger society of which they are part, although they may be the population group that inhabited the area first. As a rule, indigenous populations possess a distinctive culture that revolves around natural resources, and their way of life differs socially, culturally and/or linguistically from the dominant population. For the ILO 169 definition, see note 17.
2. The 22,720 Aboriginal people living in Nunavut in 2001 represented over 80 per cent of the territory's population. In the Northwest Territories, Aboriginal people represent 45.6 per cent of the total population according to the Canadian Council of Social Development; see http://www.ccsd.ca/factsheets/demographics/index.htm (Accessed 14 March 2007). In Karasjok and Kautokaino, the Sámi represent the absolute majority.
3. For example, Hammerfest, Alta and Loppa has a tiny minority of Sámi peoples. In Archanglesk and Murmansk, only a tiny part of the population is made up of indigenous peoples.
4. The International Covenant on Civil and Political Rights (ICCPR), Articles 1 and 47, and the International Covenant on Economic, Social and Cultural Rights (ICESCR), Articles 1 and 25.
5. The concept of permanent sovereignty over natural resources represents customary international law. Common Article 1 of the International Covenants states that: 'All peoples have the right of self-determination. By virtue of that right, they freely determine their political status and freely pursue their economic, social and cultural development. All peoples may, for their own ends, freely dispose of their natural wealth and resources without prejudice to any obligations arising out of international economic cooperation, based upon the principle of mutual benefit, and international law. In no case may a people be deprived of its own means of subsistence'. For a discussion of the close interconnection between the right of self-determination and the right of peoples/states to permanent sovereignty over natural resources, see Schrijver (1997) and for an effort to apply the principle of permanent sovereignty in the context of indigenous peoples, see Daes (2003).
6. It is too early to provide examples of how the UN Declaration on the Rights of Indigenous Peoples (UNDRIP), adopted September 13, 2007 by the UN General Assembly against the vote of Canada, USA and Australia, will eventually influence the behaviour of states and companies towards indigenous peoples.
7. For Norway, see Constitution, Article 26.
8. An Act relating to the strengthening of the status of human rights in Norwegian law (English translation available at http://www.ub.uio.no/ujur/ulovdata/lov-19990521-030-eng.pdf).
9. The list does not include the Convention on the Elimination of All Forms of Racial Discrimination (CERD) or ILO 169.
10. Some of the complexities include the distinction between self-executing and non-self-executing agreements. The categories of international agreements that fall outside the scope of the term 'treaties' include Executive Agreements (negotiated and ratified by the Executive) and Congressional Executive Agreements (negotiated by the executive but ratified by both Houses of Congress rather than just the Senate).
11. See Chapter 10 on Russia in this volume.
12. For example, Article 1(8) of the US Constitution affords to Congress the power to 'regulate Commerce with foreign Nations and among the several States, and with the

Indian tribes'. This has long been interpreted to restrict the capacity of states to regulate activities on reservations.
13 See, for example, Donihee et al. (1999) and discussions around the Finnmark Act in Chapter 9 in this volume.
14 Advisory Opinion, Western Sahara, 1975 International Court of Justice 12.
15 Mabo v. Queensland (1992), 175 Commonwealth Law Reports 1.
16 For an example, see the discussion in Awas Tingni (examined infra) of the law and practice in Nicaragua.
17 ILO 169 (Article 1(1) applies to 'peoples in independent countries who are regarded as indigenous on account of their descent from the populations which inhabited the country, or a geographical region to which the country belongs, at the time of conquest or colonisation or the establishment of present state boundaries and who, irrespective of their legal status, retain some or all of their own social, economic, cultural and political institutions'.
18 There are only 19 parties to ILO 169 as of October 2007. Norway was the first state to ratify. Both Sweden and Finland continue to explore ratification. Denmark has ratified for Greenland, but with an important and extensive interpretive declaration. Only two countries have ratified ILO 169 during the last 5 years.
19 There may be several reasons for this: (1) ILO Conventions generally have a lower profile in North America; (2) the subject matter of ILO Conventions generally creates difficult implementation problems for federal states such as the USA and Canada and (3) objections to the content of ILO 169.
20 Article 13(2) defines the term 'lands' to include 'the concept of territories, which covers the total environment of the areas which the peoples concerned occupy or otherwise use'.
21 NRS – the National Association of Norwegian Sámi – is the largest political party representing the Sámi population in Norway. All presidents to the Sámediggi come from NSR. It was established in 1968 in order to enhance the rights of the Sámi people and improve their general living conditions.
22 Available at http://www.nsr.no/website.aspx?displayid=5384 (Accessed 1 November 2006).
23 http://www.nsr.no/website.aspx?displayid=5384 (Accessed 28 August 2006).
24 Samerettsutvalget.
25 The English translation of the legal opinion was published in its entirety as Graver and Ulfstein (2004). The authors were not asked to comment on the procedural issues and they took the view that they were being asked to opine upon both whether the bill violated Sámi rights but also whether it fulfilled Norway's international obligations. The bill was also the subject of comment from the International Labour Organization (ILO; see CEACR: Individual Observation concerning [ILO 169], Norway, published 2004, ilolex document no. 062004NOR169) and from CERD (see Concluding Observations and Comments, December 10, 2003, CERD/C/63/CO/8 at para. 18).
26 Agreement of May 11, 2005. Translation available at: http://www.regjeringen.no/en/dep/ud/Documents/Reports-programmes-of-action-and plans/Reports/2005/Implementation-of-the-international-covenant-on-civil-and-political-rights.html?id=420420. See also http://www.regjeringen.no/en/dep/aid/Topics/andre/Internasjonalt_urfolksarbeid/Statement-by-the-Norwegian-Delegation-to.html?id=467701
27 See, for example, various interviews given by the Sámidiggi President Keskitalo to the media during the winter of 2006.
28 The agreed consultation procedures can be reached on: http://www.regjeringen.no/en/dep/aid/Topics/Sami-policy/PROCEDURES-FOR-CONSULTATIONS-BETWEEN-STA.html?id=450743 (Accessed 16 October 2007).
29 See *Aftenposten,* February 1, 2006, but especially various interviews in *Finnmark Dagblad* and *Norlys* during March 2006.

30 Posted at www.samiradio.org on November 25, 2005 (Accessed 11 April 2008).
31 www.samiradio.org, posted on November 25, 2005 (Accessed 5 October 2006).
32 The Human Rights Act, discussed earlier. But since Russia takes a monist approach, the Covenant is automatically part of domestic law.
33 Available at: http://www.unhchr.ch/tbs/doc.nsf/(Symbol)/fb7fb12c2fb8bb21c12563ed 004df111?Opendocument. See also the similar 1997 General Recommendation on Indigenous Peoples of the UN Committee on the Elimination of Racial Discrimination, which requires states/parties to 'recognize and protect the rights of indigenous peoples to own, develop, control and use their communal lands, territories and resources and, where they have been deprived of their lands and territories traditionally owned or otherwise inhabited or used without their free and informed consent, to take steps to return these lands and territories' (Caruso, Colchester, MacKay et al., 2003: 17).
34 http://www.unhchr.ch/tbs/doc.nsf/(Symbol)/fa24fc7cd513751bc1256fe900525608?Opendocument.
35 In petitions of this nature, the Committee is frequently faced with the unenviable task of weighing competing interpretations not only of government conduct in terms of the degree of consultation but also in terms of the assessment of the effect of the proposed resource activity on traditional land use activities; see, for example, the Lubicon case (Canada). One result is that the Committee shows some deference to governments especially where there is evidence (as there was in Lansman no. 3) that the Article 27 issue has been squarely raised and considered in litigation in the domestic courts of Finland.
36 The regional efforts include the draft Nordic Sámi Convention [an English version of the proposed Sámi Nordic Convention is reproduced in Henriksen (2006: 37–38)] and the Draft Inter American Declaration on the Rights of Indigenous Peoples. For information, see: http://www.oas.org/consejo/CAJP/Indigenousper cent20documents.asp.
37 Declaration on the Rights of Persons Belonging to National or Ethnic, Religious and Linguistic Minorities 1992, available at: http://www.ohchr.org/english/law/minorities.htm.
38. Azerbaijan, Bangladesh, Bhutan, Burundi, Columbia, Georgia, Kenya, Nigeria, Russia, Samoa and Ukraine abstained.
39 http://news.gc.ca/web/view/en/index.jsp?articleid=349759&keyword=indigenous+peoples&keyword=indigenous+peoples& (Accessed 17 October 2007).
40 Press release, September 13, 2007: *AFN National Chief Applauds Today's Passage of the UN Declaration on the Rights of Indigenous Peoples*. Available on: http://www.afn.ca/article.asp?id=3772 (Accessed 5 November 2007).
41 http://www.docip.org/declaration_last/Arctic_reg_stat.pdf (Accessed 9 October 2007).
42 Press release, September 13, 2007 for the UN General Secretary.
43 See also the efforts of some indigenous peoples to bring their concerns before the CERD Committee. For a discussion of Maori efforts to request that CERD review New Zealand's proposed foreshore legislation under its early warning and urgent action procedure, see Charters, C. and Erueti, A. (2005) 'Report from the inside: The CERD Committee's Review of the Foreshore and Seabed Act 2004', *Victoria University of Wellington Law Review*, 12.
44 For an example of an 'Arctic' petition, see the Petition brought by the Inuit Circumpolar Conference against the United States in the context of climate change and the US's failure to ratify the Kyoto Protocol, December 7, 2005. The Petition is available online at: http://inuitcircumpolar.com/files/uploads/icc-files/FINALPetition ICC.pdf.
45 Case of the Mayagna (Sumo) Awas Tingni Community v. Nicaragua Judgment of August 31, 2001. Available online at: http://www.corteidh.or.cr/docs/casos/articulos/seriec_79_ing.pdf.
46 Report No. 40/04: Case 12.053, *Merits, Maya Indigenous Communities of the Toledo District and Belize*, October 12, 2004.
47 Some caution may be appropriate. For example, the two international covenants do not protect property rights but the right to equality before the law, which forms part

of the basis for the Commission's reasoning in the Maya case, is a universally acclaimed norm.
48 Schrijver (1997) demonstrates that, as originally formulated, the concept of permanent sovereignty was very closely related to the concept of self-determination and emphasized, therefore, that it was peoples that had the right to permanent sovereignty. Over the years, however, states have succeeded in yoking the concept to states rather than to peoples. Daes' (2003) recent contribution to this debate seeks to re-connect the concept with its historical roots.
49 In Australia, commentators would refer to the Mabo case, supra note 15; and, in New Zealand, the seminal legal event would be the Treaty of Waitangi Act, creating the Waitangi Tribunal.
50 Press release, HR/4840, Permanent Forum on Indigenous Issues, May 18, 2005.
51 Document accessed on September 8, 2006 at http://www.npolar.no/ansipra/english/Index.html.
52 UNESC Commission on Human Rights, Indigenous issues, January 21, 2003 (E/CN.4/2003/90).

References

Anaya, S. J. (2004) *Indigenous Peoples in International Law*, Oxford: Oxford University Press.
Broderstad, E. G. (2001) 'Political autonomy and integration of authority: the understanding of Sami self-determination', *International Journal of Minority and Group Rights*, 8: 151–175.
Carmen, A. (2006) 'Free, prior and informed consent', in UNOHCHR (ed.) *Expert Seminar on Indigenous People's Permanent Sovereignty over Natural Resources and on Their Relationship to Land*, Geneva: United Nations OHCHR.
Caruso, E., Colchester, M., MacKay, F., et al. (2003) 'Extracting promises: indigenous peoples, extractive industries and the World Bank', in *Synthesis Report*. Baguio City: Indigenous Peoples' International Centre for Policy Research and Education.
Daes, E.-I. A. (2001) *Indigenous Peoples and Their Relationship to Land*, Geneva: United Nations Economic and Social Council, Commission on Human Rights.
Daes, E.-I. A. (2003) *Indigenous Peoples Permanent Sovereignty over Natural Resources*, Geneva: United Nations Economic and Social Council, Commission on Human Rights.
Donihee J., Gilmour J. and Burch, D. (2000) *Resource Development and the Mackenzie Valley Resource Management Act: The New Regime*, Calgary: Canadian Institute of Resources Law.
Egeland, J. (1988) *Impotent Superpower – Potent Small State: Potentials and Limitations of Human Rights Objectives in the Foreign Policies of the United States and Norway*, Oslo: Norwegian University Press.
Fjellheim, R. S. (2006) 'Arctic oil and gas – corporate social responsibility', *Journal of Indigenous Peoples Rights*, 4: 8–23.
Graver, H. P. and Ulfstein, G. (2004) 'The Sami people's right to land in Norway', *International Journal of Minority and Group Rights*, 11: 337.
Henriksen, J. B. (2005) 'Oil and gas development in the Arctic: an indigenous peoples rights perspective', *RAO/CIS Offshore*, 13–15: 655–658.
Henriksen, J. B. (2005) Oil and gas operations in indigenous peoples lands and territories in the Artic: a human rights perspective, *Galdu Cala – Journal of Indigenous Peoples Rights*, 4: 24–41.

ILO 169 (1989) *ILO Convention 169 Concerning Indigenous and Tribal Peoples in Independent Countries*, Geneva: International Labour Organization.

Norwegian Ministry of Foreign Affairs (2005) Report no 30 (2004–2005) to the Storting: *Opportunities and Challenges in the North*.

Oskal, N. (2001) 'Political inclusion of the Sami as indigenous peoples in Norway', *International Journal of Minority and Group Rights*, 8.

Royal Ministry of Labour and Social Inclusion, The (2006) *Guidelines for Consultations Between State Authorities and the Sami Parliament and Other Sami Entities (Norway)*, Oslo: The Royal Ministry of Labour and Social Inclusion.

Russian Association of Indigenous Peoples of the North (RAIPON) (2002) Policy paper, July 1, in Moscow: RAIPON.

Salim, E. (2004) *Striking a Better Balance: Extractive Industries Review Report*, Jakarta: The World Bank Group.

Schrijver, N. (1997) *Sovereignty over Natural Resources: Balancing Rights and Duties*, Cambridge: Cambridge University Press.

Stavenhagen, R. (2001) *Report of the Special Rapporteur on the Situation of Human Rights and Fundamental Freedoms of Indigenous People*, Geveva: United Nations Economic and Social Council, Commission on Human Rights.

Stavenhagen, R. (2003) *Report of the Special Rapporteur on the Situation of Human Rights and Fundamental Freedoms of Indigenous Peoples*, Geneva: United Nations Economic and Social Council, Commission on Human Rights.

Thornberry, P. (2002) *Indigenous Peoples and Human Rights*, Manchester: Manchester University Press.

Tribe, L. (2000) *Tribe's American Constitutional Law*, 3rd edn, New York: Foundation Press.

Tuhiwai Smith, L. ([1999] 2005) *Decolonizing Methodologies: Research and Indigenous Peoples*, London: Zed Books.

UNHDR (2004) 'Cultural liberty in today's diverse world', in United Nations Development Programme (UNDP; ed.) *Human Development Report*, New York: UNDP.

United Nations (2007) *United Nations Declaration on the Rights of Indigenous Peoples*, New York: United Nations. Available HTTP: http://daccessdds.un.org/doc/UNDOC/GEN/N06/512/07/PDF/N0651207.pdf?OpenElement.

World Bank (2006) *Safeguard Policies with Respect to Indigenous Peoples*, Washington: World Bank.

World Bank (2007) *ILO Convention 169 and the Private Sector*. Available HTTP: http://www.ifc.org/ifcext/enviro.nsf/AttachmentsByTitle/p_ILO169/$FILE/ILO_169.pdf (Accessed 11 April 2008).

12 Perceptions of Arctic challenges

Alaska, Canada, Norway and Russia compared

Oluf Langhelle and Ketil Fred Hansen

Introduction

As should be evident from the previous chapters, the framing of Arctic challenges varies significantly in the different countries covered in this book. In this chapter we aim to bring the material from the four countries – Alaska, Canada, Norway and Russia – together in a comparative context from the perspective of sustainable development. We want to identify, highlight and discuss differences and similarities in more detail within the overall research questions raised in the introduction: What is sustainable development believed to entail, and what does it imply for oil and gas activities in the Arctic? What are the main conflicts of interests between different groups of actors? How and why do the main concerns and conflicts vary between indigenous peoples, local people, governments and oil and gas companies in the different regions and countries?

Although differences may seem to be most visible, there are also striking similarities between different regions and countries when it comes to the framing of issues and challenges. In addition, there seems to be a common struggle, and attempts across regions and countries to reconcile different interests and concerns in the Arctic. No doubt, the work of the Arctic Council has contributed to this development, providing a common agenda for the different Arctic countries, at least within the Council. But independent of the activities in the Arctic Council, there is also an increasing interaction among different groups and actors across regions. Obviously, the oil and gas companies have for years operated in different regions, but there is now growing cooperation in the Arctic among scientists, academics, non-governmental organizations (NGOs) and indigenous peoples focusing on similar issues and concerns. This increase in cooperative efforts has arguably led to an increasing transfer of ideas and possible solutions moving from one institutional, political and cultural setting to another. The possibility of finding common solutions across regions and countries is discussed in this chapter.

From the four country chapters, it is evident that oil and gas activities in the Arctic have only partially been framed directly in the idiomatic wording of sustainable development. Despite this, however, it is apparent that much of the discussion within the four countries can be linked and interpreted within a sustainable development frame. Sustainable development has become fairly widespread in the political vocabulary in all the countries, although there are noticeable differences among them. Sustainable development is most commonly used as a frame for national policies in Norway and Canada, followed by Russia and then the United States (US). A central issue in sustainable development, however, is the balance between development and the environment. Although not always explicitly linked to sustainable development, the balancing of social, economic and environmental concerns figures prominently in all the national debates.

The rest of the chapter proceeds as follows. As our point of departure we focus first on the framing and identified challenges of oil and gas activities in the Arctic from a sustainable development perspective under the headings of developmental challenges, environmental challenges and indigenous peoples. Although these issues are addressed under separate headings, there are important linkages among them, which we also try to highlight. Second, from this comparison, we try to provide some reflections on what the global sustainable development agenda may imply for the further development of oil and gas in the Arctic.

Sustainable development and development concerns in the Arctic

It is a widely held perception in all the countries that oil and gas can benefit development in terms of job creation, profits and welfare in the Arctic.[1] Local and national politicians, indigenous peoples and other local inhabitants, together with the different oil businesses all emphasize this. The debates and story lines are first and foremost tied to the circumstances or conditions under which oil and gas is seen to be beneficial. These conditions vary somewhat among different stakeholders. At the local level, however, there seems to be general support for the conclusion in *Arctic Human Development Report* (AHDR, 2004) that the key sustainability challenge seen from the Arctic is how to ensure that more of the profits from resource extraction actually remain in the Arctic. Within the four countries, this is perhaps the perception that is most consistently advocated by local and indigenous peoples. Although the strategies followed are somewhat different, the success in capturing profits varies and the story lines that frame the message are different.

Economic challenges

Another way of framing this story line is through the question, who will benefit from increased oil and gas production in the Arctic? As we saw in Chapter 2, equity is inherent in the concept of sustainable development. How one answers this question determines to a high degree the attitudes towards oil and gas development in the Arctic. This question can also be linked to the perception of the Arctic as a storehouse of resources. As argued in Chapter 2, the study conducted by Glomsrød (2006) confirms the findings of the AHDR report. According to Duhaime and Caron (2006: 22):

> The circumpolar Arctic is exploited as a vast reservoir of natural resources that are destined for the southern, non-Arctic, parts of the countries that also

include Arctic regions, and more broadly to global markets. The Arctic is a major producer of hydrocarbons, minerals and marine resources, whose importance is confirmed by the very value of the resources produced. The economy of the Arctic is also characterized by large service industries, particularly through the role of the State. Finally, it is characterized by a limited secondary sector . . . .

These resources are petroleum, minerals, fish and forests. Of these resources, petroleum dominates the resource extraction industries. The Arctic share of global oil and gas production is significant, respectively 10.5 and 25.5 per cent (Figure 12.1). Around 97 per cent of the present Arctic oil and gas production is taking place in Alaska and Northern Russia (Lindholt, 2006).

In addition, the Arctic has 12.7 per cent of the proven petroleum reserves, and 23.9 per cent of undiscovered petroleum resources. The Arctic, therefore, has the potential to continue to supply around one-quarter of the total demand of global gas consumption (Lindholt, 2006: 29).

The Arctic region *per se*, contributes to 0.44 per cent of the world's total gross domestic product (GDP), while its population only represents 0.16 per cent (Duhaime and Caron, 2006: 17). Thus, income generation per capita in the Arctic in general is higher than for the rest of the world in general. However, for the countries having parts of their territory in the Arctic, the income per capita produced in the Arctic *per se*, represents only 80 per cent of the national per capita income. This again varies a lot from country to country and certainly from locality to locality within the Arctic zone. In the Russian Arctic, for

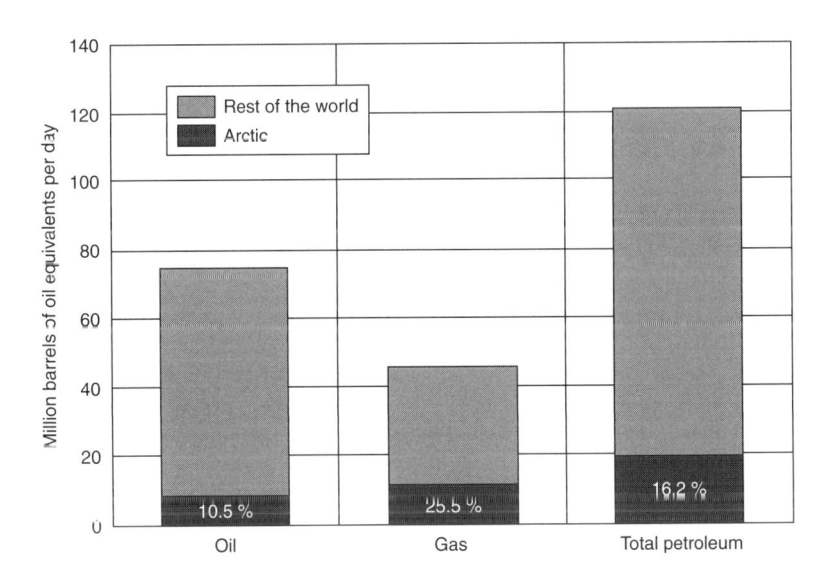

Figure 12.1 Arctic share of global petroleum production, 2002. (Source: Lindholt, 2006: 27.)

example, per capita GDP is 219 per cent of the Russian average, while in the Arctic zone in Norway the GDP per capita is only 56 per cent of Norway's national average (Duhaime and Caron, 2006: 18).

Generally speaking, the Arctic contributes to the global economy by energy and raw materials like oil, gas, diamonds, gold, wood, fish and shrimp, while it imports final goods and services (Duhaime and Caron, 2006: 20). This resembles the characteristics of a developing country. One of the major challenges, therefore, for businesses generating these large income streams and for official authorities in the Arctic countries, is to distribute the benefits in a way that the different populations in the Arctic find fair and just.

What the various populations in the different Arctic countries find fair and just however, varies significantly. At a general level, a common point is that the local inhabitants of the Arctic, indigenous peoples included, want the resources extracted from the Arctic to benefit their own region. Thus, most Arctic inhabitants agree that a fair distribution of resources implies that there should be more re-invested in the zones where it is extracted than currently takes place. Arguments for this view vary. Some local people use as their main argument the moral commitment states have to ensure equal opportunities between people. They use statistics on relative underdevelopment in their region compared to other regions in their countries to support their views.

Others, especially some indigenous peoples in Norway and Russia, refer to ILO 169 and other national laws in order to claim that the resources legally belong to them. The Russian Sámi President, Alexander Kobelev, for example, argued at the Arctic Council meeting on October 26, 2006, that:

> The Saami have watched states build a large part of their wealth on our rivers, fjords, mountains and forests. We will watch no longer. We have to enter a new phase where governments and multinational corporations stop doing the wrong things, and start doing the right. We have never given up our inherent right to our territories, however, for large parts of the Saami area our land and governance rights are still not respected.[2]

Generally speaking, indigenous groups tend to be both politically marginalized and poor in monetary terms. However, for indigenous peoples within the Arctic, the degree of marginalization and poverty varies a lot. Some groups of indigenous peoples are richer than their national average, while most still live in monetary poor conditions. For example, even if the GDP per capita in the Canadian Arctic represents 155 per cent of the Canadian national average (Duhaime and Caron, 2006: 18), indigenous peoples have a 25 per cent higher unemployment rate than other Canadians. A 65 per cent higher proportion of indigenous people than other Canadians have poor housing and 42 per cent more live on social welfare. Thus, even if the Arctic as a whole represents a larger percentage of the national economy than of the population, only a small minority of the people living there benefit from this.

Few in Norway would be classified as poor in a global monetary way of measuring it. On average, Norway disposed of 38,400 US$ (purchasing power parity; PPP) (UNDHR, 2006: 283–284). Norway has been among the ten richest countries in the world every single year for the last decade (UNHDR, 2006). However, the northernmost inhabitants in Norway lag behind the national average income (22 per cent lower than the national average income; Statistisk Sentralbyrå, 2006).

Income statistics for indigenous peoples Russia are not available. However, Russia had a per capita gross domestic product of only 9900 US$ (PPP) in 2004 (UNHDR, 2006: 283–284). In addition to a relatively low GDP per capita, the difference between the rich and the poor is greater in Russia than in Norway. While the poorest 10 per cent in Norway receive 3.9 per cent of the income, in Russia they receive only 2.4 per cent of income (UNHDR, 2006: 335–336).

To develop oil and gas activities in the Arctic further, one of the major concerns will thus be the question of distributing profits back to the localities in an equitable manner. The various states in question do this only to a certain degree. Norway, the only unitary welfare state under scrutiny, does this to a relatively high degree, while Russia redistributes resources back to the peripheries only to a limited degree.

Many indigenous groups would not measure poverty in monetary terms, but rather in terms of opportunities or possibilities to keep their identity as indigenous peoples. The identity is very much linked to language, religion and occupancy. The relationship to land is for all indigenous peoples a central factor for their identity. 'You do not share your money but you share your food. If you yourself have caught and prepared your food, sharing it represents something totally different from sharing money'. A recent study shows that 90 per cent of indigenous peoples households in the Arctic report sharing subsistence food with other community members and/or family (Poppel, 2006: 71). This also points to the fact that subsistence hunting, harvesting and fishing represent an important part of the economy in many places of the Arctic, and makes them especially vulnerable to pollution and climate change.

However, the subsistence economy is never factored into regular statistics with the result that the economic contribution to the economy of indigenous peoples is neglected, making them richer than statistics suggest. Yet, the value of land will thus also be neglected in economic terms, since subsistence economy never enters statistics. For the Inuit, for example, subsistence 'means much more than mere survival or minimum standards of living . . . It enriches and sustains Inuit communities in a manner that promotes cohesiveness, pride and sharing' (Poppel, 2006: 66).

For the indigenous peoples in Alaska and Canada, proclaiming property rights have been the most important strategy for capturing profits, a strategy that to some extent has been successful. Property rights have also become important strategies within Russia and Norway. In Russia it has been highly unsuccessful, and in Norway it remains to be seen whether or not the Sámi people will succeed in their attempts to get traditional sea rights. In all countries, however, indigenous peoples advocate a rights-based approach to resources. The story line identified in the Norwegian chapter, 'It is our right', captures the essence of this approach.

Norway has created an oil fund invested globally to keep some of the extracted resources available for future generations and not to overheat the national economy. Still, national and local politicians argue for oil and gas activities in the Arctic, using the need for job opportunities and increased welfare in Northern Norway as the arguments. These arguments are added to the main argument for expanded oil and gas exploration emphasized by national politicians in Norway, namely the global need for more energy. Helping to meet higher demands for energy in developing countries has been used by Norwegian oil company directors as well as Norwegian politicians to argue for a high level of extraction (Chapter 9). Some politicians have even argued that it is Norway's moral obligation to provide parts of this energy and that Norway has a global responsibility to contribute to poverty reduction in the world by extracting more oil and gas. Environmentalists and parts of the Sámi community argue against oil and gas activities, while a majority of the local inhabitants, the Norwegian government and the oil and gas companies are arguing for extraction now.

Russian legislation favoured devolution during the 1990s, but over the last decade we have seen an increasing re-centralization of power back to Moscow. Also, many observers and researchers doubted the application of the laws favouring Northern Russian people and indigenous peoples in Russia during the 1990s (Caulfield, 2004: 124–125). Russia's two main motives for exploiting oil and gas in the Arctic are to satisfy the steadily growing internal demand for oil and gas, and to secure their status as one of the world's most powerful countries. As Andreyeva and Kryukov (Chapter 10) have shown, the collapse of the Soviet system led to a drastic deterioration in the living standards among a large part of the population in the Russian Arctic.

Social challenges

Social issues are part of the general orientation towards development in all four countries under scrutiny. Most often social issues at stake are intrinsically linked with economic development. Which social issues are dominating the public discourse, however, varies in the different countries. Abuse of alcohol and social deprivation are particularly prominent in the stories from local communities and among indigenous peoples in Alaska, Canada and Russia, but are hardly an issue in Norway. The main concern in Norway has been linked to employment opportunities. For all four Arctic regions under study, the unemployment rate is higher than the national average, life expectancy is lower, health status worse and the level of education lower than their national average. Yet, in all the four Arctic regions discussed, statistics on different social issues like education, health and unemployment show that, in relation to their national average, Arctic populations are improving rapidly. Thus, the difference between the Arctic population and the national average is diminishing. Yet, there are still huge discrepancies.

Employment

Statistics confirm that most localities in the Arctic lack job opportunities for the population, and the region is lagging behind in social welfare and industrial modernization. In Chukotka (Russia), the entire adult population sees unemployment as a serious social problem, while 83 per cent in Alaska and 87 per cent in Canada see unemployment as a serious social problem (Poppel, Kruse, Duhaime et al., 2007: 11). In most parts of the Arctic lacking oil and gas expansion, economic opportunities have declined during the last few decades. Job prospects and investments have been reduced. The fish processing industry in Northern Norway, for example, reduced its numbers of employees by 50 per cent between 2000 and 2004 (Chapter 9). In Nunavut, unemployment is a huge issue (Chapter 8).

In general, the unemployment rate in the Arctic parts of the four countries covered in this study is higher than their national average. In Arctic Canada, there are certain regions with a 25 per cent higher unemployment rate than the Canadian average, while in Norway unemployment is only marginally higher than the national average. In Russia, the Arctic sub-regions with dominant oil and gas production have a lower level of unemployment than the national average. However, seen as a whole, the Russian Arctic has a 1 per cent higher unemployment rate than the Russian average. Yet, indigenous peoples in the Russian Arctic are four times more likely to be without employment than their Russian neighbours. In Alaska, the level of unemployment is higher in the North Slope than the state average. Thus, compared to their national average rate of unemployment, most Arctic regions face a relative disadvantage.

In most regions in the Arctic, formal unemployment is a problem. Most local inhabitants and local politicians argue that development of further oil and gas activities in their region should benefit the local population with job opportunities. However, there is a high degree of uncertainty surrounding job opportunities for the local population. Some fear that the jobs offered to the local inhabitants are temporary, short-term jobs relating mostly to the construction phase of oil and gas investments, while the more permanent jobs for experts will be given to specialist from other regions just coming to the Arctic to work, while continuing their urban living in cities outside the region. Thus, they fear that strangers will profit more than themselves. The divide between higher-paid specialist jobs taken by southerners and lower-paid unskilled workers from the Arctic will create tensions in the different Arctic localities. Michael Baffrey, working on the Arctic Council's Oil and Gas Assessment Report, argues that oil and gas employment in itself is highly variable and ultimately unsustainable.[3]

Job estimates connected to oil and gas activities vary. For Norway, estimates vary from an optimistic 10,000 new jobs to zero effect in relation to the expanded oil and gas exploration in the Norwegian Arctic (Chapter 9). However, how many of these positions will go to locals is highly uncertain. Fewer people living in Northwest Russia think increased oil and gas activities in the area will make a positive long-term impact on local employment (Hønneland, Jørgensen and Moe, 2007).

Highly educated experts from the Southern parts of Russia or from abroad are likely to take the most attractive jobs (Chapter 10). Jobs in Arctic Canada are scarce, and most locals argue that oil and gas development creates opportunities for employment.

However, people need to have specialist training to get the more attractive jobs in the oil and gas sector. Without special attention paid to adequate training, the job opportunities offered in the oil and gas sector will mostly benefit educated people from other regions in Canada than the Arctic. Where indigenous peoples have created joint venture business corporations with leading oil and gas companies, it seems that specialized training and employment possibilities for local people are better than where this has not happened (Chapter 8). In Alaska, local employment in the oil fields is low. More than 5000 non-residents commute to North Slope oil-field jobs from outside the region. Most local, indigenous people work in government, construction, service and support sectors fuelled by oil tax revenues. The unemployment rates for North Slope residents are a little higher than for the state as a whole: 7.0 per cent versus 6.7 per cent in 2006 (Chapter 7).

In the Arctic, how much of indigenous peoples' standard of living comes from subsistence farming, fishing and hunting varies a lot. In the Northern Slopes (Alaska), more than half of the population gain more than half of their food from subsistence resources. In Norway, most of the Sámi population is living in the capital city of Oslo and have regular paid urban jobs. A total of 2914 persons were connected to reindeer herding activities in Norway as of November 2005; 609 out of these classified themselves primarily as reindeer herders holding their own herd (Statistisk Sentralbyrå, 2006: 81). Only a few hundred Sámi make their direct living from reindeer herding in the Arctic parts of Norway.

In all the Arctic states, the ability for indigenous peoples to continue their main traditional occupations while working for an oil and gas operator is important. In Northern Alaska, 77 per cent of the population argue that their preferred lifestyle is to combine traditional subsistence activities with a wage job. In Russian Chukotka, 39 per cent answer they would like a waged job, while 29 per cent would prefer a combination of wages and subsistence (Poppel, Kruse, Duhaime et al., 2007: 8). The cycles followed by animals and nature make different periods of the year especially important for various indigenous occupations. To be able to survive as indigenous peoples, it is critical to reserve certain periods of the year to traditional activities. Thus, for most indigenous peoples in the Arctic, it is very important that employment in the oil and gas sector offers flexible solutions to these needs.

Health

In the various Arctic countries, health issues are of equal importance to local inhabitants. In the past 50 years, the burden of mortality and morbidity has declined in all areas of the Arctic. Average life expectancy has increased in all countries but even more in their Arctic parts. In the US, for example, life expectancy has increased by nine years since the 1950s, while it has increased by 14 years in the Arctic part of the US (Hild and Stordahl, 2004). In Canada, we find a similar pattern with an increase of 15 years for indigenous peoples during the past 50 years (Hild and Stordahl, 2004). In Norway, life expectancy for indigenous peoples has increased and is now close to the national average (Statistisk Sentralbyrå, 2006). In Russia, we notice substantially increased life expectancy up to 1990. From 1990 to 2004, life expectancy at birth declined by as much as seven years (ROSSTAT, 2004). Life expectancy at birth may be taken as an indicator of the general development and health status of the population. As another measure, 89 per cent of indigenous people in Alaska, Canada and Russia report that their health is 'good' or 'better', 8 per cent report it to be 'fair' and only a small 3 per cent report their own health to be 'poor' (Poppel, Kruse, Duhaime et al., 2007: 22).

Education

Few people in the Arctic see lack of education as a serious problem. There is little educational difference between the Arctic parts of Norway and the rest of the country. However, among the Sámi people, only 15.4 per cent (compared to the national average of 24.8 per cent) have higher education. In Russia, the level of higher education for people in the Arctic is close to the national average, but the level of education for indigenous peoples in the Arctic is three times lower. Alaska has much lower rates of high-school graduation than the nation as a whole, and within the state Alaska, Native students have the lowest rates of all: as of 2005, only 43 per cent of Alaska native youth graduate from high school (Leask and Hanna, 2005). In Canada, low graduation level among indigenous peoples is perceived as problematic, especially in Nunavut. Thus, when it comes to education, the Arctic parts of the states in this study follow their national averages closely, while the indigenous peoples in the Arctic are lagging seriously behind.

As a conclusion, we may claim that developmental concerns for the different Arctic peoples in the various states reveal similar dilemmas. First, most people living in the Arctic compare themselves in development measures to the Southern population in their own country, rather than their counterparts in other Arctic states or others within their ethnic group. Second, most people in the Arctic, whether indigenous peoples, ethnic minorities or just local nationals, are eager to see the development of oil and gas in their regions. Yet for this development to be successful, major concerns have to be addressed. Jobs have to be offered to locals, investments in collective local infrastructure have to be made, education opportunities have to be given and for the indigenous peoples, their cultural, social and legal rights have to be respected. This includes participation and influence in the planning of projects proposed on their traditional land.

Environmental concerns

Since the launch of the *Arctic Environmental Protection Strategy* (AEPS) in 1991, there has been a strong focus on environmental issues in the cooperative efforts among Arctic states, and as we saw in Chapter 2, this continued in the work of the Arctic Council and its Working Groups. At the national level, however, there are major differences in the framing of environmental issues and the way environmental issues are linked to oil and gas activities.

Looking at the different countries in general, environmental issues lost significance in Russia during the 1990s, and development concerns have to a large degree overshadowed environmental concerns (Hønneland and Jørgensen, 2006). In Canada and Alaska, conservation issues together with subsistence rights (and sustainable use) for indigenous peoples, have dominated the environmental concerns in relation to oil and gas activities. In Norway, where oil and gas activities primarily are offshore, the environmental issues in relation to oil and gas in the Arctic have had a strong focus on the vulnerability of the Barents Sea and biodiversity issues linked to fish (and fisheries), mammals and birds. The environmental issues have at times been the dominant issues in Norway in relation to oil and gas, and also figured prominently in Canada. Norway and Canada are the only countries where discourses of oil and gas production in the Arctic has been directly linked to climate change.

In the following we will elaborate and discuss some of the most striking differences, but also similarities among the regions/countries under the headings of climate change and biodiversity and sustainable use.

Global and national responses to climate change

In the Arctic Council, climate change is primarily seen as a global problem with regional effects, although the impacts in the Arctic may have serious consequences for the global climate. For the most part, the linkage between Arctic oil and gas development and global CO_2 emissions and greenhouse warming has been lifted out of the Arctic and placed at the global and national level.

When the United Nations Framework Convention on Climate Change (UNFCCC) was established in 1992, US, Canada, Norway and Russia were parties to the Convention. In 1997, all four countries signed the Kyoto Protocol. The Protocol strengthened the Convention by committing Annex I Parties to individual, legally binding targets to limit or reduce their greenhouse gas (GHG) emissions. The US, Canada, Norway and Russia are all Annex I Parties. The total of Annex I Parties commitments originally added up to a total cut in GHG emissions of at least 5 per cent from 1990 levels in the commitment period 2008–2012 with the US included.[4] Only Parties to the Convention that have also become Parties to the Protocol by ratification or approval are bound by the Protocol's commitments. Thirty-five countries and the EEC are required to reduce GHG emissions below levels specified for each of them in the treaty. The Kyoto Protocol's Annex B lists the targets for individual countries. Table 12.1 shows the Kyoto targets for US, Canada, Norway and Russia:

The Kyoto Protocol permits the use of flexible mechanisms in the implementation of national targets. The flexible mechanisms, also referred to as the Kyoto mechanisms, comprise emissions trading (ET), joint implementation (JI) and

Table 12.1 Kyoto Protocol targets for US, Canada, Norway and Russia

Country	Target (1990–2008/2012)*
US	– 7 per cent
Canada	– 6 per cent
Norway	+ 1 per cent
Russian Federation	0

Source: UNFCCC (1997), *Kyoto Protocol*.[†]
*The targets cover emissions of the six main greenhouse gases: carbon dioxide (CO_2); methane (CH_4); nitrous oxide (N_2O); hydrofluorocarbons (HFCs); perfluorocarbons (PFCs) and sulphur hexafluoride (SF_6) (see http://unfccc.int/kyoto_protocol/background/items/3145.php (accessed 5 June 2007).
[†]See http://unfccc.int/kyoto_protocol/background/items/3145.php (accessed 22 May 2007).

the Clean Development Mechanism (CDM). These were designed to help Annex I Parties cut the cost of meeting their emissions targets by taking advantage of opportunities to reduce emissions, or increase GHG removals, that cost less in other countries than at home.[5]

Any Annex I Party that has ratified the Protocol may use the mechanisms, but the Parties must provide evidence that their use of the mechanisms is 'supplemental to domestic action', which must constitute 'a significant element' of their efforts in meeting their commitments. Whether 'supplement' refers to 25 per cent or 50 per cent is still unclear. Also businesses, environmental NGOs and other 'legal entities' can participate in the mechanisms, however, under the responsibility of their governments.

While 171 Parties have ratified the Protocol to date, including Canada (2002), Norway (2002) and Russia (2004), the US has never ratified it. This decision was made, according to President Bush, because of the perceived damage to the economy of the US, the lack of commitments by the developing countries, and a lack of scientific basis for targets (UNFCCC, 2004: 4). Others have explained the decision by pressure and lobbying by industry adversely affected by mitigation policies, in particular large energy corporations in the oil and gas sector. The seemingly high mitigation costs in the US are due to a rapidly growing population and high rates of economic growth. Thus, it has been argued that the US 'would bear the lion's share of global costs in a Kyoto-like regime' (Bang, Heggelund and Vevatne, 2005: 8).

In Canada, there was strong opposition to ratification of the Protocol similar to the controversies in the US. While the reality of climate change was not equally challenged as such, the government of Alberta and the coal, oil and gas industries strongly opposed any regulation of economic activities to reduce greenhouse gases (Toner, 2000). Climate change policies are still highly controversial in Canada, and emissions have been steadily increasing. In April 2007, the Environment Minister, John Baird, delivered a drastic vision of economic breakdown if Canada were forced to comply with the Kyoto Protocol, arguing that the only way to meet Kyoto's carbon limits is to 'manufacture a recession'.[6]

In Norway, the issue of climate change was high on the political agenda in the follow-up of the Brundtland report (Langhelle, 2000). In 1989, Norway became the first country in the world to set a stabilization target for CO_2 emissions (Hovden and Lindseth, 2004). Also in Norway, however, climate change policies have been highly controversial. As pointed out in Chapter 9, Norway is probably the only country in the world where a government has resigned over the issue of climate change. This was linked to one of the main national controversies, the domestic use of natural gas and the building of gas-fired power plants (Chapter 9; Langhelle, 2000; Hovden and Lindseth, 2004). In spring 2007, the Labour Party, the largest party in the Government coalition, announced that the national target for Norway was to be climate neutral by the year 2050, reduce emissions by 30 per cent by 2020 and to over-comply the Kyoto commitments by 10 per cent. The major issue in the domestic debate, however, is how much of these reductions should be taken at home and how much should be achieved through the Kyoto mechanisms.

The case of Russia is very different. Russia's positive attitude to the Kyoto Protocol was closely related to the assumed participation of the US and the anticipation of economic gains from emissions trading between Annex I countries made possible under the Kyoto Protocol. Owing to the breakdown of the communist regime and transformation of the Russian economy, emissions of CO_2 peaked in 1989, and from 1990 to 1998, Russian CO_2 emissions decreased by 35.6 per cent, leaving Russia with substantial surplus quotas or 'hot air' for sale (Kundzewicz, Schnellnuber and Svirejeva-Hopkins, 2004; Bang, Heggelund and VeVatne, 2005).

The US is not bound by the Kyoto Protocol. This fact had several effects on Russia. It sharply reduced the outlook for big revenues since the expected value of surplus quotas was reduced. It turned many against ratification of the Protocol, and fuelled the widespread concern that emission targets could harm Russian economic growth. Since 1998, both GDP and GHG emissions have been on the rise. The non-signing of US, however, left Russia in a key role in determining whether the Protocol would be implemented or not. The Protocol would not enter into force unless Russia ratified in the absence of US participation (Sabonis-Helf, 2003; Bang, Heggelund and Vevatne, 2005).[7]

National framing of climate change in the Arctic

As we have seen, the national responses to climate change vary considerably in the four countries. Except for the US, the Kyoto commitments have been and still are important for the national follow-up and also for the framing of oil and gas discourses in the Arctic. For the US, of course, the Kyoto Protocol has no real authority at the federal level, although several of the states in the US have adopted the Kyoto target for the US. Despite the reduction targets, there has been a large increase in CO_2 emissions in all the countries, and in the later years, also in Russia (Table 12.2).

Although the above table refers to the CO_2 emissions only, and excludes land use, land-use change and forestry (LULUCF) and other GHGs,[8] the Kyoto targets

Table 12.2 Increase in CO₂ emissions from 1990 (baseline year in the Kyoto Protocol) to 2005 in percentage (without land use, land-use change and forestry), and from 2000 to 2005

Country	1990–2005	2000–2005
Canada	+ 27.1 per cent	+ 3.5 per cent
Norway	+ 24 per cent	+ 3.8 per cent
Russian Federation	– 28.6 per cent	+ 7.5 per cent
US	+ 20.3 per cent	+ 2.5 per cent

Source: United Nations Framework Convention on Climate Change (UNFCCC), Secretariat. Data compiled from http://unfccc.int/ghg_emissions_data/ghg_data_from_unfccc/ghg_profiles/items/3954.php (Accessed 11 November 2007).

remain elusive for Canada and Norway. The energy industries, which include oil and gas production, comprise a large share of total GHG emissions in all the countries, especially in Russia. As seen in Table 12.3, the energy industry's share of GHG emissions has also increased in Canada, Norway and the US, but was reduced in Russia. The energy industries include emissions from fuels combusted by fuel extraction and energy production.[9]

Within these national contexts, Arctic oil and gas has for the most part not been specifically related to climate change. The dominant story line in all the countries is linked to the Arctic as a 'storehouse of resources' and a future guarantee of 'security of supply'. In the US, this is unproblematic. There are no constraints from the Kyoto Protocol. Climate change is still disputed, although Bush acknowledged climate change in his speech to the nation January 23, 2007. Although there are currently few signs of a new policy at the federal level, many analysts believe, however, that this may change in the US with a new president. In Russia, the Kyoto commitments are still far from representing a constraint on oil and gas production in the Arctic.

The situation is different in Canada and Norway. In both countries, the Kyoto targets are more challenging, and politically challenging as well in terms of being

Table 12.3 Energy industries' sector share of total greenhouse gas emissions (without LULUCF) in 1990 and 2005 in Canada, Norway, Russia and US

Country	1990	2005	Change
Canada	31 per cent	33.2 per cent	+ 2.2 per cent
Norway	22.7 per cent	33.2 per cent	+ 10.5 per cent
Russian Federation	51.3 per cent	45.1 per cent	– 6.2 per cent
US	35 per cent	38.6 per cent	+ 3.6 per cent

Source: United Nations Framework Convention on Climate Change (UNFCCC), Secretariat. Data compiled from http://unfccc.int/ghg_emissions_data/ghg_data_from_unfccc/ghg_profiles/items/3954.php (Accessed 11 November 2007).

highly disputed (more in Canada than in Norway) and part of a wider public debate. In both countries, there is a competing story line, which links the production of oil and gas in the Arctic to climate change, which has been visible in the public debate. In Canada, Shell's investments in the exploitation of tar sands in Athabasca have received attention from environmentalists from a climate change perspective. The following quote is from a press release, dated June 8, 2006, from the Canadian environmental NGO, Ecology North, placing the Mackenzie pipeline in the global context of climate change:

> The impacts of climate change are already being felt in the NWT and around the world. These impacts are predicted to accelerate in the coming years. Natural gas that stays in the ground does not add to climate change. Extracting Mackenzie gas through the pipeline and burning it will add to climate change . . . According to a report done by the Pembina Institute on behalf of Ecology North, emissions just from building and operating the pipeline will increase the NWT's greenhouse gas emissions by 44 per cent. Burning Mackenzie gas will release at least 25 million tonnes of carbon dioxide into the atmosphere each year. This is the same increase in emissions that adding 5.5 million cars to Canada's roads would have.
>
> (Ecology North, 2006)

Ecology North's proposal is not a 'no' to the pipeline as such, however, but an attempt to 'Make the Pipeline Green!', by ensuring that the emissions caused by burning Mackenzie Gas are offset and thus 'carbon neutral'. 'To be carbon neutral, every tonne of carbon dioxide released from building and operating the pipeline and from burning the gas extracted by the pipeline would have to be offset by a reduction of carbon dioxide elsewhere'.[10] Also the Natural Resources Defense Council and the Sierra Club have linked Arctic oil and gas to climate change (Price and Bennett, 2002).

In Norway, there is a similar competing framing of oil and gas in the Arctic. Two of the most important environmental NGOs (Nature and Youth, and Friends of the Earth, Norway), and also indigenous peoples, have argued against oil and gas exploration in the Arctic from the point of view that this will contribute further to climate change. Moreover, this view is shared also by the environmental NGOs that have focused more on biodiversity issues.

Those who link oil and gas production directly to climate change argue the following way: Oil and gas production will increase domestic emissions in Norway, which is problematic given the Norwegian Kyoto targets, and the consumption of this oil and gas will further increase both domestic and foreign emissions. The target which is primarily used as the point of reference, however, is not the Kyoto targets, but the recommendations by the Intergovernmental Panel on Climate Change (IPPC) for long-term reduction and stabilization of CO_2 concentrations in the atmosphere. In the latest *Summary for Policymakers*, IPCC argues that a 50-85 per cent reduction in emissions will have to be achieved by the middle of this century in order to reach a stabilization scenario that avoids dangerous anthropogenic interference (IPCC, 2007). In this picture, the Kyoto Protocol only constitutes the first small step towards a stabilization of GHGs. Oil and gas development in the Arctic is seen as contradictory to a development path, which aims at substantial reductions in global GHGs. The following resolution adopted by the Sámi organization Norske Samers Riksforbund (NSR), in 2005, is one example of this specific framing of Arctic oil and gas in Norway:

Environment before oil and gas in the Arctic

NSR wants to draw attention to the significance of the Barents Sea for the Sámi people, for people, and for the Northern areas . . . NSR looks with great worry on the oil- and gas developments in the area . . . The impact of climate change will affect indigenous peoples especially hard, amongst other things by worsening the access to species important for traditional activities.

NSR demands that Norway shows responsibility and takes the consequences of that increased oil extraction and the use of fossil energy creates negative climate change. In addition, petroleum production constitutes a direct threat in an Arctic area. The Barents Sea must therefore not be opened up for petroleum extraction.

(Norske Samers Riksforbund, Resolution No. 9, 2005)[11]

Apart from the above linkages, the direct relation between Arctic oil and gas production and the global CO_2 emissions and greenhouse warming has not been a central part of the general sustainable development agenda in the Arctic. As argued by Glomsrød in *The Economy of the North* (2006: 10), however, the Arctic is part of the global greenhouse gas balance:

The Arctic currently supplies about 16 per cent of all oil and gas to the global economy and has reserves to keep on with this for quite some time. Hence, parts of the Arctic are seriously involved in the global greenhouse gas balance, which is subject to increasing concern related to global warming. Further, also the substantial production and reserves of minerals are indirectly involved in the large scale emission of the greenhouse gases as they are processed in coal-based and polluting smelters around the world, including some Arctic regions. Thus, the Arctic is not only affected by rapid climate change, activities in the region are also contributing their share to the global warming.

The linkages between oil and gas production in the Arctic and global climate change presented above are minority views. Understandably, the impacts of climate change have caused deep worries in the Arctic. Many of the conclusions from the Arctic Climate Impact Assessment (ACIA) project were alarming. Climate change may have dramatic impacts on biological diversity. Arctic vegetation zones are likely to shift, causing wide-ranging impacts. Animal species' diversity, ranges and distribution will be affected and many coastal communities face increasing exposure to storms. For some indigenous communities, the conclusions were dramatic:

> Changes in species' ranges and availability, access to these species, a perceived reduction in weather predictability, and travel safety in changing ice and weather conditions present serious challenges to human health and food security, and possibly even the survival of some cultures.
>
> (ACIA, 2004: 11).

The Arctic, therefore, may be more vulnerable to climate change than other areas of the world. And the Arctic may be the first region to really experience the impacts of climate change. These impacts will take place in the context of other stresses, among them chemical pollution, land-use changes, habitat fragmentation and cultural and economic changes, stresses which in many parts of the Arctic are related to oil and gas production.

Arctic oil and gas, biological diversity and sustainable use of natural resources

With the exception of the US, all countries are parties to the Convention on Biological Diversity (CBD), which was adopted at the Rio Summit in 1992. At present, there are 189 Parties to the Convention. Canada ratified the Convention in 1992, Norway in 1993 and Russia in 1995. The objectives of the CBD is threefold: the 'conservation of biological diversity, the sustainable use of its components and the fair and equitable sharing of the benefits arising out of the utilization of genetic resources' (CBD, 1993: 146). In 2002, all parties committed themselves

> to a more effective and coherent implementation of the three objectives of the Convention, to achieve by 2010 a significant reduction of the current rate of biodiversity loss at the global, regional and national level as a contribution to poverty alleviation and to the benefit of all life on earth.
>
> (CBD, 2002, *Decision VI/26*)

The 2010 Biodiversity Target was subsequently endorsed by the World Summit in Johannesburg 2002 and later by the United Nations General Assembly in 2005. In 2002, the Parties to the Convention adopted a Strategic Plan, committing them to more effective and coherent implementation of the Convention. In 2004, at its seventh meeting, an overall objective was set:

> the objective of the establishment and maintenance by 2010 for terrestrial and by 2012 for marine areas of comprehensive, effectively managed, and ecologically representative national and regional systems of protected areas that collectively, inter alia through a global network ... contribute to achieving the three objectives of the Convention and the 2010 target to significantly reduce the current rate of biodiversity loss.
>
> (CBD, 2004a, *Decision VII/28*)

In addition, a multi-year programme of work until 2010 was endorsed. In *Decision VII/30*, a preliminary framework of goals, subsidiary targets and indicators was adopted. Among them, the target that at least 10 per cent of each of the world's ecological regions should be effectively conserved, that area of particular importance to biodiversity should be protected, sustainable use promoted and that the threats to biodiversity should be addressed. The adopted goals and targets provided a flexible framework within which national and regional targets could be developed, according to national priorities and capacities (CBD, 2004b, *Decision VII/30*).

As part of the Convention requirements, Canada developed a National Biodiversity Strategy in 1995, Norway in 1997 and Russia in 2001. As opposed to climate change policies, the conservation policies in the different countries seem to have been more influenced by the work within the Arctic Council. It was at the core of the AEPS strategy from the beginning, and a core activity of the Conservation of Arctic Flora and Fauna (CAFF) Working Group.

At the AEPS Ministerial meeting in Nuuk 1993, CAFF was asked to develop a plan for a network of Arctic protected areas to ensure the necessary protection of Arctic Ecosystems. A strategy and action plan was released in 1996, and laid the foundation for the creation of the Circumpolar Protected Areas Network (CPAN). The aim of CPAN is to identify gaps in protected areas and link the different national systems 'into a comprehensive and sufficient circumpolar protected area network' (CAFF, 1996: 7).

Based on a number of earlier reports by CAFF/CPAN, the first comprehensive overview of the Arctic environment, *Arctic Flora and Fauna: Conservation and Status* was released in 2001 (CAFF, 2001). The report identified 405 protected areas (including RAMSAR-sites[12]) in the Arctic, giving formal protection to approximately 2.5 million km^2, or 17 per cent of the Arctic as defined by CAFF (2001: 77). Table 12.4 gives an overview of protected areas in Canada, Norway, Russia and Alaska as of 2000.

In 2000, in the case of Norway, most of the area was located in Svalbard. Only about 7 per cent of the Arctic mainland was protected. In the case of Russia, large marine components and several large areas designated on a regional level but not endorsed by federal authorities were included (CAFF, 2001: 78).

The CPAN Country Updates Report (CAFF, 2004) reports an increase both in number of areas and total area protected in the Arctic. Although the figures are not fully compatible, there has been a substantial increase in protected areas and proposed areas for protection since 2000. In 2004, the total percentage of the Arctic region having some type of formal protection was approximately 18 per cent. From 1997 to 2004, 124 new protected areas were created (CAFF, 2004: ii). Among them a number of areas in Canada and Russia and the creation of five new protected areas on the Arctic Archipelago in Norway, extending protected areas at Svalbard. The new protected land areas cover a total area of 4,449 km^2, or 8 per cent of Svalbard's land area (*WWF Arctic Bulletin*, No. 3, 2003).

The World Wildlife Fund's Arctic Programme, however, is still critical towards the overall implementation within the Arctic states. In the Editorial in the *WWF Arctic Bulletin* (No. 2, 2003), Samantha Smith writes:

Table 12.4 Protected areas in Arctic Canada, Norway, Russia and Alaska, classified in International Union for Conservation of Nature categories I-V, plus Ramsar international wetland sites as of 2000. Areas smaller than 10 km² are not included

Country	Number of areas	Total area (km²)	Percentage of Arctic land area of the country
Canada	61	500,842	9.5
Norway	39	41,380	25.3
Russia	110	625,518	9.9
US (Alaska)	55	296,499	50.2

Source: CAFF (2001: 78).

In 1997, Arctic countries agreed to create a circumpolar network of protected areas. Six years later, the countries are still discussing what areas should be included. Arctic governments agree that oil and gas development will be a key challenge for the region. The Council's deliverables so far include an outdated map of areas at risk, and guidelines for offshore oil and gas development that are weaker than most Arctic national regimes.

(Smith, 2003a)

According to Smith, the sizes and types of existing protected areas are inadequate to safeguard biodiversity and traditional lifestyles in the majority of the Arctic states (Smith, 2003b). A major 'gap' seems to be the lack of marine and coastal protected areas. Less than 1 per cent of the marine costal areas are protected. Only 7 per cent of the Arctic coastlines were affected by development in 2002, while 0.8 per cent had severe impacts. Hence, 'the Arctic now appears to hold the world's last remaining undeveloped costal ecosystems' (Ahlenius, Johnsen and Nellemann, 2005: 37). These ecosystems are, however, said to be critical to Arctic food chains and coastal ecosystems, and also the systems that are most at risk from petroleum exploration (Ahlenius, Johnsen and Nellemann, 2005).

Thus, oil and gas activities are still seen as a threat for the conservation of ecosystems. The conflict between conservationists and oil and gas can be found in all the countries, and is being played out in several of the cases. In Norway, the issue of 'petroleum-free zones' in the Barents Sea has been the main conflict in the development of the comprehensive management plan. In Russia, conflicts over land use have been prominent. In Canada, the Mackenzie Pipeline Project has been attacked from a conservation point of view, and in the US, the conflict over oil and gas exploration in the Arctic National Wildlife Refuge (ANWR) represents the most prominent case.

Conservation and sustainable use

The traditional conflict between conservationists and indigenous peoples, is still visible in our country studies. Canada witnessed a 'seal war' that waged for two decades between the sealers and fur industry on one side, and the environmentalists on the other, ending with a ban on the use of baby sealskins in Europe (Helander-Renvall. 2005). Similar conflicts played out in the early days of the Arctic Council, between US officials and Canadian and Permanent members regarding the harvesting of marine mammals. The US has continuously opposed any form of trade with Northern marine mammals, while Canada has supported such trade (Vanderzwaag, Huebert and Ferrara, 2003).

While conflicts over sustainable use and conservation have been important, 'conservation' is more often understood to include subsistence activities and traditional lifestyles. This is the case for CAFF's approach, and it is increasingly so among environmental organizations. Moreover, protecting traditional ways of life is in itself part of the quest for conservation. As reported by Helander-Renvall (2005: 46) legislation on Special Protected Natural Territories (SPNT) is especially important for maintaining indigenous peoples' lifestyles and for protecting biodiversity in Russia. The state ecological programme for 2002–2006 provides preservation of traditional activities of indigenous peoples. Examples are the Khatymi reserve and the Kolyma-Koren reserve.

Indigenous peoples are also increasingly part of the management of natural resources, especially in Alaska and Canada. While there is a broad range of interpretations of co-management, Helander-Renvall (2005: 46) uses Campbell's (1996) definition of co-management 'as an inclusionary, consensus-based approach to resource use and development', including the 'sharing of decision-making power' and also as 'a process of combining western scientific knowledge and traditional environmental knowledge for the purpose of improving resource management' (Campbell, 1996). The level of participation by indigenous peoples in these management boards can vary, but in the Canadian territories indigenous peoples have a legally defined role within these co-management structures, which provides them a strong voice in the management of their resources.

The role of indigenous peoples in the creation of the 20,500-km^2 Ukkusiksalik National Park in the Kivallik region, just south of the Arctic Circle, was also acknowledged by WWF-Canada. The protection of the area for subsistence hunting and ecotourism opportunities was described as 'a fundamental step in helping to achieve a well-balanced future for Nunavut', and WWF-Canada strongly supported what they saw as an expression of 'Conservation First', meaning that 'prior to industrial development, an adequate network of key cultural and ecological areas is withdrawn from industrial development, thereby protecting these critical values for future generations' (Ewins, 2003).

Given the fundamental importance of subsistence and traditional lifestyles, it is seen as an absolute necessity by indigenous peoples that the oil and gas activities do not disturb, or at least do not spoil, the possibilities of continuing traditional activities like reindeer herding, hunting, fishing and gathering. Thus, support for oil and gas is conditioned on the way oil and gas activities are conducted, and also the degree to which indigenous rights and land claims are granted and respected.

Indigenous peoples

Roughly 4 million people live in the Arctic (Bogoyavlenskiy and Siggner, 2004: 27); about 15 per cent are indigenous peoples. Compared to the 1.5 per cent of indigenous peoples in the entire world, this makes the relatively high number of indigenous peoples a common trait of the Arctic. However, within the Arctic, the proportion of indigenous peoples varies considerably. In some places in Canada and Alaska, First Nations have an absolute majority, while in parts of Russia and Norway, indigenous peoples represent an insignificant part of the numerical population. Indigenous peoples, however, have special rights, either moral or legal, or both, due to their status as indigenous peoples. Indigenous peoples' opinions about oil and gas development in the Arctic varies from country to country, but also within the countries one can find different opinions also within the same peoples (see the different country chapters in this volume).

Yet, our studies suggest that the differences in views can be traced back to differences in power accorded to the indigenous peoples by oil and gas companies and by the central governments. If local people, which in some areas are mostly indigenous peoples, are given a substantial voice and their voice is given substantial power, very few people oppose the development of oil and gas activities in their regions. The degree of influence in shaping the oil and gas development in their regions varies a lot, as does the power to negotiate pay-back and benefits for the Arctic people. As explained in Chapter 8 on Canada, the chiefs and elders representing the Deh Cho from Wrigley and Fort Simpson were very keen to get oil and gas development in their localities on condition that this development respected the natural environment when constructing, exploring and in the clean-up phase, and that they were active participants in the discussions and decisions taken by the oil and gas companies.

The Inuvialuit in Canada argue according to the same lines (Chapter 8). In Alaska, we recognize the same arguments using job opportunities, construction of schools and hospitals as desired benefits while claiming respect for traditional whaling and fishing (Chapter 7). According to Poppel, Kruse, Duhaime et al. (2007: 14), less than 15 per cent of the Arctic population in Alaska express a degree of dissatisfaction with their influence over the management of natural resource management, including oil and gas, in their localities. On the contrary, around 60 per cent professed their partial satisfaction on how the issues are handled. The Sámi in Norway are divided. Some groups argue against any oil and gas development whatsoever for environmental reasons, while the majority accept oil and gas development on condition that it takes Sámi interests into account (Chapter 9). These interests include participation in the discussions and decisions related to oil and gas development, flexible job opportunities for the Sámi, respect for sacred places and reindeer breeding areas. In Russia, data from Poppel, Kruse, Duhaime et al. (2007: 14) suggest that very few are satisfied with their influence over management of natural resources in their home areas, while as many as 83 per cent express their dissatisfaction.

State policies towards indigenous peoples and their land claims

Indigenous peoples in the Arctic live in different states that have followed various policies towards indigenous peoples during the last 150 years. Yet, we recognize similar patterns of responses by the different states towards indigenous peoples, especially how they have sought integration into the different states. A paternalistic practice of assimilation used to be the standard government response to indigenous groups all over the Arctic. Standardized procedures and common rules and regulations were the national governments' answers to all inhabitants of the state, indigenous peoples included.

If we look at the Arctic as a whole and assess the political power of indigenous peoples at the end of the Second World War compared to their political power today, there has to a certain degree been a transfer of power towards indigenous peoples (Bankes, 2004: 103; Broderstad and Dahl, 2004: 88). There is an increasing tendency by the Arctic countries to integrate indigenous affairs into regular politics and administration, be that at the local, national or regional level (Bankes, 2004; Broderstad and Dahl, 2004). Indigenous peoples have also been granted significantly better legal protection in the last 20 years, with development of different international human rights laws and specific development of various national laws (see Hansen and Bankes, Chapter 11). Still, these observations remain general, and levels of protection and power differ between the various countries.

From assimilation to formal recognition

In Norway, the Sámi were recognized by the Norwegian government as a distinct people and mostly respected as such until Norway's modernization process gained force around 1850. With modernization and emphasis on the nation-state in Norway from the mid-nineteenth century, assimilation became the new official politics towards the Sámi. The assimilation lasted for about a hundred years, up to at least the 1950s. Education systems and boarding schools with an obligation to speak the Norwegian language were forced upon the Sámi peoples. Government policy induced the Sámi to give up their nomadic lifestyle and enter into the modern monetary economy. Gradually from 1950s onwards, the Sámi of Norway have gained more rights and political power. However, Sámi representatives today still regard Norway as a colonizer. The central state still controls the legislative power, but the Finnmark Act passed in May 2005, signals a limited recognition of greater local – and thus indigenous – control of land. The Sámi themselves, however, assess the Finnmark Act as a passable compromise, but not as any acknowledgement of their land rights. In Norway, management and certain administrative tasks have been delegated to the region and to Sámi organizations; however, political authority has not been handed over from the government.

In Russia the situation is a bit different. After the Russian revolution in 1917, the indigenous peoples of the North in Russia were officially given means to modernize on their own behalf. Creation of indigenous standard written languages and the adoption of these languages in schools contributed to this modernization. Medical facilities and trading cooperatives for indigenous peoples were established. During the 1930s, however, indigenous people felt state pressure to attend state-run boarding schools and become sedentary. After the Second World War, the trend towards assimilation grew, and indigenous peoples were neglected by the central government when it came to infrastructure and modern health and education services. During the 1970s and 80s, the number of indigenous peoples declined as did the number of groups.

Up to 1991, the Moscow-based centralized power of the USSR controlled the Arctic parts of Russia without leaving much power to the regions, especially not to the indigenous peoples. Yet, as Andreyeva and Kryukov have argued in this volume (Chapter 10), the Arctic region used to receive significant economic support from the federal Soviet budget. With the breakdown of the USSR in December 1991, Russia liberalized the economy and decentralized political power. This first period of perestroika and glasnost gave more political power to the different regions, the Arctic included. However, the different indigenous peoples living in the Arctic gained little political power in this decentralization process owing to their small number in the overall population. The creation of different indigenous associations, regrouped as early as 1990 under the umbrella of RAIPON (Russian Association of Indigenous Peoples of the North), made indigenous peoples' voices heard in some regional settings. However, the central government in Russia has never listened very carefully to the voice of indigenous peoples in the Russian Arctic. Since 2001, we have seen a clear setback for indigenous peoples' political power in the Russian Arctic (Kryukov and Moe, 2006: 130–138).

At the beginning of the last century, Canada aggressively attempted to assimilate their First Nations into the dominant society by trying to wipe out their cultures and make the First Nations act and think like whites (Hughes, 2003: 30). Indigenous people used to be regarded as aimless, uncivilized and uncontrolled (Hughes, 2003: 42). In the current era, however, indigenous land claims have been settled throughout Arctic Canada since 1973. First Nations land titles have in some places included both self-government, and the right to land and its resources. Quite a few indigenous groups have established their own corporations and have become well off economically (see Chapter 8).

Indigenous peoples in Canada remained colonized far into the 1960s. Municipal governments were established during the 1950s and 1960s, but Inuit were not allowed to vote in federal elections until 1962 (Broderstad and Dahl, 2004: 90). During the 1970s, different indigenous peoples formed various associations advocating self-government or at least, greater self-determination in their regions. A greater degree of autonomy for indigenous peoples in Canada has been the result. In Canada, indigenous groups have gained exclusive titles for selected lands (Bankes, 2004: 107). Thus, in a few decades, Canada changed from being conservative and restrictive towards indigenous peoples to become one of the world's leading nations in dealing with indigenous peoples. However, with the shift in government in January 2006, Canada seems to be adopting a more restrictive position towards indigenous peoples and their rights.

In Alaska, the two most important events in the evolution of governance were the creation of the state of Alaska in 1959, with a new constitution allowing strong self-governance powers to local governments, and the Alaska Native Claims Settlement Act (ANCSA) of 1971 (see Chapter 7). These new instiutions led to the decentralization of power in the region in favour of the indigenous peoples.

States policies towards land claims

Land is primordial to indigenous peoples' cultural and physical survival. Oil and gas companies have to ensure, morally and by law, that their activities do not reduce indigenous peoples' possibility to use their traditional lands. Indigenous people do not see property rights on lands as a private business. Land should belong to the whole community, and people should use its bounty and share the benefits according to their needs. With increased interest in oil and gas resources in the Arctic, property rights have become increasingly important as well.

In the Arctic in general, most of the land is owned by the state. In all the Arctic countries, the state claims to own the sea up to 200 miles offshore. However, the general picture is not equivalent for all Arctic countries when it comes to onshore land. Property rights to land and resources have been granted to indigenous peoples in Alaska and Canada during the last three decades (Caulfield, 2004: 122). In Alaska, the US government owns about 60 per cent of the land, the state of Alaska owns 28 per cent and the Alaska Native Corporations own 12 per cent. Only 2 per cent of the land is privately owned in Alaska. In parts of Canada and Alaska, indigenous peoples engage in non-renewable resource extraction. In Russia and Norway, this has not been the case yet, and may never be. In Norway, a form of co-management of natural resources and land has been adopted, while in Russia the central government is reasserting massive control over land and resources.

Public land ownership still prevails as the dominant form of ownership in the Arctic. The various Arctic states, however, follow different policies concerning their indigenous peoples' rights to land and its resources. Canada has granted land titles with resource rights to indigenous peoples, and the central government has transferred some control and authority to regional or local bodies while keeping it public (e.g. from the federal state to Yukon territory in 1998). In the US, title to land has also been transferred to Native and state bodies in Alaska. Thus, here we can speak about a true devolution. In Russia, there has been a greater transfer of authority to the regions, but without giving any particular power to indigenous peoples. Their number being relatively insignificant, indigenous peoples in Russia have not benefited to a large extent from this decentralization of power. Yet, since 2001 we see a clear tendency of re-centralization of power from the regions towards Moscow. In Russia, special laws passed during the 1990s accorded indigenous peoples particular territorial rights (Bankes, 2004: 107).

In Norway, indigenous peoples have not figured as a prominent political issue in relation to oil and gas, mostly due to the fact that these activities for the most part are offshore. No Arctic state has been willing to discuss anything but state ownership to offshore oil and gas resources. In some states, indigenous peoples are in the process of testing this ownership claim in court.

Indigenous peoples feel the deterioration of their environment already, both by onshore and offshore oil drilling. Offshore, fisheries are disturbed and, onshore, both reindeer herding and freshwater fisheries have deteriorated. In many places, indigenous peoples realize that new projects have started without any consultation (Chapter 10). Initiatives taken by the federal government to include indigenous peoples and to compensate damage to their natural environment have seldom functioned as intended. Laws have also proven to be unclear, especially when defining the territorial limits of indigenous peoples and what constitutes a treat to their traditional culture. For the time being, indigenous peoples in the Russian Arctic have mostly suffered from oil and gas activities in the region only being economically compensated with small sums.

A common story line?

Indigenous peoples' concerns dominate the oil and gas issues in Canada and Alaska at the local and regional level. These issues are present also in Russia, especially at the local and regional level, but not high on the central political agenda. Indigenous peoples' interests and rights, however, are increasingly under pressure from oil and gas activities affecting traditional subsistence lifestyles.

How central these issues are to oil and gas development varies a lot from country to country in this study, placing the indigenous questions, claims and rights as top priority in public debates in Canada and Alaska, while leaving it to a minor question in public debates in Russia and Norway. We have noticed an increasing cooperation amongst indigenous peoples across countries and regions. Indigenous peoples and their bodies work together in different international forums, such as the Arctic Council and the UNWGIP (United Nations Working Group for Indigenous Peoples), but also in different smaller organizations like Forum for Development Cooperation with Indigenous Peoples and The Resource Center for the Rights of Indigenous Peoples. In addition, different indigenous peoples' associations every year take a number of bilateral initiatives with similar groups in the Arctic to discuss common themes. These initiatives lead increasingly to a transfer of ideas and possible solutions from one area to another. Different indigenous peoples in various areas in the Arctic are thus aware of others' successes and failures, and are thereby better able to advocate and argue for their own causes.

This development is followed by a general trend in the Arctic to devolve power towards local authorities (Bankes, 2004; Caulfield, 2004: 121). If this is the official and legal truth, it is not always perceived as such by marginalized indigenous peoples. Legal rights are not always understood the same way by the states, the oil companies and the indigenous peoples' organizations. Yet, what is clear is that most indigenous peoples welcome oil and gas activities in their region, if the industry acts as guests and respects the conventions, ideals and duties defined by their hosts, namely the indigenous peoples in the areas.

Power and politics, and the future of Arctic oil and gas

From the country case studies in this volume, there seems to be a certain inevitability about increased Arctic oil and gas exploration and production. The story lines we have identified all point more or less in the same direction. Oil and gas in the Arctic will be developed, like it or not. If we use the framework in Chapter 5 as a starting point, there are few constraints on oil and gas production. At the geopolitical level, there is hardly any resistance. The Arctic Council (with its member states), the European Union (EU) and the countries included in our study have strong and vested interests in further oil and gas exploration and production. There are many reasons for this, the most important being power and money. The increasing scarcity of oil and gas resources and the global competition for them, high oil and gas prices, increasing global demand and real worries about security of supply, all pull in the direction of going North.

The same holds for oil and gas companies. Oil and gas resources are becoming harder and harder to control. The nationalization policies in Russia, Venezuela and other countries have resulted in several oil and gas companies worrying about the ratio between reserves and production. In the case of Norway, the maturity of the North Sea oil and gas fields and declining oil production has led to an increasing focus on new fields, both nationally and internationally. As such, Statoil and also Hydro (now StatoilHydro have for several years invested in oil and gas activities in foreign countries (Ryggvik and Engen, 2005). The strong interests in the Barents Sea from these companies have also been closely linked to a wish to enter the Russian part of the Barents Sea. The Snøhvit gas field and the process plant for gas liquefaction (LNG) northwest of Hammerfest, a city in the county of Finnmark, although commercially justifiable in itself, can be seen as part of the strategy to enter the Russian parts of the Barents Sea – a strategy that has been equally important for the government. The merger between Statoil and Hydro in 2007 was more or less directly linked to these international aspirations.

The decision by state-owned Gazprom to develop the huge Shtokman gas field without foreign capital was, at first, a severe blow in this respect. Although StatoilHydro later got access to Shtokman, there is no doubt that there is a desire to retain national control over natural resources and income in the current Putin administration. This centralization can be seen partly as a response to the failed economic reforms under President Boris Yeltsin and the 'robbery' of public wealth that took place by the so-called 'oligarchs' (Goldman, 2003). Moreover, Russia seems to be following a 'patriotic' approach based on a wish to rebuild Russian status and the Russian economy, which also seems to have broad public support.

In the case of Alaska, oil and gas policies have been strongly influenced by the US focus on energy supply and energy security. The dependency on supplies from the unstable Arab world has been prominent in this regard. At the same time, there has been little focus in the current administration to address (over)consumption in the US. Canada is also strongly influenced by US energy security policies, but with a surplus of oil and gas resources, which are becoming increasingly valuable. Thus, it is hard to stay away from the Arctic.

There are, however, some caveats. Both Canada and the US are complicated political federal systems with many checks and balances (Chapter 6). These checks and balances have made further exploration in several areas like ANWR and the Mackenzie pipeline projects difficult, despite the fact that strong industrial and political interests have pushed for further developments. Issues of environment and indigenous peoples' property rights have mainly driven the resistance. The focus on energy supply and security in the US and high prices on oil and gas seem to have affected environmental groups as well. Ecology North's framing of the Mackenzie pipeline issue can be read the following way: 'It is not possible to stop it, so let's postphone it and let economy kill it'. Also in Norway, there are environmental groups who do not think it is possible to stop the expansion in the Barents Sea. WWF Norway, therefore, follows the 'conservation first' strategy in order to at least save the most valuable areas.

Russia is a strange mixture of the 'old' and 'new' political system, still undergoing transformation, but moving more towards a centralized state with control and ownership over oil and gas resources as part of the political agenda of the Putin administration. As such, there are similarities between the Norwegian and Russian oil and gas regimes. State control over the development of oil and gas resources has been strong in Norway since the discovery of oil and gas on the Norewegian continental shelf, and is increasingly so in Russia. Russia is also gaining conficence and power as money is pouring in. President Putin's resistance at the 20 October Summit in Finland 2006 towards the EU leaders' calls to ratify an international energy treaty that would liberalize trade and investment in its oil and gas sector is an example of increasing Russian power and self-cofindence.

Norway is the only unitary state included in this study. Although the unitary state model of Norway should make it easier to decide on oil and gas activities in the Arctic, cross-cutting political cleavages and a strong political opposition on energy issues have slowed down the oil and gas activities, also making Norway unpredictable. Although there is a clear majority in parliament in favour of oil and gas activities in the Barents Sea, it has been difficult to do so on a full scale, since both the current and earlier government coalitions have included one of the opposing parties. The political split over energy in Norway cuts across the traditional left–right axis in politics, and, so far, the energy cleavage has permeated all recent governments.

The adoption of the Integrated Management Plan in 2006 by parliament leaves many of the most controversial issues unsettled. The environmental NGOs lost the battle over 'petroleum-free zones', and many of the restrictions on oil and gas activities, like the areas outside Lofoten and Vesterålen (Nordland VI, Nordland VII and Troms II), are temporary. Since the Management Plan is to be updated in 2010, it means that none of the most controversial issues in the Plan have been settled. The temporary restrictions will be a major source of political controversy in years to come, and the outcomes of these battles will be settled partly in the parliamentary elections, which determine the composition of parliament and possible government coalitions, and partly by the mobilization of interests groups, either way. Already several Norwegian oil and gas companies are planning for the opening of Nordland VI, Nordland VII and Troms II in 2010. Few believe that these areas will survive the pressure from the majority in parliament, oil and gas companies and most locals.

The Comprehensive Management Plan for the Barents Sea could represent a caveat, but not even people in the Ministry of Environment believe that the plan would be able to stop anything, if a major oil and gas field was discovered in a place it should not be.

Another driving force for oil and gas development in the Arctic is the Arctic population. As we argued earlier, the key sustainability challenge seen from the Arctic is how to ensure that more of the profits from resource extraction actually remain in the Arctic. Within the four countries, this is perhaps the perception that is most consistently advocated among local and indigenous peoples. The 'development' issue is most certainly the issue that is most beneficial or unproblematic for oil and gas. This story line also reframes the challenges of oil and gas from 'if' to 'how'. Oil and gas in the Arctic is turned into a question of how you go about doing it in a proper way (see Chapter 13 on this issue).

To the degree that indigenous peoples' property rights are settled and deals are negotiated that secure subsistence livelihoods and the environment (what the Brundtland report referred to as 'sensitive development'), it is possible to imagine that most indigenous peoples will support oil and gas in the Arctic. Indigenous peoples in Alaska and Canada are already directly involved in the extraction of oil and gas. 'Sustainable use' problems are likely to increase as the activities expand. No doubt, indigenous peoples' rights will be a more central concern in the future, especially when conflicts arise. There should be ample evidence in what we have already discussed, however, that there are ways to avoid such situations (see also Chapter 13). It should be possible to find arrangements that can combine oil and gas activities with the activities of whaling, caribou hunting and reindeer breeding.

Conservation poses a 'threat' to oil and gas activities, and oil and gas activities pose a real problem for conservation. Taking conservation into account, therefore, may be both costly and difficult to reconcile with oil and gas activities. The case of ANWR in Alaska provides the evidence. Pollution from oil and gas activities may be a problem, but can be minimized through proper regulations. The occurrence of a major oil spill in the Arctic will most likely have serious consequences for the industry, but probably only temporarily.

The last caveat is climate change. This, together with biological diversity, is the most problematic environmental issue from a sustainable development perspective, and potentially the most difficult to reconcile with increased oil and gas activities in the Arctic. The focus on climate change increased dramatically in the second half of 2006, mainly due to 'strange' weather, but also due to the work of Al Gore and the release of the report from Nicholas Stern, Head of the Government Economics Service and Adviser to the Government on the economics of climate change and development in the United Kingdom (Stern, 2006). Although climate change first and foremost is a global problem, the Arctic is special in the sense that the effects of global warming will occur first in the Arctic, with feedback loops to the global climate. If the predictions are correct, climate change will undoubtedly become the major environmental issue in the Arctic. The likelihood that this in turn will reflect back on the oil and gas industry is more than probable.

Until now, the Arctic Council has primarily focused upon impacts from and adaptations to climate change. The ACIA framing was on the impacts of climate change on oil and gas activities, not the other way around. Oil and gas activities in the Arctic have never been discussed as a major source of the global CO_2 emissions and greenhouse warming, but have so far been lifted out of the Arctic and placed at the global and national level. None of the Arctic states have been particularly interested in such a discussion. It seems more or less taken for granted by the Arctic states that the oil and gas resources in the Arctic are to be explored and extracted. From a sustainable development perspective, however, it is no doubt a legitimate question to ask whether or not the discovered and undiscovered oil and gas resources should be depleted, and whether such depletion is reconcilable with a development path that many argue goes from non-renewable to renewable energy sources.

Climate change, however, may have divergent consequences in the different countries in the shorter term. For Canada and Norway, the Kyoto Protocol could restrain, or at least make it more expensive (depending on the price of CO_2), to develop already costly oil and gas resources. In Russia, there are fewer restraints from the Kyoto Protocol, but Russia may be reluctant to take on more ambitious targets. These reservations are likely to have strong support until it can be demonstrated that, in Russia, economic growth does not necessarily entail corresponding growth in energy consumption. At the federal level, the US interest seems remote at the moment, but the growing concern for climate change also in the US and the presidential elections in 2008 may change policies also at the federal level.

In the longer term, however, it is the post-Kyoto regime for greenhouse gases that most likely will have an impact on the emissions of greenhouse gases in the Arctic. In all likelihood, extraction in the Arctic will eventually have to address air emissions (through carbon capture and storage as indicated in the Program for the Norwegian Chairmanship of the Arctic Council (2006–2008) (Arctic Council, 2006). Given the last reports from IPCC, the constraints on emissions in order to avoid climate change are quite astonishing. To keep within a 2-degree increase in global mean tempartures (the target endorsed by the EU and Norway), the IPCC conclude that it is necessary to reduce global emissions from the 2000 level by 50-85 per cent by 2050 (IPCC, 2007). How to merge these ambitions with the opening of a new frontier for oil and gas in one of the most vulnerable places on Earth is not clear.

Seen from the perspective of sustainable development, therefore, it is the global sustainable development agenda that may become more and more constraining on further oil and gas activities in the Arctic and elsewhere. The consumption of fossil fuels is getting close to violating the minimum requirement for sustainable development: 'At a minimum, sustainable development must not endanger the natural systems that support life on Earth: the atmosphere, the waters, the soils, and the living beings' (WCED, 1987: 44–45).

What pulls in the other direction, however, is the growing global demand for energy, which is also linked to the global development agenda (poverty reduction and meeting human needs). Given the support for oil and gas extraction in the Arctic countries, new technology and technological developments seem to be the only feasible way politically to reconcile the above conflict at the moment. Reconciling the global need for energy (not necessarily fossil fuels) with the necessary large-scale reduction of global greenhouse gas emissions as requested by IPCC, however, poses diverging demands on the Arctic. Either way, the Arctic will be put under growing pressure in the years to come.

Notes

1 Although there may be less optimism in some parts of Russia regarding welfare and job creation for local inhabitants (Hønneland, Jørgensen. and Moe, 2007).
2 Russian Sámi President, Alexander Kobelev, at the Arctic Council Meeting, October 26, 2006.
3 See http://www.arcticpeoples.org/2006/01/24/early-findings-from-arctic-council-oil-and-gas-assessment/ (Accessed 5 June 2007).
4 The targets cover emissions of the six main greenhouse gases: carbon dioxide (CO_2); methane (CH_4); nitrous oxide (N_2O); hydrofluorocarbons (HFCs); perfluorocarbons (PFCs) and sulphur hexafluoride (SF_6). See http://unfccc.int/kyoto_protocol/background/items/3145.php (accessed 5 June 2007).
5 These mechanisms are described the following way by UNFCCC: 'Under joint implementation, an Annex I Party may implement a project that reduces emissions (e.g. an energy efficiency scheme) or increases removals by sinks (e.g. a reforestation project) in the territory of another Annex I Party, and count the resulting emission reduction units (ERUs) against its own target. While the term 'joint implementation' does not appear in Arcticle 6 of the Protocol where this mechanism is defined, it is often used as convenient shorthand. In practice, joint implementation projects are most likely to take place in economies in transition, where there tends to be more scope for cutting emissions at low cost . . . Under the clean development mechanism (CDM), Annex I Parties may implement projects in non-Annex I Parties that reduce emissions and use the resulting certified emission reductions (CERs) to help meet their own targets. The CDM also aims to help non-Annex I Parties achieve sustainable development and contribute to the ultimate objective of the Convention . . . Under emissions trading, an Annex I Party may transfer some of the emissions under its assigned amount, known as assigned amount units (AAUs), to another Annex I Party that finds it relatively more difficult to meet its emissions target'. (http://unfccc.int/kyoto_protocol/background/items/3145.php, accessed 21 May 2007).
6 See http://www.theglobeandmail.com/servlet/story/RTGAM.20070419.wclimate0419/BNStory/National/home (accessed 20 April 2007).
7 This is due to Article 25, §1 in the Protocol, which demanded the following: '1. This Protocol shall enter into force on the ninetieth day after the date on which not less than 55 Parties to the Convention, incorporating Parties included in Annex I which accounted in total for at least 55 per cent of the total carbon dioxide emissions for 1990

of the Parties included in Annex I, have deposited their instruments of ratification, acceptance, approval or accession'. The Protocol entered into force on February 16, 2005 as a consequence of ratification by Russia, which secured parties representing 55 per cent of total CO_2 emissions in industrialized countries.
8 CO_2 is, however, by far the most important greenhouse gas in developed countries.
9 See http://unfccc.int/ghg_emissions_data/information_on_data_sources/definitions/items/3817.php (accessed 23 May 2007).
10 According to Ecology North, these emissions could be offset by the following measures: 'Preserving at least 21 million acres of forests. Repairing 8 million houses (all houses in the Northern Territories, B.C., the Prairie Provinces and Ontario) to improve their energy efficiency. Building 15,660 wind turbines in Alberta. Buying carbon credits'.
11 Our translation.
12 Ramsar sites are wetlands of international importance designated under the Convention of Wetlands (Ramsar, Iran, 1971). These wetlands are commonly known as Ramsar sites.

References

Ahlenius, H., Johnsen, K. and Nellemann, C. (eds) (2005) *Vital Arctic Graphics – People and Global Heritage on Our Last Wild Shores*, UNEP/GRID-Arendal, WWF, ICC and CAFF.

Arctic Climate Impact Assessment (ACIA; 2004) *Impacts of a Warmer Arctic: Arctic Climate Impact Assessment*, Cambridge: Cambridge University Press.

Arctic Council (2006) *Programme for the Norwegian Chairmanship of the Arctic Council 2006–2008*. Online. Available HTTP: <http://arcticportal.org/en/arctic-council2> (Accessed 11 November 2007).

Arctic Environmental Protection Strategy (AEPS; 1991) *Declaration on the Protection of Arctic Environment*, June 14. Online. Available HTTP: <http://www.arctic-council.org/Archives/AEPS%20Docs/artic_environment.pdf> (Accessed 6 November 2007).

Arctic Human Development Report (2004) *Arctic Human Development Report,* Akureyri: Stefansson Arctic Institute.

Bang, G., Heggelund, G. and VeVatne, J. (2005) *Shifting Strategies in the Global Climate Negotiations*, Oslo: Cicero Report 2005: 08.

Bankes, N. (2004) 'Legal systems', in Arctic Human Development Report (2004) *Arctic Human Development Report,* Akureyri: Stefansson Arctic Institute.

Bogoyavlenskiy, D. and Siggner, A. (2004) 'Arctic demography', in *Arctic Human Development Report,* Akureyri: Stefansson Arctic Institute.

Broderstad, E. G. and Dahl, J. (2004) 'Political systems', in *Arctic Human Development Report,* Akureyri: Stefansson Arctic Institute.

CAFF (1996) *Circumpolar Protected Areas Network (CPAN) Strategy and Action Plan*, CAFF Habitat Conservation Report No. 6, Directorate for Nature Management, Trondheim, Norway.

CAFF (2001) *Arctic Flora and Fauna: Conservation and Status*, Helsinki: Edita. Online. Available HTTP: <http://arcticportal.org/arctic-council/working-groups/caff-document-library/arctic-flora-and-fauna> (Accessed 11 November 2007).

CAFF (2004). *Circumpolar Protected Areas Network (CPAN) Country Updates Report 2004*, CAFF Habitat Conservation Report No. 11, November 2004. Online. Available HTTP: <http://arcticportal.org/uploads/rQ/o5/rQo5wdkwRjc3aCyAUD4z4Q/Habitat-report-no.-11.pdf> (Accessed 11 November 2007).

Campbell, T. (1996) 'Co-management of Aboriginal resources', *Information North*, 22 (1). Arctic Institute of North America. Online. Available HTTP: <http://arcticcircle.uconn.edu/NatResources/comanagement.html> (Accessed 11 November 2007).

Caulfield, R. A. (2004) 'Resouce governance', in Arctic Human Development Report (2004) *Arctic Human Development Report,* Akureyri: Stefansson Arctic Institute.
CBD (1993) *Convention on Biological Diversity (with Annexes). Concluded at Rio de Janeiro on 5 June 1992,* United Nations Treaty Series. Online. Available HTTP: <http://www.cbd.int/doc/legal/cbd-un-en.pdf> (Accessed 12 November 2007).
CBD (2002) 'Strategic plan for the convention on biological diversity', *COP 6 Decision VI/26,* The Hague, April 7-19, 2002. Online. Available HTTP: <http://www.cbd.int/convention/cop-6-dec.shtml?m=COP-06&id=7200&lg=0> (Accessed 12 November 2007).
CBD (2004a) 'Protected areas (Articles 8 (a) to (e))', *COP 7 Decision VII/28,* Kuala Lumpur, February 9-20, 2004. Online. Available HTTP: <http://www.cbd.int/convention/cop-7-dec.shtml?m=COP-07&id=7765&lg=0> (Accessed 12 November 2007).
CBD (2004b) 'Strategic plan: future evaluation of progress', *COP 7 Decision VII/30,* Kuala Lumpur, February 9-20, 2004. Online. Available HTTP: <http://www.cbd.int/convention/cop-7-dec.shtml?m=COP-07&id=7767&lg=0> (Accessed 12 November 2007).
Duhaime, G. and Caron, A. (2006) 'The economy of the circumpolar Arctic', in Glomsrød, S. and Aslaksen, J. (eds) *The Economy of the North,* Oslo: Statistics Norway.
Ecology North (2006) Press release, June 8, 2006. Online. Available HTTP:<http://www.ecologynorth.ca/filemgmt_data/files/Carbon_Neutral_News_Release_FINAL.pdf> (Accessed 12 November 2007).
Ewins, P. (2003) 'Canada's latest national park', *WWF Arctic Bulletin,* (3): 5-6. Online. Available HTTP: <http://assets.panda.org/downloads/ab0303_o4l0.pdf> (Accessed 12 November 2007).
Glomsrød, S. (2006) 'The economy of the North: an introduction', in Glomsrød, S. and Aslaksen, J. (eds) *The Economy of the North,* Oslo: Statistics Norway.
Glomsrød, S. and Aslaksen, J. (eds) (2006) *The Economy of the North,* Oslo: Statistics Norway.
Goldman, M. I. (2003) *The Piratization of Russia. Russian Reform Goes Awry,* London: Routledge.
Helander-Renvall, E. (2005) *Biological Diversity in the Arctic. Composite Report on Status and Trends Regarding the Knowledge, Innovations and Practices of Indigenous and Local Communities,* Final Report: UNEP/CBD/WG8J/4/INF/3, September 2005. Online. Available HTTP: <http://www.cbd.int/doc/meetings/tk/wg8j-04/information/wg8j-04-inf-03-en.doc> (Accessed 12 November 2007).
Hild, C. and Stordahl, V. (2004) 'Human health and well-being', in *Arctic Human Development Report,* Akureyri: Stefansson Arctic Institute.
Hønneland, G. and Jørgensen, J. H. (2006) 'Administrativ reform in Russland', *Nordisk Østforum,* 20: 45–62.
Hønneland, G., Jørgensen, J. H. and Moe, A. (2007) 'Miljøpersepsjoner i Nordvest-Russland. Problemoppfatninger knyttet til petroleumsutbygging i Barentshavet', *Internasjonal Politikk,* 65: 7–22.
Hovden, E. and Lindseth, G. (2004) 'Discourses in Norwegian climate policy: national action or thinking globally?', *Political Studies,* 52: 63–81.
Hughes, L. (2003) *The No-Nonsense Guide to Indigenous Peoples,* London: Verso.
Intergovernmental Panel on Climate Change (IPCC; 2007) *Contribution of Working Group III to the Fourth Assessment Report of the Intergovernmental Panel on Climate Change. Summary for Policymakers.* Online. Available HTTP: <http://www.ipcc.ch/SPM040507.pdf> (Accessed 12 November 2007).

Kryukov, V. and Moe, A. (2006) 'Hydrocarbon resources and northern development', in Blakkisrud, H. and Hønneland, G. (eds) *Tackling Space. Federal Politics and the Russian North*, Lanham: University Press of America.

Kundzewicz, Z. W., Schnellnuber, H-J. and Svirejeva-Hopkins, A. (2004) 'From Kyoto via Moscow to nowhere?', *Climate Policy*, 4: 81–90.

Langhelle, O. (2000) 'Norway: reluctantly carrying the torch', in Lafferty, W. M. and Meadowcroft, J. (eds) *Implementing Sustainable Development. Strategies and Initiatives in High Consumption Societies*, Oxford: Oxford University Press.

Leask, L. and Hanna, V. (2005) *Kids Count Alaska*, Anchorage: Institute of Social and Economic Research.

Lindholt, L. (2006) 'Arctic natural resources in a global perspective', in Glomsrød, S. and Aslaksen, J. (eds) *The Economy of the North*, Oslo: Statistics Norway.

Norske Samers Riskforbund (2005) *Miljø foran olje og gass i Barentshavet*, Resolution No. 9, 2005. Online. Available HTTP: <http://www.nsr.no/website.aspx?objectid=1&displayid=10221> (Accessed 12 November 2007).

Poppel, B. (2006) 'Interdependency of subsistence and market economies in the Arctic', in Glomsrød, S. and Aslaksen, J. (eds) *The Economy of the North*, Oslo: Statistics Norway.

Poppel, B., Kruse, J., Duhaime, G. et al. (2007) *SLiCA Results*, Anchorage: Institute of Social and Economic Research, University of Alaska.

Price, M. and Bennett, J. (2002) *America's Gas Tank. The High Cost of Canada's Oil and Gas Export Strategy*, Natural Resources Defense Council and the Sierra Club.

ROSSTAT (2004) *Men and Women of Russia*, Moscow: Statistical Digest.

Ryggvik, H. and Engen, O. A. (2005) *Den Skjulte Dagsorden, Rammer for en Alternativ Oljepolitikk*, Stavanger: SAFE.

Sabonis-Helf, T. (2003) 'Catching air? Climate change policy in Russia, Ukraine and Kazakhstan', *Climate Policy*, 3: 159–170.

Smith, S. (2003a) 'The Arctic Council 12 years on: successful or out of touch?', Editorial, *WWF Arctic Bulletin*, (2). Online. Available HTTP: <http://assets.panda.org/downloads/ab0203.pdf> (Accessed 12 November 2007).

Smith, S. (2003b) 'Conservation First: achieving sustainable development in the Arctic', Editorial, *WWF Arctic Bulletin*, (3). Online. Available HTTP: <http://assets.panda.org/downloads/ab0303_o4l0.pdf> (Accessed 12 November 2007).

Statistisk Sentralbyrå (2006) *Samisk Statistikk*, Oslo: Statistisk sentralbyrå.

Stern, N. (2006) *The Economics of Climate Change*, Cambridge: Cambridge University Press. Online. Available HTTP: <http://www.hm-treasury.gov.uk/independent_reviews/stern_review_economics_climate_change/stern_review_report.cfm> (Accessed 12 November 2007).

Toner, G. (2000) 'Canada: from early frontrunner to plodding anchorman', in Lafferty, W. M. and Meadowcroft, J. (eds) *Implementing Sustainable Development. Strategies and Initiatives in High Consumption Societies*, Oxford: Oxford University Press.

UNFCCC (1997) *Kyoto Protocol. Targets*. Online. Available HTTP: <http://unfccc.int/kyoto_protocol/background/items/3145.php> (Accessed 12 November 2007).

UNFCCC (2004) *United States of America. Report on the In-Depth Review of the Third National Communication of the United States of America*, FCCC/IDR.3/USA, September 21, 2004. Online. Available HTTP: <http://unfccc.int/resource/docs/idr/usa03.pdf> (Accessed 12 November 2007).

UNHDR (2006) *Beyond Scarcity: Power, Poverty and the Global Water Crisis*, New York: United Nations Development Programme.

Vanderzwaag, D., Huebert, R. and Ferrara, S. (2003) 'The Arctic Environmental Protection Strategy, Arctic Council and multilateral environmental initiatives: tinkering while the Arctic marine environment totters', *Denver Journal of International Law and Policy*, 30 (2): 156–166.
World Commission on Environment and Development (WCED; 1987) *Our Common Future*, Oxford: Oxford University Press.

13 Managerial implications

Aslaug Mikkelsen, Ronald D. Camp II and Robert E. Anderson

Introduction

In the preceding chapters, we examined the context in which managers in the oil and gas industry must make decisions with respect to operations in the Arctic. This examination was conducted through the lens of corporate social responsibility with particular attention paid to increasing expectations around sustainable development – economic, social and environmental factors together, as the triple bottom line. In addition, we conducted this examination from the perspective of the major actors involved: communities, corporations, civil sector groups, government at all levels and supranational organizations (see Figure 13.1). Finally, this examination was done at various geographic scales: global, national, sub-national region (focusing on the Arctic within each of the subject states) and local/community.

The purpose of this chapter is to set out the major contextual forces that managers need to consider when deciding on strategy and actions with respect to oil and gas operations in the Arctic and then, more generally, make suggestions as to effective responses to these contextual forces. We do this in the four sections of this paper: (1) introduction; (2) corporations, social responsibility and sustainable development; (3) communities, oil and gas corporations and sustainable development in the Arctic and (4) managerial implications. Section 2 revisits corporate social responsibility (CSR) and proposes a matrix of possible approaches based on the corporation's view of the strategic importance of socially responsible behaviour and the nature of that behaviour. Section 3 addresses the different and often conflicting expectations of the different actor groups about appropriate corporate behaviour from the perspective of community, civil sector, state and environmental actors.

Then, in Section 4, we address the managerial implications on the oil and gas industry of the material in Sections 2 and 3, with a primary focus of the Arctic. Among other things, we explore possible approaches to CSR and sustainable development; for example: (i) whether to lead or lag behind relative to competitors and demands from other actors; (ii) whether to treat CSR as a constraint or potential source of competitive advantage and (iii) whether to treat the demands of actors as elements of a key strategic issue or as tactical issues differing by geographic scale and actors.

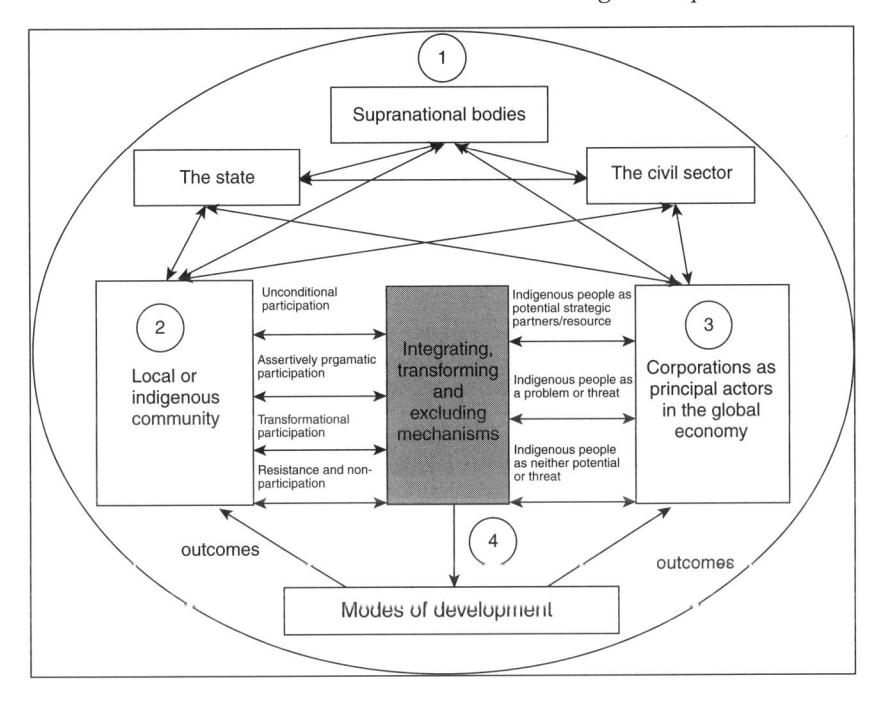

Figure 13.1 Major actors associated with the oil and gas industry in the Arctic.

We conclude that it is better to lead than lag and, by doing so, to treat CSR as a source of competitive advantage. To do this, CSR and sustainable development must be treated as a key strategic issue and not a tactical one. Although particular issues must be responded to at the national, regional and local levels, this must be done within a thoughtful corporate-wide strategic approach, not a reactive tactical location or incident specific one.

Corporations, social responsibility and sustainable development: corporate social responsibility

In the new global economy, it can be argued that directors and managers of transnational corporations (TNCs) have no particular national allegiance but rather 'are ideologically disposed to have each branch of the corporation behave like a "good corporate citizen" of the country in which it is domiciled' (Stander and Becker, 1990: 197). According to Stander and Becker, this behavioural disposition of TNCs, called 'the Doctrine of Domicile' by Sklar (1987), arises because:

> The TNC's interest in maximizing global profitability and avoiding sub optimization at a subsidiary level will be weighed against another, equally

important interest: the corporation must legitimize its host-country presence in order to obtain the stability it needs to plan the maximization of capital accumulation under its control.

(Stander and Becker, 1990: 198)

Tavis and Glade (1988) in their observations about the behaviour of TNCs in developing regions offer support for the Doctrine of Domicile. They suggest three possible approaches to interaction with developing regions available to TNCs – power-based, participative and combined strategies. The authors contend that, while power-based strategies dominated in the past, participative strategies are now necessary, if TNCs are to ensure their long-term survival. The authors attribute this shift to a number of factors including:

1   Changing expectations about what constitutes ethical and responsible TNC behaviour.
2   Increased TNC experience in developing regions and especially the partial naturalization of multinational management in overseas subsidiaries.
3   The growth of a business infrastructure in developing regions.
4   More experienced, technically competent, and self-confident organizations for negotiation and economic administration in developing regions.

This TNC behaviour (the Doctrine of Domicile) is a manifestation of the approach to CSR adopted by a firm's managers or owners. Various authors address the subject of the changing nature of CSR, as people perceive it at large and by those within the corporation. James Brummer (1991) and Ahmed Belkaoui (1984) provide particularly useful and comprehensive reviews of the subject.

James Brummer (1991) in his analysis of CSR suggests that there are four perspectives on the issue: classical, stakeholder, social demandingness and social activist. According to the classical perspective, economic performance is greatest when corporate executives respond only to the economic interests of the company's shareholders. It is argued that society's best interests are served when its specialized institutions (for-profit businesses, non-profit organizations, governments, etc.) focus exclusively on their particular functions. The social obligations of businesses are confined to satisfying legal and economic criteria. Socially responsible behaviour not mandated by law will not be pursued because it imposes unnecessary costs on the business thus reducing profitability. Proponents argue that, if a segment of society desires some goods or service, they should build a market for it or urge government to either supply it or require business to do so.

The stakeholder approach suggests that a business' responsibilities extend beyond shareholders to include all stakeholders; defined as those who: (i) are, or are likely to be, directly affected by the decisions and activities of the business; (ii) have an explicit contractual relationship with it and/or (iii) can directly affect the corporation. Corporate executives may respond to the interests of stakeholders out of corporate (shareholder) and/or individual self-interest; because such a response is demanded by law; in response to, or in anticipation of, public pressure; and/or from a genuine desire to benefit the lives of stakeholders.

According to the social demandingness approach, corporations are responsible to carry out those activities that society expects or demands of them. Managers should solicit and consider the opinions, demands and/or expectations of members of the public (not just stakeholders) regarding corporate activities and respond to the interests and needs expressed. Most theorists adopting this perspective contend that society's social and moral demands on businesses have been increasing over the past decades. It is no longer enough for firms to provide goods and services to the public for profit while satisfying the interests of stakeholders; they are expected to meet social and moral responsibilities with respect to society at large as well.

The social activist approach holds that a universal standard exists for determining responsible corporate conduct that is independent of the interests of stockholders and the claims of stakeholders. This standard demands concern for the ideal or rational interests of the public, rather than their expressed or present interests. Therefore, the standard of responsible corporate conduct is independent of current circumstances and even current interests of various groups and is absolute and universal – it has a scientific, ethical, religious and/or metaphysical basis.

In a similar vein, Belkaoui (1984), in his book *Socioeconomic Accounting*, suggests that society increasingly expects businesses to consider issues beyond profit maximization, and that these expectations have resulted in three new and related paradigms: (i) the new economic paradigm; (ii) the rectification paradigm and (iii) the ethical paradigm. The new economic paradigm seeks to rationalize two extreme views, libertarianism and radical economics, on a continuum of opinion concerning the pre-eminence of either the market or the state in solving complex economic and social problems. In this new paradigm, social and institutional economists hold that corporate behaviour can be guided by 'a clear statement of national and social goals based on a commitment to human welfare and social justice and experimentation with institutional arrangements to solve economic and social problems' (Belkaoui, 1984: 13).

The basis of the rectification paradigm is the premise that the corporation is an active market and social agent whose actions will either worsen or improve socioeconomic conditions. The continuum of rectification options available to the organization extends from complete indifference to an active role in rectifying social problems. Belkaoui suggests that indifference harms the long-term interests of the firm, while 'involvement in some forms of rectification is a way of ensuring mutual acceptance by society and a role in securing social order and affluence' (Belkaoui, 1984: 14).

The ethical paradigm also recognizes that a corporation's actions have the potential to positively or negatively impact society. The continuum of ethical options available to the organization extends a constructionist argument that a business

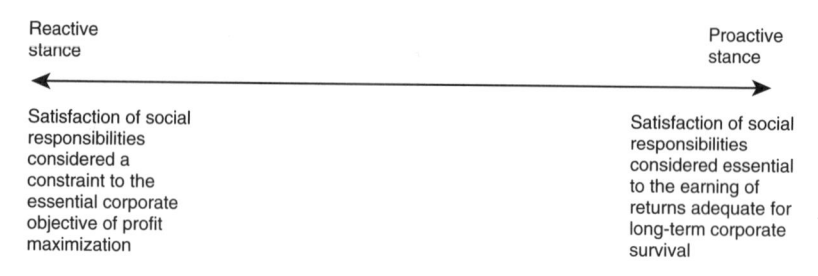

```
Reactive                                                    Proactive
stance                                                      stance
◄──────────────────────────────────────────────────────────►

Satisfaction of social                          Satisfaction of social
responsibilities                                responsibilities
considered a                                    considered essential
constraint to the                               to the earning of
essential corporate                             returns adequate for
objective of profit                             long-term corporate
maximization                                    survival
```

Figure 13.2 Corporate social responsibility and long-run survival.

serves society best when it concentrates on the objective of profit maximization, only undertaking those social activities that aid in the pursuit of profit maximization or are imposed by law. At the other end of the continuum are those who argue that an ethical corporation has an obligation to behave in a socially responsible manner, even if that behaviour interferes with profit maximization. Such a firm pursues a set of goals including social objectives and fair profits.

A synthesis of the three development paradigms described by Belkaoui and the approaches to corporate social responsibility described by Brummer, is possible. In that synthesis, corporate approaches to social responsibility and long-term survival range between two extreme stances, reactive and proactive (see Figure 13.2).

Belkaoui, Brummer and others (i.e. Armstrong, 1982; Davis, 1991; Goldman, 1995) suggest that society's expectations and the requirements of flexible or agile competition (the 'new economy') are combining to force businesses to shift their approach towards the right end of the continuum. Goldman says that:

> Management must support a proactive posture toward operations-relevant social values.... Agility [flexible competition] reflects a competitive environment that links producers and consumers far more closely than the mass production environment does. A hostile internal or external social environment precludes agility. A company cannot routinely reconfigure itself in response to changing market opportunities if it must continually fight battles ... over the environmental and community impact of its operations.
>
> (Goldman, 1995: 20)

Goldman goes on to say that it is, therefore, 'advantageous for companies to internalize social values, rather than having them constrain their operations from the outside' (Goldman, 1995: 30).

The preceding material can be combined into a 2 × 2 matrix of approaches a corporation can adopt with respect to corporate social responsibly (Figure 13.3). The horizontal continuum stretches from largely tactical, project-driven, reactive responses, which, taken together, shape a corporation's strategy at the left end, to a strategic corporate approach at the right end emerging from a belief that a corporation's approach to social responsibility is critical to its long-term success and which then shapes tactical strategies and decisions project by project. The cells

Managerial implications

	Tactical	Strategic
Triple-bottom line/social responsibility focus	(3) *Social demand:* Organizational focus on present/expressed interests of stockholders and stakeholders	(4) *Sustainable development:* Organizational focus on universal, ideal, rational social interests independent of stockholders and stakeholders
Bottom line/ profit maximization focus	(1) *Classical:* Focus on stockholder interests associated with specific project (i.e. efficiency, effectiveness & legal compliance)	(2) *Stakeholder:* Focus on stockholder and stakeholder project interests that may affect the corporation

Figure 13.3 Corporate approaches to social responsibility.

consist of the four approaches to CSR described by Brummer. So, for example, one could adopt a corporate-wide strategic approach based on the classical view of CSR (cell 1), or a corporation could react tactically on a case-to-case basis with a strong stakeholder orientation (cell 2) and so on. In general, we suggest that the global environment in which all multinational and transnational corporations operate requires a strategic (as opposed to tactical) corporate response at something beyond the level of the classic approach. This is not to say that the classic concerns are not critically important, they are. Rather, satisfying classic concerns is a necessary but not sufficient response to society's complicated and often conflicting and contradictory expectations about responsible corporate behaviour.

In the next section we turn our attention to oil and gas in the Arctic. In it we summarize the forces that need to be considered by oil and gas companies as they adopt their approach to CSR. Then in Section 4 we will return to the matrix in Figure 13.3 to discuss where we feel oil and gas companies should operate and why.

Communities, oil and gas corporations and sustainable development in the Arctic

In the first parts of this section, we consider the rights, power, expectations and needs of three groups – Indigenous communities, the civil sector and states – as they relate to oil and gas companies operating in the Arctic and elsewhere. We look at the civil sector and the state focusing on aspects of their rights, power, expectations and needs that may conflict with or be supplementary to those of Indigenous communities. Then, in the final part of this section, we draw together these arguments under the umbrella of sustainable development (see also the model for major actors associated with oil and gas development in the Arctic in Figure 13.1).

Communities, rights and development

According to Sayre Shatz (1987b), people in developing regions exhibit one of three approaches to economic development and interaction with multinational enterprises (MNEs): acceptance, rejection or pragmatic. He characterizes the approaches as follows: In the acceptance approach, local or Indigenous communities view relations with MNEs as economically rational, with benefits generally outweighing costs for the community (Shatz, 1987b). This view leads to a decision to optin and unconditionally participate in business interactions with MNEs (see Figure 13.1).

Under the rejection approach, local or Indigenous communities view MNE activities as essentially baleful. Harm is seen to outweigh benefit for the community (Shatz, 1987b). This view leads to a decision to opt out and resist participation in business interactions with MNEs.

> Between these two is the pragmatic approach which views MNE operations as mixed in effect; which believes that feasible government pressures can produce a worthwhile improvement in the benefit-cost mix; and which therefore advocates intelligent bargaining.
>
> (Shatz, 1987b: 108-109)

For the local or Indigenous community, this means a decision to opt in to interaction with an MNE on a case-by-case basis, where participation may be transformed to best meet the needs of the community.

Shatz and others contend that the pragmatic approach has become the strategy of choice for most developing people, including the First Nations in Canada.[1] The assertively pragmatic approach of the Tahltan people of British Columbia and the Yukon towards companies wishing to develop natural resources on their traditional lands is typical.

> We wish to make it very clear that the Tahltan People and the Talhtan Tribal Council are not inherently opposed to any specific type of business or resource development within our country. However, we do feel strongly that any development within our tribal territory must adhere to some basic principles.
>
> Before a resource development project can commence within Tahltan territory, it will be necessary for the developer and the Tahltan Tribal Council to enter into a project participation agreement that encompasses the following elements and basic principles:
>
> 1   assurance that the development will not pose a threat of irreparable environmental damage;

2   assurance that the development will not jeopardize, prejudice or otherwise compromise the outstanding Tahltan Aboriginal rights claim;
3   assurance that the project will provide more positive than negative social impacts on the Tahltan people;
4   provision for the widest possible opportunity for education and direct employment-related training for Tahltan people in connection with the project;
5   provision of the widest possible employment opportunities for the Tahltan people with respect to all phases of the development;
6   provision for substantial equity participation by Tahltans in the total project;
7   provision for the widest possible development of Tahltan business opportunities over which the developer may have control or influence;
8   provision for the developer to assist the Tahltans to accomplish the objectives stated above by providing financial and managerial assistance and advice where deemed necessary. (Notzke, 1994: 215)

Transnational bodies and rights of Indigenous communities

Transnational organizations are working to shape the interactions between Indigenous communities, governments, civil organizations and corporations. Events during the final decades of the twentieth century and the opening decade of the twenty-first resulted in Indigenous people emerging as a group of some consequence to all players in the global economy. We will briefly cover these events by focusing on three transnational statements pertaining to the rights of Indigenous peoples: (i) ILO 169 of 1989 of the International Labour Organization; (ii) the United Nations Declaration on the Rights of Indigenous People that went before the General Assembly in the fall of 2006 for a ratification vote; (ii) and the policy of the World Bank towards Indigenous people revised in 2005. These policy decisions address four areas of rights for Indigenous peoples: the right to control over their own land and resources; the right to economic benefit from traditional lands; the right to self-determination in development and the right to maintain and strengthen institutions, cultures and traditions.

The International Labour Organization Convention 169

In 1989, the International Labour Organization (ILO), revised its Convention 107 of 1957 regarding Indigenous peoples and created the new ILO Convention 169. In response to pressure from Indigenous people, Article 8 of the 1989 convention dropped the 'integration' or 'assimilation' philosophy and recognized the rights of Indigenous people 'to retain their own customs and institutions, where these are not incompatible with fundamental rights defined by the national legal system and with internationally recognized human rights'.

Two additional articles are relevant for companies wishing to do business with Indigenous communities. ILO 169 Article 14 addresses land rights and Article 15 addresses resource rights. In particular, Article 14 calls for the recognition of Indigenous peoples' ownership rights over the lands, which they have traditionally occupied and usage rights over lands 'to which they have traditionally had access for their subsistence and traditional activities'. It also calls upon governments to identify these traditional lands and effectively protect Indigenous peoples' rights of ownership and possession.

Article 15 addresses the rights of Indigenous people to control over the natural resources associated with their lands, including participation in the use, management and conservation of these resources. In the case where the State retains ownership of mineral or sub-surface resources, Indigenous peoples have the rights to consultation over the use of these resources, to participate in the benefits from these activities, and to be fairly compensated for any damages which they may sustain from such activities.

United Nations Declaration of the Rights of Indigenous People [2]

For more than two decades, efforts have been under way in the United Nations (UN) to develop international standards to address the widespread discrimination and marginalization that has forced Indigenous peoples worldwide into situations of extreme poverty and cultural destruction. On June 29, 2006, the Human Rights Council of the United Nations adopted a resolution on the Declaration on the Rights of Indigenous Peoples 'that offers both an inspiring affirmation of the rights of Indigenous peoples and an assurance that "the human rights . . . of all shall be respected"' (Amnesty International Canada, 2006).

In the preamble, the Declaration recognizes the importance of lands, territories and resources to Indigenous people, that they have the right to develop in their own way (PP5), that these rights are inherent to their existence as peoples (PP6), and that these rights are essential rebuilding Indigenous communities as Indigenous people wish to rebuild them (PP8). Furthermore, these rights include control by Indigenous peoples over developments affecting them and their lands, territories and resources in order to maintain and strengthen their institutions, cultures and traditions, and to promote development in accordance with their aspirations and needs as a people.

A series of statements about specific rights follow the preamble. Statements most relevant to land and resource rights and development on their own terms address the rights to maintain and develop their political, economic and social systems or institutions, their own means of subsistence and development, and the right to engage freely in traditional activities. These statements establish the notion of traditional ownership, occupation and use of land. They also establish a right to fair and equitable redress, restitution or compensation for the use of lands, territories and resources, which they have traditionally owned, occupied or used.

The World Bank policy on Indigenous people

In 2005, the World Bank instituted a new policy with respect to Indigenous people, BP4.10, replacing OD4.2 dated September 1991. While only pertaining to the activities of the Bank itself and the projects it funds, the new policy recognizes Indigenous rights, including those related to land and resources. One of the centrepieces of the new policy is the concept of 'free, prior and informed consultation'.

The Bank's policy states that borrowers are responsible to ensure that the consultation process with Indigenous peoples is free from external manipulation, interference or coercion, and that the Indigenous peoples concerned have 'prior access to information on the intent and scope of the proposed project in a culturally appropriate manner, form, and language'. This notion of being culturally appropriate extends further. Borrowers need to recognize existing Indigenous institutions, including councils of elders, headmen and tribal leaders. They should also start the consultation process early enough to allow 'for adequate lead time to fully understand and incorporate concerns and recommendations of Indigenous Peoples into the project design'.

In addition to free, prior and informed consultation, the Bank's policy requires a social assessment of positive and negative project effects. These effects may be unique to Indigenous communities in that their distinct circumstances and close ties to land and natural resources affect their relative vulnerability. Furthermore, these unique risks require diligence in planning the measures necessary to avoid, minimize, mitigate or compensate for adverse effects, and to ensure that the Indigenous peoples receive culturally appropriate benefits under the project.

Finally, the policy requires the development of an Indigenous Peoples Plan. This plan incorporates a framework for ensuring free, prior and informed consultation during project implementation. It also incorporates an action plan for ensuring that Indigenous peoples receive culturally appropriate social and economic benefits. Where necessary, this includes measures to increase the capacity of the borrowing agency to provide these benefits. It also includes measures to avoid, minimize, mitigate or compensate for potential adverse effects for the Indigenous communities involved.

Concluding comments on statements on Indigenous peoples' rights

The ILO Convention and the UN Declaration are not binding on states unless they ratify them (Canada, the United States and Russia have not done so) and the World Bank policy applies only to projects in which the organization is involved. However, according to the UN's International Working Group on Indigenous Affairs (IWGIA), the three statements put pressure on governments to live up to their objectives and 'serve to reinforce such universal principles as justice, democracy, respect for human rights, equality, non-discrimination, good governance and good faith' (IWGIA, 2006).

These expectations are not limited to governments; corporations are subject to them as well. This is especially true for multinational enterprises, including oil and gas companies, whose activities take place on remote, often traditional Indigenous, lands.

The four cases studied in this book (Chapters 7-10, see also Chapter 11) clearly show that the Indigenous people of the Arctic expect the principles of ILO 169, the UN Declaration and the World Bank policy to be respected and followed by governments, corporations and others. Their expectations are expressed in demands for long-term, sustainable economic and social benefits as an outcome of oil and gas developments, and the inter-related environmental concerns tied to place-specific matters, such as whale and caribou migration, and, more generally, their deep concerns about the impact of global climate change on their traditional lands and activities.

The civil sector

As could be seen in the previous section on transnational bodies, within Arctic communities, two groups of civil institutions are important for our discussion: those focusing on environment, and those focusing on human rights. Many civil sector organizations promote interests and values that conflict with Indigenous groups' rights as set out in ILO 169, the UN Declaration and the World Bank policy. Some groups, focused on conservation and preservation, are in open conflict with Indigenous groups about what constitutes acceptable activities on Indigenous lands. Other groups, concerned with human rights, find aspects of traditional (often male-dominated) governance models and an emphasis on collective over individual rights, objectionable. The discussion that follows will highlight both the convergence and divergence of interests between these groups of civil institutions and Indigenous communities in the Arctic.

Environmental issues

There is considerable common ground between the people of the communities in the Arctic and environmental groups in two general areas – climate change, and local regional impacts of oil and gas exploration, extraction and transportation. The people of the Arctic see first hand the impact of climate change on their traditional lands and activities and the need for economic development, as do environmental groups. The range of responses regarding possible solutions to these issues is similar for both groups. Some favour continued development of hydrocarbon resources as essential to economic sustainability, and are looking to methods of controlling and reducing greenhouse emissions; for example, the underground storage of CO_2 and more efficient combustion. Others see conservation and alternative fuels as the answer to energy problems and question the wisdom of ongoing reliance on hydrocarbons, and hence the wisdom of oil and gas activities in the Arctic. This is part of the larger challenge facing oil and gas corporations, and is not unique to the Arctic.

The local impacts of oil and gas activities have also drawn considerable attention. Clearly, the communities expect companies to respect Indigenous and other rights, including the right to pursue traditional activities on the land. This concern is common across all the cases, as is evident from the concern for whale migration among the people of the North Slope of Alaska, and concerns about caribou among the Indigenous people of the Mackenzie in Canada. Similar concerns are apparent in Norway and Russia.

However, while communities and environmental groups generally demand similar things from companies with respect to these rights, some environmental groups have agendas that extend beyond that of the people of the communities. This often finds voice in demands for conservation and protection of areas or species, even at the expense of local traditional practices and economic development, bringing the conservation/preservation-oriented environmental groups into conflict with the communities. This, in turn, places corporations in a challenging position. For example, taking a proactive approach to supporting the whaling activities of Indigenous groups would demonstrate respect for Indigenous rights to traditional practices. However, it would also bring the corporation into conflict with many environmental groups that want a complete ban on whaling. Similarly, there is clearly a disagreement between Indigenous people and conservationists about what should or should not be allowed in protected areas, in Alaska and elsewhere.

Human rights issues

Human rights organizations also impact the operations of oil and gas companies in the Arctic and elsewhere. In general, their concerns centre on equality rights. Many organizations seek redress for Indigenous people and support them in their drive to reacquire their traditional lands and resources and rebuild their communities. However, other organizations see concerns about gender, individual rights and democracy (one person, one vote, majority rules) as critical issues. These latter organizations operate at two levels, within Indigenous communities, and between Indigenous and non-Indigenous communities. At the within-communities level, many organizations express concerns about involvement, governance, accountability and equitable treatment. At the Indigenous/non-Indigenous level, some organizations argue that preferential treatment towards Indigenous peoples is appropriate.

However, conflict exists between Indigenous and non-Indigenous communities where there are significant numbers of the latter (e.g. Yellowknife, Canada) in the Arctic. This conflict is usually expressed as resentment towards preferential treatment for Indigenous people, and is an issue with which the four Arctic governments must wrestle. One partial solution has been to mandate special treatment beyond the Indigenous community, for example, hiring and procurement targets for people in the region of the project, rather than just Indigenous people. This has worked to some extent, though employees often object to diversity programmes that give preferential treatment to specific groups of employees. Objections to special rights for Indigenous people by others affected by these projects have found some support with organizations whose focus is on individual rights and equality.

Government issues affecting business activities in Indigenous communities

We think three conclusions emerge from our case studies with respect to government. First, governments are generally in accord with the notion of Indigenous rights and what constitutes appropriate social responsibility with respect to Indigenous communities in the Arctic, as expressed in ILO 169, the UN Declaration and the policies of the World Bank. This is not to say that all four states accept all that is in these three pronouncements. Indeed, Canada voted against taking the UN Declaration on Indigenous Rights to the General Assembly in its present form, but for reasons that do not disavow most of the principles in the Declaration. If a corporation was to espouse these general principles, it would find itself in general accord with policies of the four states we studied.

Second, the four states leave no doubt as to their position regarding the continued use of oil and gas. All see development of these resources in their Arctic regions as essential. They vary in their approach to the issue of greenhouse gases but none seem about to adopt a radical shift away from oil and gas as a source of energy. This settles the issue for oil and gas corporations. The states have declared that oil and gas development in the Arctic is socially responsible and critical for economic sustainability.

Third, the states have made a series of pronouncements on various issues that might lead to conflict between communities and civil groups that relieve corporations of some of the challenges of clearly conflicting expectations as to social responsibilities. Alaskan authorities accept limited whaling. Canadian authorities permit the trade in furs. Both require preferential hiring and procurement schemes and so on. In spite of objections from other stakeholders, firms must incorporate these things into their approach to CSR.

Contradictory demands on sustainable development in the Arctic

In the previous sections on transnational bodies and on the civil sector we have looked at the context in which oil and gas companies must operate in the Arctic and elsewhere. From these sections, we can see that the demands from communities, the civil sector and governments are myriad and often contain contradictory elements. If a corporation seeks to satisfy the expectations about Indigenous rights expressed in ILO 160, the UN Declaration and the policies of the World Bank, it may make other important stakeholders unhappy. For example, the principle related to Indigenous rights clearly supports ongoing whaling by the Indigenous people of the North Slope. Socially responsible corporations working in the North Slope should, therefore, support communities in this traditional endeavour, and minimize and mitigate negative impacts of their operations on whales and whaling. However, the anti-whaling movement would likely not be satisfied with actions by these corporations that assisted the whale hunt.

Similar issues arise in regard to protected areas and interaction with legitimate community leaders. While companies and communities likely could reach agreement about the careful and sustainable use of special areas, some civil sector groups want absolutely no activity in these areas. Likewise, corporations must work with the leaders that are there and yet that leadership may not be 'proper' by either Indigenous or civil sector standards. It will often fall to the state to specify legally recognized leaders.

Similar arguments can be made from the perspective of a corporation trying to meet the needs of other major stakeholder groups. None of the states involved subscribes completely to the principles expressed in ILO 160, the UN Declaration and the policies of the World Bank. One obvious example in Canada is the environmental assessment process. It treats Indigenous people as one group, equal among many, whose interests need to be considered when activities will occur on traditional lands. Furthermore, the Canadian government considers the oil and gas under traditional land to be the property of all Canadians with royalties from its use accruing to all Canadians. Yet Indigenous people feel that these are their lands and their resources; that they should decide what can be done, where and how; and that they should benefit from the resources of the lands. Grand Chief Matthew Coon Come of the Cree of Northern Quebec captured the essence of this issue, saying 'immense wealth – several billion dollars a year – is being taken [out] of land in Eeyou Astchee . . . and that we Crees get no share in the wealth, either in the form of royalties, business opportunities or work' (Grand Council of the Crees, 1998: 1).

In reality, the Indigenous people in all four jurisdictions do possess separate, special rights recognized by the states involved and by other bodies such as the UN and the World Bank. Corporations must address and accommodate these special rights, because, in spite of objections by some, these rights will not disappear. Another reality is the lack of consensus on these rights and their status relative to other values and concerns, which can make negotiations difficult and good agreements, acceptable to all, difficult to reach and sustain.

So what can a corporation do? Clearly the stakeholders are demanding socially responsible behaviour. And equally obviously these demands extend beyond those found in the classic view of CSR. We think these demands can best be viewed as a package captured by the concept of sustainable development. In the next section we briefly examine current practices, and offer suggestions we feel might be useful to oil and gas corporations trying to promote sustainable development in the Arctic.

Managerial implications

In this section, we go back to the concepts of corporate social responsibility presented in Section 2. In particular, we argue that the approach to date has tended to operate towards the left of the continuum presented in Figure 13.2, that is treating CSR issues as a constraint and a cost to be minimized in pursuit of profit and shareholder wealth maximization, and in the lower left corner of the matrix presented in Figure 13.3 and modified earlier as Figure 13.4 (cell 1).

Based on the analysis of the cases presented in this volume, we argue that, in order for CSR activities to promote sustainable development in local and

	Tactical/reactive	Strategic/proactive
Triple-bottom line/ social responsibility focus	(3) Social demand	(4) Sustainable development
Bottom line/profit maximization focus	(1) Classical	(2) Stakeholder

Figure 13.4 Corporate approaches to social responsibility.

Indigenous Arctic communities, corporations must operate under a more strategic, proactive orientation (see cell 4 in Figure 13.4). This orientation implies two things. First, the identification and satisfaction of social responsibilities must be considered a critical, strategic corporate objective essential to long-run corporate survival and profitability. Second, the corporation must be committed to developing an economy that is socially, environmentally and economically sustainable.

In the remainder of this section we offer some observations as to: (i) why we think the current approaches to CSR activities of oil and gas companies in the Arctic are closer to point 1 than 4 in Figure 13.4; (ii) why we think a company should operate at point 4 in Figure 13.4; (iii) what operating at this position means to a company and (iv) how we think a company might move to cell 4, an enhanced stakeholder approach, should it choose to.

Current approaches to corporate social responsibility

The strategy of the oil and gas companies is to extract oil and gas for refinement and sale. To pursue this strategy, they need access to oil and gas deposits, and permits for exploration and development. Historic, adverse impacts on the environment and communities have damaged the reputations of oil and gas companies. This, in turn, has made it essential for oil and gas companies to create goodwill with governments and communities controlling rights to these resources. In the current approach by oil and gas companies, social engagement and dialogue appear to be seen as prerequisites for gaining access to resources, and to obtaining and maintaining licences to operate.

This current approach has several shortcomings. Engagement with communities tends to be fragmented and ad hoc. Consequently, there is a lack of long-term and cumulative impact planning, and a lack of an integrated management framework. This in turn seems to lead to the practice of dealing with negative spill-overs through monetary compensation, even when it is difficult to put a monetary figure on cultural impacts. This holds true even though there appears to be a general consensus that it would be more effective to create an alternative strategy for sustainable communities. Often efforts to be more proactive are hindered by a lack of strategic focus, which allows a lack of monitoring, enforcement and follow-up on strategic impact assessments, and the practice of ticking boxes by external affairs officers and other managers with limited power within the companies. Keeping the stakeholders satisfied enough to maintain access to resources seems to be the overall aim of the processes, and the success factor that the activities are measured against.

Classical approach

In practice we have identified four different models for how a company evaluates or takes responsibility in the situation. From the first to the fourth, the movement is up and to the right in Figure 13.4. The first approach to CSR, classical, is simply to run the permitting processes and the operations as long as the company can, by dealing with the authorities and upper decision-makers of the system. The guiding strategic vision appears to be limited, with management thought more focused on the tactical issues of securing important natural resources for a given project. Likewise, the underlying idea of CSR seems to be that 'if it is legal, it is right'. This practice corresponds to the ideas of Friedman outlined in Chapter 4.

Stakeholder approach

The second, and more widespread approach, is the stakeholder approach, in which the company's behaviour and thinking is more strategic. This more proactive approach encourages the company's managers to give limited attention to the community representatives and other stakeholders. Recognizing the potentially negative effect on future projects, managers appear to be concerned with ameliorating the harm caused by oil and gas projects through handing out money or monetary support in order to avoid further damage to a company's reputation, and to secure access to resources and licenses. In this model, corporate representatives go straight from issue identification to compensation in order to minimize long-term resistance to the company's projects and strategic initiatives.

Social demand approach

In the third approach, social demand, the company deals directly with project concerns identified by the community and then physically mitigates those concerns to the extent they can. The 'community' is listened to and the company is flexible in providing a solution. While this approach provides a kind of symptomatic therapy, where the companies address legitimate needs identified by the stakeholders, there is little or no effort made to get beyond immediate project concerns. Again, corporate thinking appears to be focused on tactical issues associated with specific projects and securing specific natural resources. The underlying causes of problems are neither identified nor dealt with. In addition, no effort is made to identify further economic benefits associated with the community surrounding the project, either as a source of labour with unique knowledge or as a potential market.

Enhanced approach to CSR

To maximize their impact for both stockholders and stakeholders, CSR activities should be strategically focused beyond the corporate bottom line. As was explained in the previous sections, current approaches to CSR tend to be limited. While the social demand approach does focus on being socially responsible, and attending to social and environmental issues in addition to bottom-line fiscal responsibilities, it is not strategically focused and does not generally go beyond mitigating current problems. CSR activities based in this approach are, therefore, prone to be unfocused and subsequently diffuse in their impact. For external stakeholders, this means that programmes attempt to satisfy rather than optimize. For corporate decision-makers and stockholders, this means that corporate actions are constrained by the often conflicting demands of stakeholders. It also means that the company has a limited ability to leverage these activities across the corporation to promote long-term corporate success or sustainable development.

The direction of the preceding arguments suggests that attaining a strategic focus in CSR entails pursuing normative behaviours as they pertain to sustainability and relationships with communities (Indigenous and otherwise). Strategic CSR and sustainable development have a moral and ethical basis, and need to play a central role in guiding corporate strategy, tactics and operations. As a component of strategic vision and decision-making, these moral and ethical bases can potentially free the company from the vagaries of conflicting societal expectations to pursue its own long-term agenda. This increases the probability of an oil and gas company maximizing its own value from operations in the Arctic and elsewhere. Even if this strategic, sustainable development approach does not allow companies to eliminate stakeholder conflicts, it does help corporate decision-makers to manage or at least externalize these conflicts in a way that supports their efforts to meet the classic requirements regarding shareholders and other close stakeholders (employees, etc.) so as to ensure ongoing corporate survival.

Sustainable development approach

The fourth approach to CSR is collaborative and strategic, where the aim is to contribute to sustainable development on all levels. To solve social problems, the industry needs to participate in providing education, employment and business opportunities. At the same time, the community should take part in developing the strategy and see some value in what the industry is doing. In identifying and developing the solutions, industry and government need to respect local knowledge and accept an incremental development process. If the aim is to preserve and give local people the opportunity to develop on their own terms, disregarding their local and traditional knowledge may be as fatal as disregarding the existence of a global economy and modern technology. This means that oil and gas companies have to balance between the local and traditional knowledge and concerns, and using good science and technology to mitigate the effects of their operations on local communities. The key is an enhanced stakeholder approach to social responsibility with a sustainable development approach to CSR as the core business value.

Establishing a corporate sustainable development approach

Personal relevance is one of the defining characteristics of values as they help individuals distinguish between what is good and bad or benevolent and threatening, when scanning their environment (Rokeach, 1973; Rohan, 2000). What is personally relevant also goes for managers at the top and middle level in companies, where salaries, bonuses and fringe benefits are related to profitability, productivity and turnover, but not always to the resilience of local communities. However, the fear of damage to their reputation and the consequence for branding seems to get also hard-nailed managers to the negotiation table. Many managers come to realize that a proactive problem-solving strategy is better than reactive negotiation about compensations.

Local and Indigenous people in Arctic communities want to modernize yet preserve their traditions. To accomplish this, they need an explicit discussion of what they wish to preserve and what they are willing to change. To assist in this effort to attain sustainable development, oil and gas companies working in the Arctic need to help these communities balance concerns for economic development and progress against the carrying capacity of the environment and the developmental opportunities and resilience of the society. If people do not succeed in these balancing activities, their attitudes towards oil and gas exploration and development are more likely to be polarized, making social dialogue and problem-solving more difficult for the companies involved. Surfacing all the necessary information from the start of a field development in an area and envisioning desirable end states for all involved participants and affected people is the only way to anticipate and resolve this type of conflict in a way that promotes sustainable development. In order to accomplish this goal, companies pursing a sustainable development approach to social responsibility need to establish a high level of stakeholder engagement and social dialogue.

Stakeholder engagement and social dialogue

Stakeholder engagement involves: identifying and consulting stakeholders; building relationships that foster external and internal understanding and trust and improving design and management so that activities are integrated into the local social, economic and environmental context and accepted by communities. The analysis of the stakeholders and their concerns are used as input for strategies on creating positive effects and mitigating negative effects in the communities and societies where the companies are active. Since creating positive effects and avoiding negative effects requires movable targets, the tools for stakeholder engagement should be dynamic, containing a continuous correction of the direction. Stakeholder dialogue and participative tools like partnership are used for this purpose. Good communication to most people is not only a means but also an end. A key pitfall in all stakeholder engagement processes is that the meetings become the end value. From here it is easy to 'tick boxes' and to develop an excessively narrow scope and input on what is going on. Therefore, feedback mechanisms and assessments of issues and processes by outsiders are key success factors in stakeholder management.

A major cause of conflict between oil and gas companies and Arctic communities comes from non-strategic funding for community projects, which tend to be limited in scope. Internally, different managers have to fight for limited capital available for their CSR projects. Funding is usually provided for CSR projects, rather than programmes, which means that it is only provided for a specific time period. Given that these decisions are usually not tied to strategic organizational issues, the size of this funding is not always realistic and the chances of extra money are limited – at least not always available on short notice. Furthermore, as the budget period proceeds, funds may be redirected towards other, more strategic, emerging priorities. This in turn encourages public relations officers and company managers to focus on social dialogue aimed at 'educating' local people on 'how oil and gas companies are operated', rather than looking at mutually appropriate ways to attain strategic goals associated with CSR projects. While a two-way understanding for companies and communities about how each operates is a prerequisite for social dialogue and sustainable development, it is not sufficient. In the context of sustainable development, the social dialogue needs to also explore the expectations in the communities on what constitutes the CSR – or the legitimacy of CSR as discussed in Chapter 4.

To be proactive and strategic in their approach, oil and gas companies must use participatory tools, and stakeholder engagement and dialogue with the local communities. Currently, the more strategically minded efforts start consultation after securing a lease. Company representatives report that this consultation process consists of providing information to the community about what the plans are and after that maintaining ongoing communication with different stakeholders at different levels. One government spokesperson in the state of Alaska stated that:

> The bigger companies that have the resources, have whole teams of people that are regularly travelling to the slope, certainly providing, you know, information, listening to concerns. Even the medium sized companies do it.

However, the issue is more than just increasing the level of dialogue. In this stakeholder engagement, the industry has observed that, owing to the limited organizational and management capacity in these Arctic communities, the community representatives are wearing many hats: the same person acting as mayor, tribal leader, corporate employee and whaling captain or crew member. The villages on, for example, the North Slope are small, ranging from under 600 in six of the eight villages, from Atqasuk, being the smallest with only 250 people, to 4429 people in Barrow.[3] Based upon current Nuiqsut data, the village expects 10-12 annual public meetings with industry dealing with NPR-A issues, 6-8 BLM workshops for National Petroleum Reserve – Alaska (NPR-A)

NE, NW, SW and Subsistence Oversight Panel, and reviewing of over 200 oil industry-related pieces of incoming correspondence.[4] This does not include the meetings with other industries and government departments. Nuiqsut has had an enormous increase in paperwork. They receive approximately 70 e-mails a month related to the oil and gas industry, and requests to attend many meetings. The city administration has only one city administrator, one book-keeper/treasurer, one city clerk and one cultural coordinator for all the work tasks. In this and similar situations on other locations, the local authorities are overwhelmed by the number of requests for community meetings. One mayor says:

> There's no coordinated planning, so they get asked, you know, maybe five times a week for community meetings, and there's no coordination between the people who are asking for the meetings. So that's some source of frustration, but each company, each agency feels like they're meeting their required mandates to get community participation and involvement.

If sustainable development is the desired end state, an integrated management plan for how to get there should be developed for the regions in the Arctic and, if possible, also across regions and national borders in the Arctic. This management plan needs to include better designs for making the engagement process more efficient as well as more effective. One way to do this is to look more strategically at local employment as a way to develop local capacity for communication and decision-making around critical development issues associated with Arctic communities and oil and gas companies.

Although it might not be the companies' job to provide new visions for local communities, local communities might not be able to foresee the full scope of the impact the oil and gas activities. In the Mackenzie Valley situation, for instance, Justice Berger provided the Native peoples in Canada with a ten-year break to enable them to organize, and get the necessary competence to evaluate the situation and the solutions. Strategic corporate efforts to develop local human resources might be a way to accelerate this process in similar Arctic communities. Part of this process would be an effort to develop employee skills and mental frameworks that would facilitate coordinating between the major actors associated with oil and gas in the Arctic (see Figure 13.1) aimed at strategic coordination of planning and action in the Arctic.

In Alaska, for example, some people are working to change the situation for the local communities. On behalf of the North Slope Borough, the consultancy URS has developed a comprehensive plan for the North Slope and they hope to succeed in getting this as a steering document for the region. A goal of the work has been to encourage discussion and assignment of responsibility between the Borough, communities, state and federal governments and the industry in tackling oil and gas issues. The work aims to get the different authorities collaborating. By first developing it as a technical report, this would be less threatening to the different agencies involved. A technical report is advisory, but when and if it is adopted as part of a comprehensive plan, it has regulatory power. Commenting on their own work, URS underlines that one advantage a comprehensive plan offers is moving away from project-by-project planning, thus highlighting the cumulative effects of oil and gas activities. The question for corporate executives is what kind of efforts the oil and gas companies will make to fit into and be complementary with their activities in regard to this plan.

Issue management

Oil and gas companies have a diverse set of management tools they use in different phases of their operations. Stakeholder engagement, environmental and social impact assessment and issue management are used to detect matters of actual or potential controversy. If not acted upon or acted upon in the wrong manner, issues can develop into crises or in other ways have negative effects on business. In doing or expanding business, key areas of interest for issue management are typically job opportunities, education or skills training, business opportunities (with a local content), environmental, and social or health impacts. However, other issues, not as easily observed or identified, may be relevant as well.[5] When defining the social aspects of sustainability, Waage et al. (2005: 1149) indicate that access to both human and social resources (health, education, social networks), and also natural and financial resources, like education and jobs, are important.

In addition to capacity issues, oil and gas companies operating in Arctic communities also face issues unique to Indigenous communities, such as values and norms associated with social and environmental impacts. As shown in the case studies, the follow-up of social and environmental impact assessments is often lacking or of low quality. These impact assessments might also have built-in taboos or story lines that are not questioned. For example, in most of the impact assessments available to our research group, major impacts on subsistence activities are listed in Alaska, Canada and Russia. In most of the corresponding mitigation plans, efforts to maintain the subsistence activities are suggested. However, none of these plans contained a discussion of the realistic barriers to trying to maintain subsistence activities, such as negative impacts on the environment, changes to a Western lifestyle or negative impact on secondary school attendance associated with preparing children to take over responsibility for hunting. Consequently, there is a great risk of underestimating the local and regional social, economic and cultural impacts of the operations.

Structural barriers to corporate sustainable development

Discussions around the core of the subsistence activities may be limited by current international standards (i.e. ILO 169 concerning Indigenous and Tribal Peoples and the World Bank Operational Policy and Bank Procedure 4.10)[6] of how to deal with Indigenous people. These standards are formulated in a way that prevents or makes the companies and governments go around the long-term consequences of the activities. The outline of participatory principles are also formulated in a way that encourages 'tick boxes practice' – accepting the wishes

Managerial implications 371

of local people and perhaps not informing them that the consequences of accepting this may have long-term negative consequences to their core activities.

Government policies towards mitigation encoded in Environmental or Social Impact Assessments (EIA/SIA) indicating measures to avoid, minimize or remedy impacts, may also work against taking a more proactive, strategic orientation towards CSR. Mitigation consists of the actions taken to proactively or reactively solve the issues identified by stakeholder engagement, environmental or social impact assessment or other tools of issue management. The main problem identified with oil and gas exploration and development activities is insufficient effort to avoid negative impacts combined with an idea that it is possible to use monetary compensation to 'pay your way out of the problems' when they arise. At an early stage of area and project planning, avoidance or impact minimization can create sustainable development over time, whereas monetary compensation later in the process usually does not give this result.

According to a list published by the State of Alaska, typical mitigation actions included in lease sales are regulations on the use of explosives, off-road travel (the use of ice roads), specifications for infrastructure design so as to avoid impact to wildlife, requirements concerning respect for archaeological sites, sufficient training of all employees about environmental, social and cultural concerns, local hire 'to the extent that they are available and qualified', partnerships with the communities to recruit and hire locals, discussion of potential disruption to subsistence activities with the local communities, minimization of impact to wetlands, location of pipeline structures that facilitates containment and clean-up of spills and protection of pipelines from environmental hazards such as currents and storms.[7] While these actions are positive, their wording confounds mitigation, which tends to be tactical and reactive, with activities that should be proactive and strategic, such as community partnering and local human resource development. This, therefore, encourages stakeholders to view these activities in terms more consistent with either cell 1 of Figures 13.3 and 13.4, classical CSR, or cell 3, social demand, both of which are tactical and reactive, concerned with immediate rather than long-term issues.

The focus that results from these impact statements and subsequent mitigation activities works against sustainable development within the communities and in the regions. Most actions are not part of an integrated comprehensive management plan for the whole area. Creating sustainable development means carefully choosing mitigation acts that have a long-term effect. The scope of the chosen actions is often within the budget limits set for each project. Using sustainable development as the core value means that 'the business case for sustainable development' should not become a sufficient guiding principle for company behaviour. It is nice that CSR often pays, but sometimes short-time profitable behaviour has to be changed because it is in conflict with creating long-term viable communities and a sustainable environment.

Representatives of the industry in Alaska and Canada may argue that the inhabitants in some parts of the Arctic are fully compensated for loss of livelihood and other negative impacts. Improvements in living standards and technology

(Western standards) are used as evidence. Looking at our data, this argument and way of thinking might be questioned. Without a comprehensive plan for sustainable development for present and future generations, it is hard to conclude that the strategies used so far have been very successful. Financial compensation is necessary, but seldom sufficient as an adequate compensation measure.

Attitudes of local managers and supervisors in the industry may also interfere with taking a more strategic approach to CSR in Arctic communities. One industry manager says: 'Local residents want jobs that pay wages, but it's not necessarily good for them', and: 'The attention of the leaders (in the communities) is directed outwards instead of towards internal problems. The adaptive ability is limited'.

These attitudes are areas of conflict but are not necessarily dealt with in the dialogues between the Natives and the companies. Most often the managers are seeing the social issues separated from the business and the commercial side of the activities. One manager says: 'The community members want to be valued. Our company wants to extract values'. To the Native people, being valued seems to be linked to respect for their subsistence lifestyle and traditions, and the wish not to have to fight for this kind of human right. In the *Arctic Human Development Report* (Arctic Council, 2004) this is called 'fate control', cultural continuity and ties to land.

According to our informants, the most successful oil and gas companies in the Arctic have been those most willing to enter serious dialogue with the Indigenous peoples. This sounds easy, but many industry managers feel trapped in a situation where they are paying their way out instead of finding good and long-term solutions. One external affairs manager says:

> The ideal way to handle this would be (1) to avoid conflicts, (2) to reduce them if they occur and then (3) to compensate. The reverse order happens however more often. The most common conflict has to do with subsistence and caribou. Avoidance would mean to not put structures in a way that prevents the animals from moving where they used to move. Reduction could mean to make the pipelines we use taller, use different colour coating, combine pipelines so that there is only one obstacle, separate roads and pipelines, consult with guides etc. Mitigation means monetary compensation from cash funds to the Natives that have to travel further to hunt. This is a common tool on the North Slope.

Some of the informants in our Arctic project presented 'Good Neighbour Policy' as a mitigation tool. A Good Neighbour Policy can be a voluntary arrangement between a company and a stakeholder group to establish an insurance policy and a monetary fund. They commit to pay a large amount of money if a catastrophic event (spill) occurs that prevents people from, for example, doing their traditional hunting (like whaling).[8] The payment can cover extra gas money and compensation for more dangerous working conditions. For example, if the whalers have to go further out in the ocean to hunt the whales.

Going beyond mitigation, an emerging trend is industry and community representatives sitting down to find solutions in conflicts. The jointly created solutions often go beyond what the governments require because the industry sees it as good business investment to have 'goodwill' within the community. 'Good neighbour policy' may be seen as a sort of private arrangements between a company and a community with the government on the sideline. It is a business strategy for influencing and shaping the relationship between the community and the company, and may be seen as a result of weakened positions of the governments. Making such policies and agreements transparent and just is a challenge.

After decades of oil and gas activities in the Arctic and elsewhere, the industry seems to accept that the companies have to contribute to give a 'local content' of their activities. Often, however, representatives of the industry argue that possibilities for doing this are limited, and attribute the lack of possibilities to factors outside the industry. For example, in Canada, unions in the companies that get the orders hire using the seniority of tenure principle. Since the locals, especially in Indigenous communities, lack tenure and necessary skills, they cannot be hired. Similarly, in Alaska US anti-discrimination laws make it difficult to give priority to people from the North Slope or Alaska before people from the lower 48 states. Furthermore, lack of skills and drug abuse are problems in Arctic communities. At the same time, the government is seen as being responsible for providing education, treatment and infrastructure in these communities.

Chapter 4 discussed the CSR principle of public responsibility and corporate citizenship, and how hard it is to limit the scope of corporate responsibility when the states or the authorities have not developed a sufficient infrastructure to guarantee the civil rights of the population. The corporate responsibility is to decide on how to integrate and balance values and issues of value creation, social equity and environmental protection in the core economic decisions and actions. There is no clear recipe for how this should be done, except that the concept of 'public responsibility' (Preston and Post, 1975) underlines the importance of politics and the public process rather than the personal morality view of different stakeholder groups defining the corporate responsibility in the societal arena and justice in the stakeholder engagement process.

A stakeholder engagement process aims to achieve a fair distribution of benefits and burdens associated with oil and gas projects based on justice and efficiency, taking into account their respective risks and vulnerabilities. Part of this balance is taking advantage of the new pattern of opportunities that the oil and gas activities may create. One problem with this process is that the community might have limited capacity to see and utilize these advantages and that pre-existing power relations may determine on whose premises the advantages shall be taken. This influences the perceptions of justice associated with oil and gas projects in the Arctic.

Members of communities and organizations focus on fairness when forming their opinions about change, also expansion of oil and gas activities in the Arctic. Fairness can be considered both as a desirable end state and as a desirable state of conduct. Thibaut and Walker (1975) suggest that people prefer fair procedures because such procedures imply that participants have larger degrees of outcome control than in processes that are perceived as unjust (e.g. low levels of participation and little opportunity for a voice). Leventhal, Karuza and Fry (1980) suggest that individuals for whom procedural justice is an important value will focus their sense-making activities on issues, such as: (a) procedural consistency (i.e. the degree to which the same procedures are used across persons, groups and issues); (b) without self-interest; (c) on the basis of accurate information; (d) with opportunities to correct the decisions; (e) with the interest of all concerned parties represented and (f) following moral and ethical standards. Oil and gas companies pursing sustainable development goals for the Arctic therefore need to develop practices that are seen to be procedurally fair.

Transparency

Transparency and accountability are key tenets of CSR. Oil and gas companies need to communicate their credibility in this area. We focus on the lack of transparency that follows from main mitigation practices, the lack of integrated management plans and coordinated activities from the industry, and the use of gifts and entertainment in 'paying your way'. When the gifts consist of money, they are often called 'speed money' – a name that speaks either of the intention of the donor or the interpretation of the recipient. Bribes generally involve payments to someone to pervert the course of business by taking improper or illegal action. Facility payments are small payments usually to low-level public officials to enable or speed up a process that it is the official's job to arrange so business can be conducted effectively. Typically this involves issuing licences, permits or conflict avoidance agreements.

As we have seen in the earlier chapters, the lists of stakeholders in the different Arctic regions are long. The number of external affairs persons employed by the different companies also mirrors this situation. Companies often mitigate and compensate for impacts their activities have on the communities. However, when companies' operations are uncoordinated and when there is no public overview over the support the companies are giving to the local communities, transparency is reduced. Transparency is the number one principle of building an anti-corruption culture. Where there is transparency, concerns can be reported, recorded and discussed, and improvements suggested. As long as there is no overview of what is given and used by the companies, the split-and-rule culture is fuelled and short-term facilitation may be chosen instead of long-term business relations and development processes. This system also focuses on 'keeping them satisfied' – not a very respectful way of thinking in a balanced business relationship. Lack of overview of social investment and how the companies sponsor local activities means lack of transparency.

Corporate/community partnerships as CSR

Some CSR company programmes work better than others in the pursuit of sustainable development. This part of the chapter explores two issues with strong potential to increase the strategic nature of CSR, and move oil and gas companies in the Arctic towards cell 4 of the corporate approaches to CSR model (see Figure 13.4). The first issue concerns the approach taken by oil and gas companies regarding negotiation with local communities. The second concerns initiatives by the industry, government or non-governmental organizations (NGOs) to build local capacities through community-based intermediary organizations.

Approaches to negotiations with Arctic communities

In approaching communities to negotiate access to strategic resources, companies have two behavioural strategies available: power-based/distributive, and interest-based/integrative (Walton and McKersie, 1965). Power-based negotiations are consistent with a short-term, tactical focus on identifying and accessing potential oil and gas reservoirs in the shortest time possible. From this perspective, time is money and long negotiations lead to frustration for company employees, management and shareholders. However, renegotiation means that issues already settled are up for debate again, and may require more time. From a strategic perspective, negotiations are a legitimate activity to identify different groups' interests within, outside and in contact with the company (Buchanan and Badham, 1999).

Most negotiations involve mixed motives. In the context of oil and gas exploration and development in Arctic communities, these interests include: preserving the traditional subsistence activities and culture of Indigenous peoples taking part in the oil and gas activities; promoting a green image of oil and gas companies; contributing to sustainable development and at the same time choosing cheaper technological solutions that satisfy both environmental concerns and cost limits set by market mechanisms. The main challenge in the social dialogue is, therefore, simultaneously to handle the cooperative and the competitive dynamics (Rognes, 1995). An integrative approach appears to be most appropriate for sustainable development and, therefore, most appropriate for a proactive, strategic approach to CSR.

Assuming an interest-based strategy for negotiating with Arctic communities, oil and gas companies should focus primarily on the cooperative element in negotiations and stakeholder engagement where the aim is to create a win–win solution. To succeed in these negotiations, the underlying interests, stakes or concerns of each partner should be identified. Scoping, issue management and impact assessments involving information exchange, active listening and problem-solving, are useful tools in this. However, if the companies are not open to alternative approaches in their operations, the value of the tools will be limited.

Consistent with earlier research into negotiation (Pruitt, 1981; Pruitt and Carnevalle, 1993) and observations of negotiations between Arctic communities and oil and gas companies, a coercive style – perceived by the partners as coming to the stakeholder meetings with fixed solutions – seems to block further dialogue and problem-solving. This blockage may result from either the problem-solving activities embedded in this kind of negotiation or from the subjective experience of unfairness associated with coercive behaviours.

Alternatively, relationships based on mutual exchange of information and trust give a kind of satisfaction and in themselves provide motivation to continue the relationship. For example, in Alaska, a reduction in trust may have led to a reduction in attendance at public hearings. Power-based strategies lead to suspiciousness or worse. The use of ultimatums is the extreme of this strategy.[9] To promote an approach to negotiating with communication that is consistent with strategic CSR, appropriate institutions and arenas for negotiation need to be constructed. Many of the partnership agreements between oil and gas companies are more like signals of intention. Many of the meetings between representatives of the NGOs/local communities and the companies are not explicitly seen and defined as negotiations, and thus not planned. (In fact, the people attending the meetings may not even realize these are negotiations.) Redefining all kinds of contact between stakeholders and the companies might make the collaboration 'cleaner'. It is, therefore, important for companies pursuing strategic CSR to label these contacts explicitly as part of the negotiating process. This also means that the partnerships have to be expanded from being short-term contact with NGOs and authorities to include long-term planning and development actions.

Continuous engagement with all relevant stakeholders is a prerequisite for effective reviewing of impacts and options. The oil and gas companies have already taken partnership into use, but the practice seems fragmented, dominated by external affairs thinking and only marginally linked to the core business activities. Partnership is a tool for transforming stakeholder engagement and dialogue into joint actions, as 'everyone does what they do best'. These partnerships are instrumental in finding solutions to defined problems. They also provide learning opportunities for all five major actors associated with community development (see Figure 13.1). In addition, deliberation between the five major actors can provide cultural change, new perspectives, trust and respect that may enhance sustainable development on the community level.

Community-based intermediary organizations

Community-based intermediary organizations (CBIO) are community-led, apolitical, non-profit organizations seeded by industry, government or development organizations. CBIOs are designed to fulfil some or all of the following four roles: First, they provide communities impacted by industry operations with the institutional and technical competence to engage effectively with company operations. Second, they act as the voice of the community in relation to operational concerns and ways to mitigate negative impacts and monitor industry commitments. Third, CBIOs build an agreement in the community on developmental concerns and priorities, including opportunities to benefit from the industry's operation. Fourth, they participate in addressing broader community concerns and local development priorities through direct delivery of projects and services.

The main advantage of CBIOs lies in their long-term capacity building, strategic and independent nature. Leveraging this advantage, CBIOs are a mechanism by which oil and gas companies can proactively and systematically build long-term relations with local communities, respond to their operational concerns and contribute to local development priorities. For example, one of the ideas behind stakeholder engagement is to use existing organizational and technical capacity within communities to engage with existing institutions like NGOs through community liaison officers, advisory panels, foundations and company-managed social investment programmes. However, some communities have low- or non-existent technical and organizational capacity to address the potential impacts and benefits of oil and gas projects effectively. In these cases, companies focused on strategic CSR should consider supporting the establishment of a new engagement structure that links corporate planning and monitoring systems, strategies, goals and objectives with CBIO involvement.

Community-based intermediary organizations come in different forms and may serve a wide range of issues according to local conditions, priorities and existing capacities in a given community. Impact-oriented CBIOs (often referred to as Industry Relations Committees), for example, focus principally on enhancing community understanding and the ability to participate in decision-making processes concerning the environmental and social impacts of the industry, including development of mitigating strategies and monitoring of any commitment. One example of an impact-oriented CBIO is the subsistence Advisory Panel at the North Slope. The weakness of this type of CBIO is that the discussion can easily be very short-sighted and, instead of moving activities towards sustainable development, may focus instead on short-term wishes for goods, equipment or infrastructure in the communities, which are easier to identify and implement. Benefits-oriented CBIOs are sometimes referred to as Community Development Corporations. They focus principally on assisting communities in building their capacity to address local community development priorities, such as health, education, housing, employment and income generation and community safety.

Many communities lack the experience time, expertise or resources to participate and influence engagement on issues that affect them directly, or to take advantage of potential benefits related to the industry's activity. CBIOs are in this way a first step. However, their main focus is not business development, and there seems to be a growing consensus amongst external stakeholders to different kinds of industry that the private sector and market mechanisms should play a greater role in poverty reduction and environmental stewardship alongside more traditional aid and grant-oriented approaches to development. As a consequence of this, the Shell Foundation in 2002 launched a social entrepreneurship initiative with a new way of thinking by looking for long-term business opportunities linked to their core business and to needs in the communities where they operate.[10] The idea is to work on environmental and social impacts of their own activities at the same time as they are developing business, the third element of sustainable development. By doing this, it is possible to design investment partnerships to: (1) provide business development support to small and medium enterprises (SMEs) to increase their financial performance; (2) leverage local capital into investment funds for enterprise by offsetting technology and financial risks and (3) apply commercial credit appraisals to SMEs in order to ensure higher levels of financial viability.

Conclusions

To be effective at promoting sustainable development in Arctic communities, corporate social responsibility needs to be proactive/strategic and intended to contribute to the company's triple bottom line. The success of these programmes, in turn, is a function of the quality of the stakeholder engagement, the effectiveness of the social performance programme, the delivery and the progress on social investment and social impact management. Without a comprehensive plan that is communicated to the public, it is difficult to measure the effectiveness of the social performance of the companies and transparency is halted, increasing the risk of choosing short-term facilitation over long-term business relations and development processes such as education and skill development. Our case studies have shown that most stakeholders see education as the key for development for local communities. The lack of such education and the lack of development of skills needed in the oil and gas industry is the number one challenge of the future for oil and gas companies. To reach the target of sustainable development, the companies need to be proactive in developing a normative core of sustainable development, building environmental and social factors into their economic models, rather than doing this only when they are forced to by the possibility of damage to their reputation. Developing a normative core of sustainable development as business value means taking the lead and proactively using CSR as a competitive advantage through participatory management tools.

Notes

1. For example, see Anderson and Bone (1995a, b), Anderson (1995) and Notzke (1994).
2. A plain text version of the Declarations is available at: http://www.iwgia.org/sw1592.asp.
3. http://www.co.north-slope.ak.us/nsb/default.htm.
4. NPR-A Grant document.
5. How each of areas are linked to the normative core of sustainable development are discussed in Chapter 4.
6. http://www.unhchr.ch/htlm/menu3/b/62.htm and http://www.worldbank.org.
7. http://www.dog.dnr.state.ak.us/oil/products/publications/northslope/nsaw2004/NS_2004%20Mits.pdf and http://www.dog.dnr.state.ak.us/oil/products/publications/beaufortsea/bsaw2004/BS_2004Mits.pdf.
8. At the North Slope of Alaska there are negotiations between the Natives, their Alaskan Eskimo Whaling Commission (AEWC) and the companies about a 'Good Neighbor Policy'.
9. In Nigeria, local people in situations of poverty, resignation and hopelessness have used sabotage and other illegal means to acquire benefits from oil resources.
10. An example of this is Shell Foundation's social entrepreneurship that was launched in June 2000 (UK-registered charity with an endowment of US$250 million). This foundation is part of the Shell Group commitment to long-term sustainable development. It emerged from external stakeholder engagement combined with a review of social investment across the group carried out in the 1990s. Given the Shell Foundation's mission to be a catalyst for partnerships that deliver sustainable solutions to those social and environmental challenges in which the energy industry and multinational corporations have a particular role and responsibility, two core programmes were developed. The first was a Sustainable Energy Programme to tackle the social

and environmental issues associated with energy production and consumption both globally and locally (access to modern energy service; household energy and health; sustainable transport and biodiversity). The second was a Sustainable Communities Programme to address the problems and challenges faced by vulnerable communities in the process of globalization. This programme comprised two main goals. The first goal was to maximize market access and supply-chain opportunities by assisting local, community-linked enterprises to produce and supply goods, services and inputs for sale or purchase by or through larger enterprises nationally and internationally. The second goal was to develop local markets and supply chains by piloting innovative ways for small and medium enterprises, NGOs and others to work with established, larger-scale enterprises, including transnational companies, to stimulate local markets and develop products, supply chains and distribution systems. The Shell Foundation does not provide funding to operating units within the Shell Group to address local issues related to operations or to mitigate negative impacts. This type of investment falls under social impact management. Similarly, the Shell Foundation is not involved in philanthropy.

References

Amnesty International Canada and the Native Women's Association of Canada (2006) *Draft United Nations Declaration on the Rights of Indigenous Peoples: Canada's Delaying Tactics Not in the Interest of Human Rights*, June 23. Online. Available HTTP: http://www.iwgia.org/graphics/Synkron-Library/Documents/Noticeboard/News/International/PublicstatementAmnesty.htm (Accessed 19 November 2007).

Anderson, R. B. (1995) 'The business economy of the First Nations in Saskatchewan: a contingency perspective', *The Canadian Journal of Native Studies*, 15: 310–346.

Anderson, R. B. and Bone, R. M. (1995a) 'First Nations economic development: a contingency perspective', *The Canadian Geographer*, 39: 120–130.

Anderson, R. B. and Bone, R. M. (1995b) 'Aboriginal people and forestry in the Churchill River Basin: the case of the Meadow Lake Tribal Council', *The Proceedings of the Churchill Heritage River Conference*, Saskatoon: University of Saskatchewan.

Arctic Council (2004) *Arctic Human Development Report*, Akureyri: Stefansson Arctic Institute.

Armstrong, G. R. (1982) 'The firm and society', in Duerr, W. (ed.) *Forest Resource Management*, Corvallis, OR: OSU Book Stores, Inc.

Belkaoui, A. (1984) *Socioeconomic Accounting*, Westport, CT: Quorum Books.

Brummer, J. (1991) *Corporate Responsibility and Legitimacy*, New York: Greenwood Press.

Buchanan, D. and Badham, R. (1999) 'Politics and organizational change: the lived experience', *Human Relations*, 52: 609–629.

Davis, J. (1991) *Greening Business: Managing for Sustainable Development*, London: Basil Blackwell.

Goldman, S. (1995) *Agile Competition: The Emergence of a New Industrial Order*, Hamilton, Ontario: The Society of Management Accountants.

Grand Council of the Crees (1998) *Growing Crisis Over Wealth-sharing from Resources of Cree Lands Concerns Annual General Assembly*, www.gcc.ca.

International Working Group on Indigenous Affairs (2006) *Declaration on the Rights of Indigenous Peoples*. Online. Available HTTP: http://www.iwgia.org/sw248.asp (Accessed 19 November 2007).

Leventhal, G. S., Karuza, J. and Fry, W. R. (1980) 'Beyond fairness: a theory of allocation preferences', in Mikula, M. G. (ed.) *Justice and Social Interaction*, New York: Springer Verlag.

Notzke, C. (1994) *Aboriginal Peoples and Natural Resources in Canada*, North York: Captus Press.

Preston, L. E. and Post, J. E. (1975) *Private Management and Public Policy: The Principle of Public Responsibility*, Englewood, NJ: Prentice-Hall.

Pruitt, D. G. (1981) *Negotiation Behaviour*, New York: Academic Press.

Pruitt, D. G. and Carnevalle, P. J. (1993) *Negotiation in Social Conflict*, Buckingham: Open University Press.

Rohan, M. J. (2000) 'A rose by any name? The values construct', *Personality and Social Psychology Review*, 4: 255–277.

Rognes, J. K. (1995) 'Negotiating cooperative suppliers relationships: a planning framework', *The International Journal of Purchasing and Material Management*, 31: 19–26.

Rokeach, M. (1973) *The Nature of Human Values*, New York: The Free Press.

Shatz, S. (1987a) 'Socializing adaptation: a perspective on world capitalism', in Becker, D. G. and Sklar, R. (eds) *Postimperialism: International Capitalism and Development in the Late Twentieth Century*, Boulder, CO: Lynne Rienner Publishers, Inc., 161–177.

Sklar, R. L. (1987) 'A class analysis of multinational corporate expansion', in Becker, D. G. and Sklar, R. (eds) *Postimperialism: International Capitalism and Development in the Late Twentieth Century*, Boulder, CO: Lynne Rienner Publishers, Inc., 19–39.

Stander, H. and Becker, D. (1990) 'Postimperialism revisited: the Venezuelan wheat import controversy of 1986', *World Development*, 18: 197–213.

Tavis, L. and Glade, W. (1988) 'Implications for corporate strategies', in Tavis, L. (ed.) *Multinational Managers and Host Government Interactions*, Notre Dame, IN: University of Notre Dame Press.

Thibaut, J. and Walker, L. (1975) *Procedural Justice: A Psychological Analysis*, Hillsdale, NJ: Erlbaum.

Waage, S. A., Geiser, K., Irwin, F., et al. (2005) 'Fitting together the building blocks for sustainability: a revised model for integrating ecological, social, and financial factors into business decision-making', *Journal of Cleaner Production*, 13: 1145–1163.

Walton, R. E. and McKersie, R. B. (1965) *A Behavioral Theory of Labor Negotiations. An Analysis of a Social Interaction System*, New York: McGraw-Hill.

Index

(Figures in **bold**, tables in *italics*)

Aboriginal Pipeline Group (APG), 181–86, 194
Agenda 21, 15, 17, 22, 24, 64, 85
Alaska
   Alpine Bay, 2, 129, 140, 148, 151, 153, 155, 158, 165, 170
   Coastal Zone Management Act in, 145
   Constitution, 147
   economy of, 141
   government in, 117
   indigenous peoples' rights in, 129
      (*see also* indigenous peoples: legal rights of)
   leasing and permitting in, 144
   National Petroleum Reserve–Alaska (NPR-A), 140, 145, 155, 160–61, 368
   natural resource ownership in, 121
   North Slope (*see* North Slope of Alaska)
   North Slope Borough (NSB), 117, 144, 154–57, 163–65, 170
   Permanent Fund, 147, 157
   planning and zoning in, 145–46
   property rights in, 121, 143, 294, 321
      (*see also* Alaska Native Claims Settlement Act; indigenous peoples: legal rights of)
   revenues, 147–48, **148**
   Treaty of Cession (1867), 128
Alaska Coalition, 150–51, 165, 168
Alaska Eskimo Whaling Commission (AEWC), 142, 150, 154, 163–64, 378
   (*see also* whaling)
Alaska National Interest Lands Conservation Act (ANILCA), 129, 143, 163
Alaska Native Claims Settlement Act (ANCSA, 1971), 121, 123, 128–29, 133, 135, 141–44, 148–49, 157–58, 339

Alaska Permanent Fund, 147, 157
Alberta, 8, 175, 181, 188, 327
Alberta tar sands, 181
alcohol and drug abuse, 140, 152, 159, 188, 191, 322
Aleut International Association (AIA), 40
Alpine Bay, 2, 129, 140, 148, 151, 153, 155, 158, 165, 170
Alpine Satellite Development Plan, 153, 158
American Convention on Human Rights, 306
American Declaration on the Rights and Duties of Man, *292*, 306
Antarctica, 50–51, 54, 235
Arctic
   cooperative agreements, 94 (*see also specific agreement*)
   economy, 32, 319
   environmental characteristics, 47, 51, 209–12
   ice sheet, 45–46, 50–54, **52**
   income, 319
   oil reserves, 2, **3**, 227
   petroleum industry in, 88, **88**, **351**
   petroleum production in, **319**
   regulatory framework for, 89–95
   sustainable development in, 25–30
   warming of, 50–53, 194
      (*See also* Canada; Norway; Russia; United States)
Arctic Athabaskan Council (AAC), 30, 40
Arctic Climate Impact Assessment (ACIA), 32–33, 38, 331, 344
Arctic Council
   Arctic Oil & Gas Guidelines (Arctic Council), 79
   chairmanship, 1, 15, 344
   climate change and, 326

Arctic Council *(Continued)*
 Framework Document (Chapeau) for the Sustainable Development Program (Arctic Council), 28
 history, 1, 27
 Policy Document, 33–34
 petroleum exploration and production and, 341
 Sustainable Development Action Plan (SDAP), 28–29
 sustainable development and, 4, 15
 Working Group on Sustainable Development (WGSD) of, 15, 28
Arctic Environmental Protection Strategy (AEPS), 25–27, 325, 333
Arctic Flora and Fauna: Conservation and Status (CAFF), 36, 333
Arctic Gas consortium, 173, 177
Arctic Human Development Report (AHDR), 25, 29–30, 87, 318–22
Arctic Institute of North America, 282
Arctic Monitoring and Assessment Program (AMAP), 26, 30, 35–36
Arctic National Petroleum Reserve, 140, 145, 148, 166, 368
Arctic National Wildlife Refuge (ANWR), 140, 143, 155, 161, 165–68, 334, 342–43
Arctic Slope Regional Corporation (ASRC), 129, 144, 148, 150–51, 158,
Assessment of Oil and Gas Activities in the Arctic (AMAP), 38
Awas Tingni case, 306–7

Barents Sea
 environmental organizations and, 210
 fisheries in, 283
 foreign policies and, 228–30
 geography, 207, **209**
 ice cover, 279
 integrated management plan for, 208
 management of, 343
 petroleum activities in, 207–8, 215, 279
Beaufort Sea, 8, 153, 163, 174–76, 188
Bellona, 210, 213, 215, 221, 282
bioacoustics, 154
biodiversity, 1, 4, 22, 29, 35–36, 97, 210, 283, 326, 330, 333–35
bribes, 374
British Petroleum Ltd., 211, 247–49
Brundtland Commission, 6, 16, 18, 19, 58, 64–65, 75, 343

business
 community-based intermediary organizations and, 376–77
 corporate social responsibility and (*see* corporate social responsibility)
 developmental activities of, 74
 economic paradigm and, 353
 economic responsibilities of, 70–72, *76*
 ethical paradigm and, 353–54
 global economy and, **88**
 government and, 362
 institutional capacity building, 72, 74–75, *76*
 institutional (public) responsibilities of, 70, 72–76, *76*
 issue management by, 370
 negotiations with indigenous peoples by, 375–77
 philanthropy by, 164, 368
 rectification paradigm and, 353
 societal responsibilities of, 76, 352–53, 372
 support centers, 190
 sustainable development and, 74

Calgary, University of, 282
Canada
 Calder decision and, 123, 176–77
 CO_2 emissions by, *329*
 co-management boards in, 124
 federal government in, 113–15
 indigenous peoples' rights in, 123–25, *124*, 173, 187, *292*
 Kyoto Protocol and, 327
 local government in, 116
 National Energy Board (NEB), 184
 natural resource ownership in, 120, 184–87
 in Nunavik, 2
 in Nunavut (*see* Nunavut)
 property rights in, 120, *124*, 177–78, 184–87, 294, 339, 363
 provincial authority and rights in, 115
 territories of, 116 (*see also* Northwest Territories; Nunavut; Yukon)
 transfer payments in, 124
 Treaty 8 in, 123, 179
 Treaty 11 in, 123, 179
Canada Petroleum Resources Act (Canada, 1985), 117, 121
Canadian Arctic Resources Committee, 192–94
Canadian Association of Petroleum Producers (CAPP), 185

Canadian Wildlife Federation, 193
carbon dioxide
  atmospheric, 45, 47–50, **50**, **53**
  emissions, *329*, 344
  geological reservoir for, 48
  in soil, 48
  tax of Norway, 205
  vegetation content of, 48–49
Carbon Sequestration Leadership Forum (CSLF), 34
cash economy, 8, 142, 150, 157–60, 165
Circumpolar Biodiversity Monitoring Program (CBMP), 37. (*See also* biodiversity)
Circumpolar Protected Areas Network (CPAN), 333
climate change
  assessment, 29, 45
  corporate social responsibility and, 80
  definition, 22
  evidence for, 47, 51, 193, 215, 329 (*see also* ice sheet)
  fossil fuel consumption and, 33
  global warming and, 6, 46–48, 51–53, 151, 261, 331, 334
  governmental responses to, 326–28, 344
  human health and, 46
  indigenous peoples and, 193, 281, 321, 360
  land clearing and, 33
  modeling, 45, 52
  petroleum industry and, 32–34, 39, 93, 214, 326, 330–31, 344
  sea level, 45, 47, 50–54
  sustainable development and, 343
Climate and Environment Changes in the Arctic (CECA), 51
Community-based intermediary organizations (CBIOs), 376–77
Conflict Avoidance Agreements (CAA), 163–64. *See also* Good Neighbor Policies
ConocoPhillips Oil Company, 129, 155, 163–64, 181, 249
Conservation of Arctic Flora and Fauna (CAFF), 26, 36–37, 333, 335
Convention on Biological Diversity (CBD), 23, 332
Convention Concerning Indigenous and Tribal Peoples in Independent Countries. *See* International Labour Organization Convention 169 on Indigenous and Tribal Peoples

Convention on the Elimination of All Forms of Racial Discrimination (CERD), 299
corals, 45–46, 210
corporate social responsibility (CSR)
  business forces and, 68
  business responsibility in, 60–61, **60**, 70-77, 311–12, 352–53, **354**, **355**
  capitalism and, 68–69, 77
  classical approach to, 59, **60**, 70–71, 352, **355**, 363, **363**, 365
  climate change and, 80
  corporate/community partnerships and, 374–77
  corporate managers and, 67
  definition of, 57–58, 62
  development issues and, 62
  ethical and moral basis for, 366
  governance model for, 67, 69
  governmental regulation and, 65–66
  history of, 59, 66–67
  institutional capacity building and, 72, 74–75
  institutional transformation and, 5
  management and, 364–78
  social activist approach to, **60**, **60**, 62–63, 70, 352–53, **355**, **363**
  social demanding approach to, **60**, **60**, 352, **355**, **363**, 365
  stakeholder approach to, 59, 352–53, **355**, **363**, 365, 367–70, 373
  sustainable development and, 5, 64–70, 80, 351–55, **355**, 366–67
  transparency in, 374
  See also business

Deh Cho people, 175, 180–81, 183–86, 188–90, 194–95, 336
dependency theory, 89
discourse analysis
  data analysis in, 103–4
  definition of, 95
  discourse coalitions in, 97–98
  informants in, 101, 103
  interview guide in, 100–101, *102*
  limitations of, 104–5
  story lines in, 96–98
Doctrine of Domicile, 351–52
drug and alcohol abuse, 140, 152, 159, 188, 191, 322

earthquakes
  glacial, 51–54, 262
  technogenic, 262

education of indigenous people, 29, 55, 63, 140, 175, 325
Emergency Preparedness and Response (EPPR), 26, 30
employment, 31, 62, 142, 159, 175, 188–89, 216–18, 245, 277, 320, 322–24, 369, 373
energy consumption, 23, 33, 230–32, 244
ENI Norge, 211, 215, 218–19
environment
	Arctic, 47, 51, 209–212 (*see also* Canada; Norway; Russia; United States*)*
	Arctic communities and, 360–61
	climate change and (*see* climate change)
	culture and, 192
	disruption of, 37–38, 261 (*see also* oil spills; pipeline(s): leakage from)
	impact statements and, 371
	organizations for (*see* environmental organizations)
	pollutants in, 26–27, 35–36, 47, 99
	protection of, 191–92, *334*
	threats to, 192
	See also climate change
environmental organizations, 150
	indigenous peoples and, 167, 194, 361
	perspective of, 168–69, 210
	Sierra Club, 93, 194, 330
European Union, 2, 5, 58, 118, 232, 235–36, 341–342
extinctions, 46, 153. *See also* climate change
Exxon Valdez accident, 5, 34, 170, 211. *See also* oil spills; pipeline(s): leakage from

facility payments, 374
Finnmark Act, 113, 126, 128, 223, 234, 299–300, 337
fisheries
	Arctic, 31, 210, 216, 222, 277, 280–84, 216, 222
	environment and, 51, 208
	petroleum industry and, 282–84, 340
	rights involving, 297
Framework Document (Chapeau) for the Sustainable Development Program (Arctic Council), 28

Gazflot, 277
General Agreement on Tariffs and Trade (GATT), 89
glaciers
	earthquakes from, 51–52, 54
	melting of, 45–46, 52–53, 55

global economy, **88**, 90–95, 320, 331, **351**, 351, 357, 366
global warming, 6, 46–48, 51–53, 151, 261, 331, 334. *See also* climate change
Good Neighbor Policies (GNP), 163, 372. *See also* Conflict Avoidance Agreements
greenhouse gases (GHG)
	composition of, 45, 47, 345
	effects of, 45, 48, 51
	emissions, 33–34, 51, 261, *329*
	offshore oil and gas and, 214
	See also carbon dioxide; climate change; Kyoto Protocol; methane
Greenland, 2, **3**, 12, 46–47, 51–54, 293
Gulf Canada Resources Ltd., 181
Gwich'in Council International (GCI), 40
Gwich'in people, 8, 40, *124*, 167, 175, 179–81, 184, 186–87, 193–94
GX Technology Corporation, 163

human rights
	ILO 169 and (*see* International Labour Organization Convention 169 on Indigenous and Tribal Peoples)
	indigenous rights and, 292, 294, 308
	instruments, *292*, 306
	law, 291, 295–305, 337
	organizations for, 361
	terra nullius doctrine and, 295
hunting, 132, 141, 150, 153, 158, 160, 169, 223, 280, 297, 321, 335, 343. *See also* mammals; whaling

ice sheet, 45–46, 50–54, **52**. *See also* climate change
ILO 169. *See* International Labour Organization Convention 169 on Indigenous and Tribal Peoples
Imperial Oil Company, 175, 181, 184, 192
Indian and Northern Affairs Canada (INAC), 185
indigenous peoples
	Canadian territorial government and, 117
	cash economy of, 8, 142, 150, 157–60, 165
	climate change and (*see* climate change)
	colonial settlement and, 309–10, 338
	consultation with, 301–302, 308–309, 368
	corporate social responsibility and (*see* corporate social responsibility)
	cultural rights of, 302–303
	demographics of, 291, 336

indigenous peoples *(Continued)*
  drug and alcohol abuse in, 140, 152, 159, 188, 191, 322
  economic state of, 272–73, 320
  education of, 29, 55, 63, 140, 175, 325
  employment, 31, 62, 142, 159, 175, 188–89, 216–18, 245, 277, 320, 322–24, 369, 373
  environmental organizations and, 167, 334–35, 361
  fishing by, 277, 280–81 *(see also* fisheries)
  foods of, 193 *(see also* fisheries; hunting; whaling)
  health of, 152, 258, 324–25
  hunting by *(see* hunting; whaling)
  legal rights of, 112, 123–132, 176–78, 184–85, 269, 276, 306–308, 310–11, 321, 337, 343, 356, 361–63 *(see also* International Labour Organization Convention 169 on Indigenous and Tribal Peoples; *specific group of indigenous people)*
  mixed economy of, 157
  modernization of, 89, 150, 155–6, 158, 177, 337, 367
  natural resource management by, 335
  negotiations with, 375–76
  Permanent Participants status of, 12, 27, 40
  power and, 164–65, 187, 341–45
  prior informed consultation and, 308–309
  reindeer herding by *(see* reindeer herding)
  relationships between, 194
  revenue distribution and, 320–21
  resource claims of, 112, 225–26
  self-determination for, 160, 173, 226, 275–76, 304–305
  social aid for, 157
  socioeconomic development of, 187, 258, 322
  subsistence activities of, 47, 157, 275–76 *(see also* fisheries; hunting; reindeer herding; whaling)
  sustainable development and, 24, 155–56, 369 *(see also* sustainable development)
  time concept and, 309
  whaling by, 144, 150, 153–54, 168, 361–62 *(see also* Alaska Eskimo Whaling Commission)
  See also specific group of indigenous peoples

institutional capacity building, 72, 74–75, 76
integrated management plans, 11, 99, 152, 208, 212–13, 220–24, 342, 369
Intergovernmental Panel on Climate Change (IPCC), 38, 45, 330, 344–45,
International Covenant on Civil and Political Rights; ICCPR), 291, 292, 296, 302
International Energy Agency (IEA), 2, 230–31, 235
International Labour Organization Convention 169 on Indigenous and Tribal Peoples (ILO 169), 126, 128, 223, 292, 292, 294–302, 305, 310, 320, 357–58, 360, 362, 370
International Monetary Fund (IMF), 89
International Partnership for Hydrogen Economy (IPHE), 34
International Union for Conservation of Nature and Natural Resources (IUCN), 18
International Whaling Commission (IWC), 153. See also whaling
Inuit Circumpolar Council (ICC), 26, 40
Inupiat Commission of the Arctic Slope, 142
Inupiat people, 118, 139, 141–42, 150, 153–54, 157–58, 165, 169–70
Inuvialuit people, 124, 178–79, 184–85, 187–88
Inuvialuit Regional Corporation (IRC), 184–85, 189–90, 194

JSC Gazprom, 229, 245, 247–50, 264–65, 267, 273–74, 280, 341
JSC LUKOIL, 247, 249–50, 254, 276
JSC Rosneft, 247, 250, 257, 284
JSC Surgutneftegas, 250
JSC Transneft, 247, 256

Khantyt-Mansii Autonomous Okrugs, 2, 75. See also Russia
KTK Pipeline, 247. See also pipeline(s)
Kyoto Protocol, 47, 205, 214, 326–28, 327, 329–31, 344–45

land claims. *See* indigenous peoples: legal rights of
liquefied natural gas (LNG), 3, 208, 215–17, 341
local governments, 31–32, 118, 120, 339
Lofoten, 2, 207, 210, 214–16, 218, 230–31, 234–36

Mackenzie Valley Gas Pipeline, 124, 175, **177**, 181–194 (see also pipeline(s))
malaria, 46
mammals
  distribution of, 37–38
  marine, 37, 139, 153, 210, 335 (see also whaling)
  petroleum industry and, 153
  See also hunting; reindeer herding; whaling
Maya case, 306, 314
Melkøya, 2
methane, 45, 48–50, **50**, 215. See also greenhouse gases
Métis people, *124*, 179–80, 184, 188,
mining, 31, 37, 128, 191, 198, 267–69, 281
Mobil Oil Canada, 181
modernization theory, 89. See also indigenous peoples: modernization of
Murmansk region, 263
  demographics of, 278
  ecological assessment of, *278*
  pollution in, 262, 278–79
  resource division in, 282
  socio-economic development in, 263–65, 279–80

NadymGazprom, 267, 277
National Association of Norwegian Sámi (NSR), 224, 226, 298, 331
National Petroleum Reserve–Alaska (NPR-A), 140, 145, 155, 160–61, 368
Natural Resources Defense Council, 330
Niger Delta, 62, 69, 75
non-governmental organizations (NGOs), 5, 165, 330
North Slope of Alaska
  employment in, 142 (see also employment)
  environmental sustainability and, 151–55
  history of, 140
  map of, **141**
  offshore petroleum activities, 139
  Prudhoe Bay (see Prudhoe Bay)
  social problems in, 140
  stakeholder discourses in, 150
North Slope Borough (NSB) (Alaska), 117, 144, 154–57, 163–65, 170
Northwest Territories, 116
  demographics, 175–76, 312
  geography, 174
  governance of, 116
  mining in, 31 (see also mining)
  oil and gas activities in, 8, 174–75

Norway
  Barents Sea and (see Barents Sea)
  climate change and, 328 (see also climate change)
  CO_2 emissions by, *329*
  demographics, 175–76, 206
  economy in, 217, 322
  Ekofisk field in, 201, 236
  employment in, 216–18
  environmental movement in, 201–202, 210, 220–21
  federal government in, 112, 118, 203–204 (see also Norway: Storting)
  Finnmark, 118–19, 125–267, 216, 219–220, 233
  Finnmark Act, 113,126, 128, 223, 234, 299–300, 337
  fisheries in, 216, 222
  geography, 205, **206**
  Goliat oil field in, 9, 208, 211, 213, 215, 218
  Government Pension Fund, 205
  Hammerfest, **206**, 207, 215–17, 341
  Human Rights Act (1999), 118, 293
  indigenous peoples' rights in, 125–27, 292, *292*
  industrial structure in, 219
  integrated management plan of, 208
  Kyoto Protocol and, 214, 327
  land ownership in, 120, 340 (see also Finnmark Act)
  licensing in, 203
  Lofoten (see Lofoten)
  Melkøya, 2
  Mining Act, 128
  Ministry of Petroleum and Energy, 118–19, 203–4
  natural resource ownership in, 120
  Norwegian Petroleum Act of, 121, 202
  Norwegian Petroleum Directorate (NPD), 204, 235
  petroleum-free zones in, 212–14, 334–332
  Petroleum Tax Act (1975) of, 205
  policy making in, 227
  politics of, 202, 219, 342
  regional and local government in, 118
  resource management by, 202
  Russian relations with, 232
  Sameting of (see Sámi people: governance of)
  Shtokman field and, 215, 229, 341, 248, 251, 264, 280

Norway *(Continued)*
  Snøhvit gas field, 9, 207, 216–18, 341
  Soria Moria Declaration, 213, 215–217, 233
  State's Direct Financial Interest (SDFI), 204–5
  Storting, 118–19, 126, 201, 203–4, 214, 227
Norwegian Continental Shelf (NCS), 201–2, 204, **206**, 207, 235–36
Norwegian Pollution Control Authority (SFT), 210, 212, 225, 229
Nuiqsut people, 140, 152, 158–60, 164, 368–69
Nunavik, 2
Nunavut, 2, 8, 116–17, *124*, 312, 325, 335
Nuuk Declaration, 26–27

Offshore Oil and Gas Guidelines (PAME), 34. *See also* petroleum industry: offshore
oil and gas. *See* petroleum industry
oil spills, 5, 34–35, 154, 168, 170–71, 210–11, 283, 291, 343. *See also* pipeline(s): leakage from
Organisation for Economic Co-operation and Development (OECD), 15, 73, 230
Organization of Petroleum Exporting Countries (OPEC), 2
Ottawa Declaration, 1–2, 27
Our Common Future (WCED), 4, 15–16, 21–22, 24–25

Pan-Canadian Innovations Initiative for the Northern Women in Mining and Oil and Gas Project, 191
permafrost
  as carbon reservoir, 48–50, 215
  damage to, 33, 193
  definition of, 49, 193
petroleum industry
  Arctic share of, 319
  climate change and, 32–34, 39, 93, 214, 326, 330–31, 344 (*see also* climate change)
  cumulative effects of, 152
  employment by, 216–218, 323–4 (*see also* employment)
  fisheries and, 282–84, 340
  negative influences of, 33, 37–38
  offshore, 8, 33–35, 80, 139, 147, 150, 154, 164, 169, 207, 214, 229, 242–44, **242**, *247*, 256, 26–69, 340,
  permafrost (*see* permafrost)
  pollution from, 35–36, 47
  positive influences of, 33
  production, 11, 35, 38, 134, 140, 147, **148**, 166, 170, 175, 181, 201, 203–4, 208, 221, 235, 243, 244–48, 251, 256, *259*, **262, 267**, 278, 318–19, **319**, 330–31, 341 (*see also* production sharing contracts)
  regulatory framework for, 89–95
  reserves, 2, **3**, 5, 12, 206, 243, 265, 319
  seismic testing for, 37, 154, 163–64, 203, 262
  sustainable development and, 30–37, 332–34, 345
  transport and, 10, 33, 35–36, 47, **49**, 154, 170, 256–57, 280 (*see also* pipeline(s))
  wildlife refuges and, 149 (*see also* Arctic National Wildlife Refuge)
Phillips Petroleum Company Norway, 201
pipeline(s)
  costs, 148
  environmental effects of, 153, 183, 192–93, 330
  leakage from, 170–71, 193 (*see also* oil spills)
  Mackenzie Valley Gas Pipeline and, 124, 175, 177, 181–94
  property rights and, 141, 144, 178, 180–81
  Russian, 247–48, 266, 280
  socioeconomic development and, 187–91
  Trans Alaska 145, 166
Pipelines Operations Training Committee (POTC), 188
pollutants, 26–27, 35–36, 47, 99
production sharing contracts, 251, 253
Protection of the Arctic Marine Environment (PAME), 26. *See also* *Offshore Oil and Gas Guidelines*
Prudhoe Bay
  history, 8, 129, 140–41
  large-scale development in, 2
  oil production of, 166, 170
  pipeline leak and, 170

reindeer herding, 37, 125–27, 135, 223, 270, 273, 275–77, 280–81, 324, 336
regime of accumulation, 91–93, 95, 174, 195

regulation
  corporate social responsibility and, 65–68, *76*,
  environmental protection and, 146, 193, 229
  implementation, 310–11
  indigenous peoples and, 224
  of petroleum industry, 231, 284
  social, 92–95, 173
  theory, 88–91
  See also specific laws
resource claims. *See* indigenous peoples: resource claims of
Rio Accords
  Agenda 21, 17, 22, 24, 64, 85
  Convention on Biodiversity, 17
  Framework Convention on Climate Change, 17, 23, 38, 326
  Rio Declaration on Environment and Development, 17, 27
  Statement of Forest Principles, 17
Rio Earth Summit, 15
Russia
  Arkhangelsk region of, 100, 243, 263–64, *278*
  Barents Sea oil and gas exploration by, 228–30 (*see also* Barents Sea: petroleum activities in)
  centralism in, 246
  Code of Indigenous Administration (1822), 130
  constitution, 120, 293, 310
  continental shelf, 243, 248–49
  corporatism, 246, 284
  economic transition of, 241–42, 257–58, 263, 268, 270, 342
  employment in Northern regions of, 245, 263 (*see also* employment)
  environmental problems in, 261
  federal government in, 113–16
  fisheries in (*see* fisheries: Arctic)
  greenhouse gas emissions by, 261, *329*
  indigenous peoples of, 270–72, *292*, *293* (*see also* indigenous peoples)
  judicial system of, 246
  Khantyt-Mansii Autonomous Okrugs, 2, 75 (*see also* Russia)
  Kyoto Protocol and, *327*, 328
  land ownership in, 120–21, 339
  Law on the Protection of the Environment, 122
  Law on the Subsoil (1992), 116, 121, 250–52
  legislation in, 249, 250–52, 341
  Ministry of Natural Resources (MNR), 272
  motivations of, 322
  Murmansk region of (*see* Murmansk region*)*,
  natural resource ownership in, 120–21
  obshchinas in, 121, 131–32
  oil spills in, 283 (*see also* oil spills)
  paternalism in, 246, 284
  petroleum companies in (*see specific companies*)
  petroleum export by, 256
  petroleum history in, 243
  petroleum production by, 229, **244,** 245, *259*, *260*, **267**
  petroleum reserves in, 241–42, **242**, *243*, 245, 249, **250**
  petroleum revenues in, *247*, **267**
  pipelines in, 247–48 (*see also* pipeline(s))
  power engineering in, 265
  principle of joint jurisdiction in, 252–53
  production sharing contracts (PSCs) in, 251, 253
  resettlement from the North in, 268
  Shtokman field and, 215, 229, 245, 248, 251, 257–58, 263–64, 280, 341
  socioeconomic problems in, 258
  as stakeholder, 247
  sustainable development in, 257–60, 284, 318
  Territories of Traditional Natural Use (TTNU) in, 131, 269, 271–72
  Underground Resources Law, 254–56
  Yamal-Nenets (*see* Yamal-Nenets)
Russian Agency of Sub-soil Resources (Rosnedra), 249
Russian Association of Indigenous Peoples of the North (RAIPON), 40, 270–72, 278, 310, 338
Russian Joint Stock Company, 264

Sahtu Dene people, *124*, 174, 179–80, 184–85, 190, 193
Sakha Republic, 31, 114, 132, 242, 248, 250, *259*, *260*
Sakhalin Association of Indigenous Peoples, 272
Sámi Council, 40, 223–24, 301, 305
Sámi people
  assimilation of, 337
  culture of, 126–28, 223–24
  demographics of, 118, 126, 223, 324

Sámi people *(Continued)*
  industrial development and, 280–82
  governance of, 113, 119, 126–27, 224, 226, 301 (*see also* Sámi Council)
  national association of, 224
  property rights of, 123, 125–27, 224, 226, 297–300
  reindeer herding by, 125–26, 128, 281 (*see also* reindeer herding)
  resource claims of, 225–26, 297
  Russian, 280–81
  Sameting (*see* Sámi people: governance of)
  self-determination of, 226–27
sea level, 45, 47, 50–54. *See also* climate change
seismic monitoring and testing, 37, 154, 163–64, 203, 262
Shell Canada Ltd, 163, 181
Shell Oil Company, 12, 150, 163–64, 248, 273, 330, 377–79
Shell Salym Development (SSD), 273
Siberia, 2–4, 31, 34, 48–49, 131–32, 242–43, 250, 265, 269, 277
Siberia–Energy (SE) Ltd., 273
Sierra Club, 93, 194, 330. *See also* environmental organizations
stakeholder theory, 59–60. *See also* corporate social responsibility
Statoil (now StatoilHydro), 62, 75, 78–79, 201, 204–5, 229, 245, 250, 264, 341
Status of Women's Council (SWC), 190–91
Storting. *See* Norway: Storting
sustainable development
  in Arctic, 25–30
  Arctic Council and, 4, 15 (*see also* Arctic Council)
  Arctic petroleum and, 30–37, 332–34
  Brundtland Commission (*see* Brundtland Commission)
  business and, 74
  in Canada, 318
  conservation and, 324–25
  corporate barriers to, 370–74
  corporate social responsibility and (*see* corporate social responsibility: sustainable development and)
  definition of, 4, 20
  economics of, 29–30, 38, 192, 257
  energy consumption and, 23 (*see also* energy consumption)
  environment and, 29

  environmental sustainability and, 21–22, 38, 192
  global framing of, 16–25
  history of, 18
  indigenous peoples and, 24, 155–56, 369 (*see also* indigenous peoples)
  institutional condition for, 64–70
  national policies for, 318
  in Norway, 318
  petroleum exploration and, 30–37, 332–34, 345
  requirements for, 195
  revenue distribution and, 318
  in Russia, 257–60, 284, 318
  social aspects of, 29–30, 38, 192
  in United States, 318
  World Commission on Environment and Development (WCED) and, 19

Task Force on Sustainable Development and Utilization (TFSDU), 26
tax revenues, 31, 170, 324
technology
  corporate social responsibility and, 366–67
  environment and, 35, 154–55, 366
  indigenous peoples and, 8
  petroleum industry and, 200, 215, 221, 229–30, 235, 366
terra nullius doctrine, 133, 295
Teshekpuk Lake, 160–61, **162**
Trans Alaska Pipeline System, 141, 145, 166
TransCanada Pipelines Ltd., 181–82
transfer payments and, 32, 124, 187
transport of petroleum. *See* petroleum industry: transport and; pipeline(s)
tundra, 37, 47, 49, **49**, 158, 170. *See also* permafrost

unemployment. *See* employment
United Nations
  Conference on Environment and Development (UNCED), 4, 17–18, 22, 25–26
  corporate social responsibility and, 58 (*see also* corporate social responsibility)
  Declaration of the Rights of Indigenous Peoples (UNDRIP), 291, 294, 303–5, 358–59
  Development Program (UNDP), 74
  Environmental Programme (UNEP), 18, 212

United Nations *(Continued)*
   Framework Convention on Climate Change (UNFCCC), 17, 23, 38, 326
   Millennium Development Goals, 46
United States
   Alaska (*see* Alaska)
   Bureau of Land Management (BLM), 145, 160–61, 368
   CO_2 emissions by, *329* (*see also* greenhouse gases)
   federal government, 113
   human rights in, *292*, 293
   Kyoto Protocol and, 327, 328 (*see also* Kyoto Protocol)
   local government in, 116
   Minerals Management Service (MMS) of, 122, 145, 147–49
   National Environmental Policy Act (NEPA), 145, 149
   National Petroleum Reserve–Alaska (NPR-A), 140, 145, 153, 160–61, 167, 368
   natural resource ownership in, 120
   Outer Continental Shelf (OCS) Lands Act, 147–49
   Working Group for Indigenous Peoples (UNWGIP), 301, 340
University of Calgary, 282

water storage, 46
Wescoast Energy Inc., 181
whaling, 144, 150, 153–54, 168, 361–62. *See also* Alaska Eskimo Whaling Commission
Working Group on the Conservation of Arctic Flora and Fauna (CAFF), 26, 36–37, 39, 333, 335,

Working Group on Emergency Preparedness and Response (EPPR), 26, 30
Working Group on the Protection of the Arctic Marine Environment (PAME), 26, 30, 34–35, 79
World Bank, 15, 58, 62, 89, 93, 261, 295, 311, 357, 359–60, 362–63
World Bank Country Assistance Strategy (CAS), 261
World Commission on Environment and Development (WCED), 4, 19–20. *See also* Brundtland Commission; *Our Common Future*
World Conservation Strategy (1980), 18–19
World Council of Churches, 18, 93
World Health Organization (WHO), 46
World Trade Organization (WTO), 15
World-Wide Fund for Nature (WWF), 18, 210, 213–15, 230, 282, 335
World Wildlife Federation, 183
World Wildlife Fund Norway, 210, 342

Yamal-Nenets
   development indicators, *259*, *260*
   economics of, 241, 245, 267–69
   gas fields and pipelines, 245, **266**, 274
   gas production, 245, 265
   gas revenues, **267**
   geography, 265
   indigenous peoples and, 272–75
   large-scale development in, 214
yedoma, 48–49
Yukon, 40, 116–17, 121, 123, *124*, 124, 179–81, 339
Yukon Oil and Gas Act (1998), 121